Lecture Notes in Mathematics
A collection of informal reports and seminars
Edited by A. Dold, Heidelberg and B. Eckmann, Zürich

Series: Forschungsinstitut für Mathematik, ETH, Zürich · Adviser: K. Chandrasekharan

T0226008

80

H. Appelgate, M. Barr, J. Beck, F. W. Lawvere, F. E. J. Linton, E. Manes, M. Tierney, F. Ulmer

Seminar on Triples and Categorical Homology Theory

ETH 1966/67

Edited by B. Eckmann

1969

Springer-Verlag Berlin · Heidelberg · New York

Library of Congress Catalog Card Number 68-59303 Printed in Germany. Title No. 3686

P r e f a c e

During the academic year 1966/67 a seminar on various aspects of category theory and its applications was held at the Forschungsinstitut für Mathematik, ETH, Zürich. This volume is a report on those lectures and discussions which concentrated on two closely related topics of special interest; namely a) on the concept of "triple" or standard construction with special reference to the associated "algebras", and b) on homology theories in general categories, based upon triples and simplicial methods. In some respects this report is unfinished and to be continued in later volumes; thus in particular the interpretation of the general homology concept on the functor level (as satellites of Kan extensions), is only sketched in a short survey.

I wish to thank all those who have contributed to the seminar; the authors for their lectures and papers, and the many participants for their active part in the discussions. Special thanks are due to Myles Tierney and Jon Beck for their efforts in collecting and coordinating the material for this volume.

B. Eckmann

Table of Contents

Introduction

The papers in this volume were presented to the seminar on category theory held during the academic year 1966-67 at the Forschungsinstitut für Mathematik of the Eidgenössische Technische Hochschule, Zürich. The material ranges from structural descriptions of categories to homology theory, and all of the papers use the method of standard constructions or "triples."

It will be useful to collect the basic definitions and background in the subject here, and indicate how the various papers fit in. References are to the bibliography at the end of the Introduction.

Before beginning, one must waste a word on terminological confusion. The expression "standard construction" is the one originally introduced by Godement [1]. Eilenberg-Moore substituted "triple," for brevity [4]. The term "monad" has also come into use. As for the authors of this volume, they all write of:

1. Triples. $\mathbf{T} = (T,\eta,\mu)$ is a triple in a category $\underline{A}$ if $T : \underline{A} \to \underline{A}$ is a functor, and $\eta : id_{\underline{A}} \to T$, $\mu : TT \to T$ are natural transformations such that the diagrams

$$\begin{array}{ccc} T & \xrightarrow{T\eta} & TT \\ & \searrow\!\!\!\searrow & \downarrow{\scriptstyle \mu} \\ & & T \end{array} \qquad \begin{array}{ccc} T & \xrightarrow{\eta T} & TT \\ & \searrow\!\!\!\searrow & \downarrow{\scriptstyle \mu} \\ & & T \end{array} \qquad \begin{array}{ccc} TTT & \xrightarrow{T\mu} & TT \\ {\scriptstyle \mu T}\downarrow & & \downarrow{\scriptstyle \mu} \\ TT & \xrightarrow[\mu]{} & T \end{array}$$

commute. η is known as the unit of the triple, μ as the multiplication, and the diagrams state that η,μ obey right and left unitary and associative laws.

(Notation: In the Introduction morphisms will be composed in the order of following arrows. In particular, functors are evaluated by being written to the right of their arguments.)

As for natural transformations, if $\varphi : S \to T$ is a natural transformation of functors $S,T : \underline{A} \to \underline{B}$, and $\psi : U \to V$ where $U,V : \underline{B} \to \underline{C}$, then $\varphi U : SU \to TU$, $S\psi : SU \to SV$ are the natural transformations whose values on an object $A \in \underline{A}$ are

$$A\varphi U = (A\varphi)U : (AS)U \to (AT)U ,$$

$$AS\psi = (AS)\psi : (AS)U \to (AS)V .$$

Other common notations are $(\varphi U)_A$, $(\varphi * U)_A$ as in [1], as well as $(U\varphi)_A$, ... This should make clear what is meant by writing $T\eta$, $\eta T : T \to TT$, transformations which are in general distinct.)

The original examples which were of interest to Godement were: (a) the triple in the category of sheaves over a space X whose unit is $\mathcal{F} \to \mathcal{C}^0(X,\mathcal{F})$, the canonical flasque embedding, and (b) the triple $(\ ) \otimes R$ generated in the category of abelian groups $\underline{A}$ by tensoring with a fixed ring R; the unit and multiplication in this triple are derived from the ring structure:

$$A \xrightarrow{a\otimes 1} A \otimes R , \qquad A \otimes R \otimes R \xrightarrow{a\otimes r_0 r_1} A \otimes R .$$

It was also Godement's idea that by iterating the triple simplicial "resolutions" could be built up and homology theories obtained. For example, the complex of sheaves

$$0 \to \mathcal{F} \to \mathcal{C}^0(X,\mathcal{F}) \rightrightarrows \mathcal{C}^0(X,\mathcal{C}^0(X,\mathcal{F})) \mathrel{\substack{\to\\\to\\\to}} \ldots$$

gives rise to sheaf cohomology. Although restricted to abelian categories, this was the prototype of the general homology theories to which triples lead.

Note that the situation dualizes. A <u>cotriple</u> in a category $\underline{B}$ is a triple $\mathbf{G} = (G,\varepsilon,\delta)$ where $G : \underline{B} \to \underline{B}$, $\varepsilon : G \to id_{\underline{B}}$, $\delta : G \to GG$ and counitary and coassociative axioms are satisfied.

2. <u>Algebras over a triple</u>. A $\mathbf{T}$-<u>algebra</u> [4] is a pair (X,ξ) where $X \in \underline{A}$ and $\xi : AT \to A$ is a unitary, associative map called the $\mathbf{T}$-<u>structure</u> of the algebra:

$$\begin{array}{ccc} X & \xrightarrow{X\eta} & XT \\ & \searrow^{=} & \downarrow \xi \\ & & X \end{array} \qquad \begin{array}{ccc} XTT & \xrightarrow{X\mu} & XT \\ \xi T \downarrow & & \downarrow \xi \\ XT & \xrightarrow[\xi]{} & X \end{array}$$

$f : (X,\xi) \to (Y,\theta)$ is a <u>map</u> of $\mathbf{T}$-algebras if $f : X \to Y$ in $\underline{A}$ and is compatible with $\mathbf{T}$-structures: $fT.\theta = \xi f$.

The category of $\mathbf{T}$-algebras is denoted by $\underline{A}^{\mathbf{T}}$.

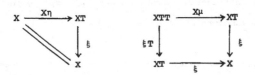

For example, if $\underline{A}$ is the category of abelian groups and $\mathbf{T}$ is the triple $(\ ) \otimes R$, then a $\mathbf{T}$-structure on an abelian group A is a unitary, associative operation $A \otimes R \to A$. Thus $\underline{A}^{\mathbf{T}}$ is the category of R-modules.

Many other intuitive examples will soon appear. An example of a dual, less obvious kind arises when a functor $\underline{M} \to \underline{C}$ is given. By taking the direct limit of all maps $M \to X$ where $M \in \underline{M}$, one obtains a value of a so-called singular cotriple, XG. The corresponding coalgebras, that is, objects equipped with costructures $X \to XG$, have interesting local (neighborhood) structures. Appelgate-Tierney study this construction in this volume ("Categories with models") taking for $\underline{M} \to \underline{C}$ such model subcategories as standard simplices, open sets in euclidean space, spectra of commutative rings,...

3. Relationship between adjoint functors and triples. Recall that an adjoint pair of functors [2] consists of functors $F : \underline{A} \to \underline{B}$, $U : \underline{B} \to \underline{A}$, together with a natural isomorphism

$$\mathrm{Hom}_{\underline{A}}(A, BU) \cong \mathrm{Hom}_{\underline{B}}(AF, B)$$

for all objects $A \in \underline{A}$, $B \in \underline{B}$.

Putting $B = AF$ we get a natural transformation $\eta : id_{\underline{A}} \to FU$ called the unit or front adjunction. Putting $A = BU$, we get $\varepsilon : UF \to id_{\underline{B}}$, the counit or back adjunction. These natural transformations satisfy

$$\begin{array}{ccc} F & \xrightarrow{\eta F} & FUF \\ & \searrow^{=} & \downarrow F\varepsilon \\ & & F \end{array} \qquad\qquad \begin{array}{ccc} U & \xrightarrow{U\eta} & UFU \\ & \searrow^{=} & \downarrow \varepsilon U \\ & & U \end{array}$$

P. Huber [3] observed that

$$\mathbf{T} = \begin{cases} T = FU : \underline{A} \to \underline{A} \\ \eta : id_{\underline{A}} \to T \\ \mu = F\varepsilon U : TT \to T \end{cases} \qquad\qquad \mathbf{G} = \begin{cases} G = UF : \underline{B} \to \underline{B} \\ \varepsilon : G \to id_{\underline{B}} \\ \delta = U\eta F : G \to GG \end{cases}$$

are then triple and cotriple in $\underline{A}$ and $\underline{B}$, respectively. This remark simplifies the task of constructing triples. For example, Godement's example (a) above is induced by the adjoint pair of functors

$$\text{Sheaves}(X_0) \underset{f^*}{\overset{f_*}{\rightleftarrows}} \text{Sheaves}(X) \ ,$$

where X_0 is X with the discrete topology and $f : X_0 \to X$ is the identity on points.

Conversely, Eilenberg-Moore showed [4] that via the $\underline{A}^{\mathbb{T}}$ construction triples give rise to adjoint functors. There is an obvious <u>forgetful</u> or <u>underlying</u> <u>A-object</u> <u>functor</u> $U^{\mathbb{T}} : \underline{A}^{\mathbb{T}} \to \underline{A}$, and left adjoint to $U^{\mathbb{T}}$ is the <u>free</u> <u>$\mathbb{T}$-algebra</u> <u>functor</u> $F^{\mathbb{T}} : \underline{A} \to \underline{A}^{\mathbb{T}}$ given by $AF^{\mathbb{T}} = (AT, A\mu)$. The natural equivalence

$$\mathrm{Hom}_{\underline{A}}(A,(X,\xi)U^{\mathbb{T}}) \tilde{\to} \mathrm{Hom}_{\underline{A}^{\mathbb{T}}}(AF^{\mathbb{T}},(X,\xi))$$

is easily established.

Thus, granted an adjoint pair $\underline{A} \to \underline{B} \to \underline{A}$, we get a triple $\mathbb{T} = (T,\eta,\mu)$ in $\underline{A}$, and we use the $A^{\mathbb{T}}$ construction to form another adjoint pair $\underline{A} \to \underline{A}^{\mathbb{T}} \to \underline{A}$. To relate these adjoint pairs we resort to a canonical functor

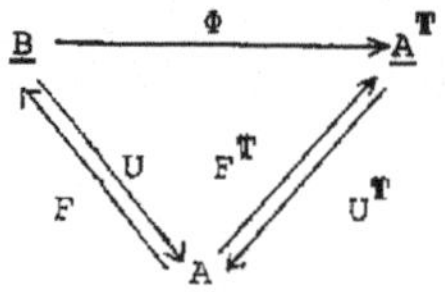

with the properties $F\Phi = F^{\mathbb{T}}$, $U = \Phi U^{\mathbb{T}}$. Φ is defined by $B\Phi = (BU, B\varepsilon U)$. Its values are easily verified to be $\mathbb{T}$-algebras. Intuitively, $B\varepsilon : BUF \to B$ is the canonical map of the free object generated by B "onto" B, and the $\mathbb{T}$-structure of $B\Phi$ is just the $\underline{A}$-map underlying that.

4. <u>Tripleability</u>. The adjoint pair (F,U) is <u>tripleable</u> [5] if $\Phi : \underline{B} \to \underline{A}^{\mathbb{T}}$ is an equivalence of categories.

Sometimes Φ is actually required to be an isomorphism of categories. This is particularly the case when the base category $\underline{A}$ is the category of sets.

Readers who replace "triple" with "monad" will replace "tripleable" with "<u>monadic</u>".

Intuitively, tripleableness of (F,U) means that the category $\underline{B}$ is definable in terms of data in $\underline{A}$, and that $U : \underline{B} \to \underline{A}$ is equivalent to a particularly simple sort of forgetful functor.

Examples. (a) Let $\mathcal{V}$ be an equational category of universal algebras (variety), that is, the objects of $\mathcal{V}$ are sets with algebraic operations subject to equational conditions (groups, rings, Lie algebras,..., but not fields, whose definition requires mention of the inequality $x \neq 0$). The adjoint pair $\underline{A} \to \mathcal{V} \to \underline{A}$ is tripleable, where $\mathcal{V} \to \underline{A}$ is the underlying set functor and $\underline{A} \to \mathcal{V}$ is the free $\mathcal{V}$-algebra functor ([5], and see [6] for the introduction of universal algebra into category theory). In fact, if the base category $\underline{A}$ is that of sets, F.E.J. Linton shows that triples and equational theories (admitting a just amount of infinitary operations) are entirely equivalent concepts ([7], and "Outline of functorial semantics", this volume). From the practical standpoint, formulations in terms of triples tend to be concise, those in terms of theories more explicit. The components T, η, μ of the triple absorb all of the operational and equational complications in the variety, and the structure map $XT \to X$ of an algebra never obeys any axiom more involved than associativity.

(b) In general, tripleableness implies a measure of algebraicity. The adjoint pair Sets $\to$ Topological spaces $\to$ Sets (obvious functors) is not tripleable. But the paper "A triple theoretic construction of compact algebras" by E. Manes (this volume) shows that compactness is in this sense an "algebraic" concept.

(c) Let $\underline{A}$ be the category of modules over a commutative ring. Linear algebras are often viewed as objects $A \in \underline{A}$ equipped with multiplicative structure. But here the universal-algebra description of structure is inappropriate, a binary multiplication, for example, not being a K-linear map $A \times A \to A$, but rather a K-bilinear map. It was precisely this example which motivated the intervention of triples. Let $\mathcal{A}$ be any known category of linear algebras (associative, commutative, Lie, ...). Then if the free algebra functor exists, the adjoint pair $\underline{A} \to \mathcal{A} \to \underline{A}$ is tripleable.

In view of the applicability of the tripleableness concept in algebra and in geometry (descent theory), it is useful to have manageable tests for tripleableness. Such tests are discussed and applied by F.E.J. Linton in his paper "Applied functorial semantics, II" (this volume).

5. Homology. Let $\underline{A} \xrightarrow{F} \underline{B} \xrightarrow{U} \underline{A}$ be an adjoint pair, $\varepsilon : UF \to id_{\underline{B}}$ the counit, and let

$X \in \underline{B}$. Iterating the composition UF and using ε to construct face operators, we construct a simplicial "resolution" of X:

$$X \leftarrow XUF \leftleftarrows X(UF)^2 \Lleftarrow \ldots$$

If appropriate coefficient functors are applied to this resolution, very general homology and cohomology theories arise. These theories are available whenever underlying pairs of adjoint functors exist. When the adjoint pairs are <u>tripleable</u> these theories enjoy desirable properties, notably classification of extensions and principal homogeneous objects [5].

A lengthy study of such homology theories is given in this volume by Barr-Beck, "Homology and standard constructions." Cotriple homology is now well known to encompass many classical algebraic homology theories, and agrees with general theories recently set forth in these Lecture Notes by M. André [8] and D.G. Quillen [9].

In "Composite cotriples and derived functors," Barr studies the influence on homology of so-called "distributive laws" between cotriples. Such distributive laws are discussed elsewhere in this volume by Beck, in a paper which is more in the spirit of universal algebra.

The classical "obstruction" theory for algebra extensions has not yet been carried over to triple cohomology. In his paper "Cohomology and obstructions: Commutative algebras", Barr works out an important special case, obtaining the expected role for H^2 (the dimension indices in triple cohomology being naturally one less than usual).

Finally,one has to wonder what the relationship between this adjoint-functor "simplicial" homology and classical derived-functor theories is. In the final paper in this volume, "On cotriple and André (co)homology, their relationship with classical homological algebra", F. Ulmer shows that on an appropriate functor category level, triple cohomology appears as the satellite theory - in the abelian-category sense - of the not-so-classical "Kan extension" of functors. Incidentally, as triple cohomology, that is to say, general algebra cohomology, must not vanish on injective coefficients, it cannot be referred to categories of modules after the fashion of Cartan-Eilenberg-Mac Lane.

This then summarizes the volume - apart from mention of F.W.Lawvere's paper

"ordinal sums and equational doctrines", which treats in a speculative vein of triples in the category of categories itself - the hope is that these papers will supply a needed and somewhat coherent exposition of the theory of triples. All of the participants in the seminar must express their gratitude to the E.T.H., Zürich, and the Director of the Forschungsinstitut, Professor B. Eckmann, for the hospitality and convenient facilities of the Forschungsinstitut in which this work was done.

References

[1] Godement,R., Théorie des faisceaux, Paris, Hermann, 1957.

[2] Kan,D.M., Adjoint functors, Trans. A.M.S. 87 (1958), 294-329.

[3] Huber,P., Homotopy theory in general categories, Math. Ann. 444 (1961), 361-385.

[4] Eilenberg,S. and J.C.Moore, Adjoint functors and triples, Ill. J.Math. 9 (1965), 381-398.

[5] Beck,J., Triples, algebras and cohomology, Dissertation, Columbia University, 1967.

[6] Lawvere,F.W., Functorial semantics of algebraic theories, Dissertation, Columbia University, 1963.

[7] Linton,F.E.J., Some aspects of equational categories, Proc. of the Conf. on Categorical Algebra, Springer Verlag, 1966.

[8] André,M., Méthode simpliciale en algèbre homologique et algèbre commutative, Lecture Notes in Mathematics, No. 32, Springer Verlag, 1967.

[9] Quillen,D.G., Homotopical algebra, Lecture Notes in Mathematics, No. 43, Springer Verlag, 1967.

AN OUTLINE OF FUNCTORIAL SEMANTICS *

by

F. E. J. Linton

This paper is devoted to the elucidation of a very general structure-semantics adjointness theorem (Theorem 4.1), out of which follow all other structure semantics adjointness theorems currently known to the author. Its reduction, in § 10, to the classical theorem in the context of triples requires a representation theorem (see § 9) asserting that the categories of algebras, in a category $\mathcal{A}$, tentatively described in § 1 (and used in [20] in the special case $\mathcal{A} = \mathcal{S}$) , "coincide" with the categories of algebras over suitably related triples, if such exist.

§§ 7 and 8 pave the way for this representation theorem. A table of contents appears on the next page; a more detailed outline of the contents of §§ 3-11 is sketched in § 2. Portions of this paper fulfill the promises made in [18] and at the close of § 6 of [19].

*The research here incorporated, carried out largely during the author's tenure of an N.A.S.-N.R.C. Postdoctoral Research Fellowship at the Research Institute for Mathematics, E.T.H., Zurich, while on leave from Wesleyan University, Middletown, Connecticut, was supported in its early stages by a Faculty Research Grant from the latter institution.

Contents. page

§ 1. Introduction to algebras in general categories.

Functorial semantics generalizes to arbitrary categories the classical notion [6] of an abstract algebra. This notion is usually [7], [24] defined, in terms of a set Ω of "operations", a set-valued "arity" function[1] n defined on Ω, and a collection E of "laws[2] governing the operations of Ω", as a system (A,λ) consisting of a set A so equipped with an Ω-indexed family $\lambda = \{\lambda(\theta) \mid \theta \in \Omega\}$ of $n(\theta)$-ary operations

$$\lambda(\theta) : A^{n(\theta)} \longrightarrow A \qquad (\theta \in \Omega)$$

---

[1] Its values are often constrained to be ordinals, or cardinals, or positive integers.

[2] E.g., associativity laws, unit laws, commutativity laws, Jacobi identities, idempotence laws, etc.

that the body of laws is upheld. An algebra homomorphism from $(A,\mathfrak{A})$ to $(B,\mathfrak{B})$ is then, of course, any function $g : A \to B$ commuting with all the operations, i.e., rendering commutative all the diagrams

$$\begin{array}{ccc} A^{n(\theta)} & \xrightarrow{\quad g^{n(\theta)} \quad} & B^{n(\theta)} \\ {\scriptstyle \mathfrak{A}(\theta)}\downarrow & & \downarrow{\scriptstyle \mathfrak{B}(\theta)} \\ A & \xrightarrow{\quad g \quad} & B \end{array} \qquad (\theta \in \Omega) .$$

Among the algebras of greatest interest in functorial semantics are those arising by a very similar procedure from a functor

$$U : \mathcal{X} \longrightarrow \mathcal{A} .$$

To spotlight the analogy, we first introduce some suggestive notation and terminology, writing $\mathcal{S}$ for a category of sets (in the sense either of universes [25] or of Lawvere's axiomatic foundations [16], [17] in which $\mathcal{A}$'s hom functor takes values.

Given the functor $U : \mathcal{X} \longrightarrow \mathcal{A}$, we define, for each $\mathcal{A}$-morphism $f : k \to n$, a natural transformation

$$U^f : U^n \longrightarrow U^k ,$$

from the functor

$$U^n = \mathcal{A}(n,U(-)) : \mathcal{X} \longrightarrow \mathcal{S}$$

to the functor

$$U^k = \mathcal{A}(k,U(-)) : \mathcal{X} \longrightarrow \mathcal{S} ,$$

by posing

$$(U^f)_X = \mathcal{A}(f,UX) : U^n X \to U^k X \qquad (X \in |\mathcal{X}|) .$$

Moreover, whenever, A, B, n, k are objects of $\mathcal{A}$ and $f : k \to n$, $g : A \to B$ are $\mathcal{A}$-morphisms, we set

$$A^n = \mathcal{A}(n,A) \in |\mathcal{S}| ,$$

$$A^f = \mathcal{A}(f,A) : A^n \to A^k , \quad \text{and}$$

$$g^n = \mathcal{A}(n,g) : A^n \to B^n .$$

The class

$$\text{n.t.}(U^n,U^k) \qquad (\text{resp.} \quad \mathcal{S}(A^n,A^k))$$

is to be thought of as consisting of all natural k-tuples of (or all k-tuple-valued) n-ary operations on U (resp. on A) .

A U-algebra is then defined to be a system (A,λ) consisting of an object $A \in |\mathcal{A}|$ and a family

$$\lambda = \{\lambda_{n,k} | \ n \in |\mathcal{A}| , \ k \in |\mathcal{A}|\}$$

of functions

$$\lambda_{n,k} : \text{n.t.}(U^n,U^k) \longrightarrow \mathcal{S}(A^n,A^k)$$

satisfying the identities

(1.1) $\quad \lambda_{n,k}(U^f) = A^f \qquad (f \in \mathcal{A}(k,n))$

(1.2) $\quad \lambda_{n,m}(\theta'\circ\theta) = \lambda_{k,m}(\theta')\circ\lambda_{n,k}(\theta) \qquad (\theta : U^n \to U^k , \ \theta' : U^k \to U^m)$.

Writing (compare [10])

(1.3) $\quad \{\lambda_{n,k}(\theta)\}(a) = \theta * a$ ($= \theta *_\lambda a$, when one must remember λ)

whenever $\theta : U^n \to U^k$ and $a \in A^n$, the identities (1.1) and (1.2) become

ALG 1) $\quad U^f * a = a \circ f \qquad (f \in \mathcal{A}(k,n) , \ a \in A^n)$

ALG 2) $\quad (\theta'\circ\theta) * a = \theta' * (\theta * a) \qquad (\theta : U^n \to U^k , \ \theta' : U^k \to U^m , \ a \in A^n)$.

As U-algebra homomorphisms from (A,λ) to $(B,\mathfrak{B})$ we admit all $\mathcal{A}$-morphisms $g : A \to B$ making the diagrams

(1.4)
$$\begin{array}{ccc} A^n & \xrightarrow{g^n} & B^n \\ {\scriptstyle \mathcal{A}_{n,k}(\theta)}\downarrow & & \downarrow{\scriptstyle \mathcal{B}_{n,k}(\theta)} \\ A^k & \xrightarrow[g^k]{} & B^k \end{array}$$

commute, for each natural operation θ on U ; in the notation of (1.3), this boils down to the requirement

ALG 3) $g \circ (\theta * a) = \theta * (g \circ a)$ $\qquad (\theta : U^n \to U^k, \ a \in A^n)$.

We write U-Alg for the resulting category of U-algebras. The prime examples of U-algebras are the U-algebras $\Phi_U(X)$, available for each object $X \in |\mathcal{X}|$, given by the data

(1.5)
$$\Phi_U(X) = (UX, \lambda_U(X)) ,$$

where, in the notation of (1.3), $\lambda_U(X)$ is specified by

$$\theta * a = \theta_X(a) .$$

It is a trivial consequence of the defining property of a natural transformation that each $\mathcal{A}$-morphism $U(\xi)$ $(\xi : X \to X')$ is a U-algebra homomorphism

$$U(\xi) : \Phi_U(X) \longrightarrow \Phi_U(X') ,$$

and that these passages provide a functor

$$\Phi_U : \mathcal{X} \longrightarrow \text{U-Alg} : \begin{cases} X \longmapsto \Phi_U(X) , \\ \xi \longmapsto U(\xi) , \end{cases}$$

called the <u>semantical comparison functor for</u> U .

We must not ignore the <u>underlying $\mathcal{A}$-object functor</u>

$$|\ |_U : \text{U-Alg} \longrightarrow \mathcal{A} : \begin{cases} (A,\lambda) \longmapsto A , \\ g \longmapsto g . \end{cases}$$

For one thing, the triangle

$$\begin{array}{ccc} \mathcal{X} & \xrightarrow{\Phi_U} & \underset{\sim}{\text{U-Alg}} \\ & {\scriptstyle U}\searrow \quad \swarrow{\scriptstyle |\ |_U} & \\ & \mathcal{A} & \end{array}$$

commutes. For another, awareness of $|\ |_U$ is the first prerequisite for a much more concise description of $\underset{\sim}{\text{U-Alg}}$ as a certain pullback. The only other prerequisite for this is the recognition that the system $\mathfrak{A}$ in a U-algebra $(A,\mathfrak{A})$ is nothing but the effects on morphisms of a certain set-valued functor, again to be denoted $\mathfrak{A}$, defined on the following category $\mathfrak{T}_U$, the (full) clone of operations on U : the objects and maps of $\mathfrak{T}_U$ are given by

$$|\mathfrak{T}_U| = |\mathcal{A}| ,$$

$$\mathfrak{T}_U(n,k) = \text{n.t.}(U^n,U^k) ;$$

the composition in $\mathfrak{T}_U$ is the usual composition of natural transformations. We point out the functor

$$\exp_U : \mathcal{A}^* \longrightarrow \mathfrak{T}_U : \begin{cases} n \longmapsto n , \\ f \longmapsto U^f , \end{cases}$$

and remark that the functions $\mathfrak{A}_{n,k}$ are obviously the effects on morphisms of a functor (necessarily unique)

$$\mathfrak{A} : \mathfrak{T}_U \longrightarrow \mathfrak{S}$$

whose effect on objects is simply

$$\mathfrak{A}(n) = A^n .$$

(Proof: (1.1) and (1.2).) Likewise, given a U-algebra homomorphism $g : (A,\mathfrak{A}) \to (B,\mathfrak{B})$, the commutativity of (1.4) makes the system $\{g^n : A^n \to B^n \mid n \in |\mathcal{A}|\}$ a natural transformation from $\mathfrak{A}$ to $\mathfrak{B}$. In this way, we obtain a functor

$$\underset{\sim}{\text{U-Alg}} \longrightarrow (\mathfrak{T}_U, \mathfrak{S})$$

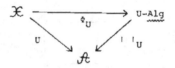

making the square

$$(1.6)\qquad \begin{array}{ccc} \underset{\sim}{\text{U-Alg}} & \longrightarrow & (\mathcal{T}_U, \mathbb{S}) \\ {\scriptstyle |\ |_U}\downarrow & & \downarrow{\scriptstyle (\exp_U, \mathbb{S})} \\ \mathcal{A} & \underset{Y}{\longrightarrow} & (\mathcal{A}^*, \mathbb{S}) \end{array}$$

commute; here functor categories and induced functors between them are denoted by parentheses, and Y is the Yoneda embedding $A \longmapsto \mathcal{A}(-,A)$. In § 5 we shall see (it can be proved right away, with virtually no effort)

Observation 1.1. Diagram (1.6) is a pullback diagram.

With this introduction to algebras in general categories behind us, we turn to a description of what lies ahead.

§ 2. General plan of the paper.

Motivated both by Observation 1.1 and the desire to recapture the structure-semantics adjointness of [15], we spend the next two sections, with a fixed functor

$$j : \mathcal{A}_0 \longrightarrow \mathcal{A} ,$$

studying the passage from

$$V : \mathcal{A}_0^* \longrightarrow \mathcal{C}$$

to the $\mathcal{A}$-valued functor $\mathfrak{m}^{(j)}(V)$ defined on the pullback of the pullback diagram

$$\begin{array}{ccccc} \text{pullback} & & \longrightarrow & & (\mathcal{C}, \mathbb{S}) \\ \downarrow{\scriptstyle \mathfrak{m}^{(j)}(V)} & & & & \downarrow{\scriptstyle (V,\mathbb{S})} \\ \mathcal{A} & \underset{Y}{\longrightarrow} & (\mathcal{A}^*, \mathbb{S}) & \underset{(j^*,\mathbb{S})}{\longrightarrow} & (\mathcal{A}_0^*, \mathbb{S}) \end{array} ,$$

the passage from

$$U : \mathcal{X} \longrightarrow \mathcal{A}$$

to the composition

$$\mathfrak{s}^{(j)}(U) : \mathcal{A}_o^* \xrightarrow[j^*]{} \mathcal{A}^* \xrightarrow[Y]{} (\mathcal{A},\mathcal{S}) \xrightarrow[(U,\mathcal{S})]{} (\mathcal{X},\mathcal{S}) \quad ,$$

the adjointness relation between $\mathfrak{m}^{(j)}$ and $\mathfrak{s}^{(j)}$, and the modification of this adjointness that results from consideration of the full image cotriple on the (comma) category $(\mathcal{A}_o^*, \underset{\sim}{Cat})$. In § 5 we present some remarks on the constructions of §§ 3-4, including a proof of Observation 1.1 and an indication of the manner in which the structure-semantics adjointnesses of [5], [15] and [19] are recaptured by specializing the functor j .

The next three sections digress from the main line of thought, to present tangential results, without which, however, the main line of thought cannot easily continue. In § 6, the least important of these digressions, we present two completeness properties of the categories of algebras arising in § 4 (slightly less satisfying results along the same line can be achieved also for those arising in § 3 -- we forego them here). The material of §§ 7-8 is necessitated by the frequent possibility of associating a triple [3] $\mathbf{T} = (T,\eta,\mu)$ to an $\mathcal{A}$-valued functor $U : \mathcal{X} \to \mathcal{A}$, in the manner of [1], [14], or [26]. This can be done, for example, when U has a left adjoint $F : \mathcal{A} \to \mathcal{X}$, with front and back adjunctions $\eta : id_{\mathcal{A}} \to UF$, $\beta : FU \to id_{\mathcal{X}}$, by setting

$$T = UF \ , \quad \eta = \eta \ , \quad \mu = U\beta F \quad .$$

It will be seen in § 9 that if $\mathbf{T}$ is a triple suitably associated with $U : \mathcal{X} \to \mathcal{A}$, the category of U-algebras and the category $\mathcal{A}^{\mathbf{T}}$ (constructed in [3 , Th. 2.2], for example) of $\mathbf{T}$-algebras are canonically isomorphic. To this end, § 7 reviews the definitions of triples, of the categories $\mathcal{A}^{\mathbf{T}}$, and of the construction [13] of the Kleisli category of a triple, while § 8 is devoted to a full elucidation of the manner in which a triple on $\mathcal{A}$ can be associated to an $\mathcal{A}$-valued functor.

The above mentioned isomorphism theorem in § 9 is proved there in two ways: once by appeal to a general criterion, which depends on a result of § 6 and on the availability of a left adjoint to $| \ |_U : U\text{-}\underset{\sim}{Alg} \to \mathcal{A}$, and once (sketchily) by a somewhat more involved argument that constructs the isomorphism explicitly, still using, of course, the left adjoint just mentioned.

In § 10, the result of § 9 is used to recover the structure-semantics adjointness for the context of triples from that of § 4. Finally, in § 11, we give a proof of the isomorphism theorem of § 9 that is entirely elementary -- in particular, that is quite independent of the knowledge that $|\ |_U$ has a left adjoint, and from which that fact follows. The exposition of this last § is so arranged that it can be read immediately after § 1, without bothering about §§ 3-10.

§ 3. Preliminary structure-semantics adjointness relation.

The granddaddy of all the structure-semantics adjointness theorems is the humble canonical isomorphism

$$(\mathcal{X},(\mathcal{Z},\mathcal{S})) \cong (\mathcal{Z},(\mathcal{X},\mathcal{S}))$$

expressing the symmetry [8] of the closed category $\underset{\sim}{\mathrm{Cat}}$ of categories. Here we are using $(\mathcal{X},\mathcal{Y})$ to denote the category of all functors from $\mathcal{X}$ to $\mathcal{Y}$, with natural transformations as morphisms.

Until further notice, fix a functor $j : \mathcal{A}_o \to \mathcal{A}$.

The first prototype of structure and semantics (rel. j) will be functors passing from the category $(\underset{\sim}{\mathrm{Cat}},\mathcal{A})$ of all $\mathcal{A}$-valued functors $U : \mathcal{X} \to \mathcal{A}$, with domain $\mathcal{X} \in |\underset{\sim}{\mathrm{Cat}}|$, to the category $(\mathcal{A}^*_o,\underset{\sim}{\mathrm{Cat}})$ of all functors $V : \mathcal{A}^*_o \to \mathcal{C}$ with codomain $\mathcal{C} \in |\underset{\sim}{\mathrm{Cat}}|$, and back again, as outlined in § 2. Of course, we think of $(\underset{\sim}{\mathrm{Cat}},\mathcal{A})$ and $(\mathcal{A}^*_o,\underset{\sim}{\mathrm{Cat}})$ as comma categories [15], so that the $(\underset{\sim}{\mathrm{Cat}},\mathcal{A})$-morphisms from $U' : \mathcal{X}' \to \mathcal{A}$ to $U : \mathcal{X} \to \mathcal{A}$ are those functors $x : \mathcal{X}' \to \mathcal{X}$ satisfying $U' = U \circ x$, while the $(\mathcal{A}^*_o,\underset{\sim}{\mathrm{Cat}})$-morphisms from $V : \mathcal{A}^*_o \to \mathcal{C}$ to $V' : \mathcal{A}^*_o \to \mathcal{C}'$ are those functors $c : \mathcal{C} \to \mathcal{C}'$ satisfying $V' = c \circ V$.

The proof of the basic lemma below is so completely elementary that it will be omitted. To find it, just follow your nose.

Lemma 3.1. For each pair of functors $U : \mathcal{X} \to \mathcal{A}$, $V : \mathcal{A}_o^* \to \mathcal{C}$, the canonical isomorphism

$$(\mathcal{X},(\mathcal{C},\mathbb{S})) \cong (\mathcal{C},(\mathcal{X},\mathbb{S}))$$

(where $\mathbb{S}$ is a category of sets receiving $\mathcal{A}$'s hom functor) mediates an isomorphism

(3.1) $$M(j;U,V) \cong S(j;U,V)$$

between the full subcategory $M(j;U,V) \subset (\mathcal{X},(\mathcal{C},\mathbb{S}))$ whose objects are those functors $F : \mathcal{X} \to (\mathcal{C},\mathbb{S})$ making the diagram

(3.2) $$\begin{array}{ccccc} \mathcal{X} & & \xrightarrow{\quad F \quad} & & (\mathcal{C},\mathbb{S}) \\ {\scriptstyle U}\downarrow & & & & \downarrow{\scriptstyle (V,\mathbb{S})} \\ \mathcal{A} & \xrightarrow[Y]{} & (\mathcal{A}^*,\mathbb{S}) & \xrightarrow[(j^*,\mathbb{S})]{} & (\mathcal{A}_o^*,\mathbb{S}) \end{array}$$

commute, and the full subcategory $S(j;U,V) \subset (\mathcal{C},(\mathcal{X},\mathbb{S}))$ whose objects are those functors $G : \mathcal{C} \to (\mathcal{X},\mathbb{S})$ making the diagram

(3.3) $$\begin{array}{ccccc} \mathcal{C} & & \xrightarrow{\quad G \quad} & & (\mathcal{X},\mathbb{S}) \\ {\scriptstyle V}\uparrow & & & & \uparrow{\scriptstyle (U,\mathbb{S})} \\ \mathcal{A}_o^* & \xrightarrow[j^*]{} & \mathcal{A}^* & \xrightarrow[Y]{} & (\mathcal{A},\mathbb{S}) \end{array}$$

commute. Moreover, the isomorphisms (3.1) are natural in the variables $U \in (\underset{\sim}{Cat},\mathcal{A})$ and $V \in (\mathcal{A}_o^*,\underset{\sim}{Cat})$.

The crudest structure and semantics functors (rel. j), to be denoted $\mathfrak{s}^{(j)}$ and $\mathfrak{m}^{(j)}$, respectively, are defined as follows.

Given $U : \mathcal{X} \to \mathcal{A}$ in $(\underset{\sim}{Cat},\mathcal{A})$, $\mathfrak{s}^{(j)}(U)$ is the composition

$$\mathfrak{s}^{(j)}(U) = (U,\mathbb{S}) \circ Y \circ j^* : \mathcal{A}_o^* \to \mathcal{A}^* \to (\mathcal{A},\mathbb{S}) \to (\mathcal{X},\mathbb{S}) .$$

It is clear that $(x,\mathbb{S}) : (\mathcal{X},\mathbb{S}) \to (\mathcal{X}',\mathbb{S})$ is an $(\mathcal{A}_o^*,\underset{\sim}{Cat})$-morphism $\mathfrak{s}^{(j)}(U) \to \mathfrak{s}^{(j)}(U')$ whenever $x : \mathcal{X}' \to \mathcal{X}$ is a $(\underset{\sim}{Cat},\mathcal{A})$-morphism from $U' : \mathcal{X}' \to \mathcal{A}$

to $U : \mathcal{X} \to \mathcal{A}$.

In the other direction, given $V : \mathcal{A}_o^* \to \mathcal{C}$ in $(\mathcal{A}_o^*, \underline{\mathrm{Cat}})$, define $\mathfrak{m}^{(j)}(V)$ to be the $\mathcal{A}$-valued functor from the pullback $\mathcal{P}_V^j$ in the pullback diagram

$$(3.4) \qquad \begin{array}{ccccc} \mathcal{P}_V^j & & \longrightarrow & & (\mathcal{C}, \mathfrak{S}) \\ {\scriptstyle \mathfrak{m}^{(j)}(V)} \downarrow & & & & \downarrow {\scriptstyle (V, \mathfrak{S})} \\ \mathcal{A} & \xrightarrow[Y]{} & (\mathcal{A}^*, \mathfrak{S}) & \xrightarrow[(j^*, \mathfrak{S})]{} & (\mathcal{A}_o^*, \mathfrak{S}) \end{array} .$$

It is clear, whenever $c : \mathcal{C} \to \mathcal{C}'$ is an $(\mathcal{A}_o^*, \underline{\mathrm{Cat}})$-morphism from $V : \mathcal{A}_o^* \to \mathcal{C}$ to $V' : \mathcal{A}_o^* \to \mathcal{C}'$, that

$$(c, \mathfrak{S}) : (\mathcal{C}', \mathfrak{S}) \to (\mathcal{C}, \mathfrak{S})$$

induces a functor $\mathcal{P}_{V'}^j \to \mathcal{P}_V^j$ between the pullbacks that is actually a $(\underline{\mathrm{Cat}}, \mathcal{A})$-morphism $\mathfrak{m}^{(j)}(V') \to \mathfrak{m}^{(j)}(V)$.

With these observations, it is virtually automatic that $\mathfrak{s}^{(j)}$ and $\mathfrak{m}^{(j)}$ are functors

$$\mathfrak{s}^{(j)} : (\underline{\mathrm{Cat}}, \mathcal{A}) \to (\mathcal{A}_o^*, \underline{\mathrm{Cat}})^* ,$$
$$\mathfrak{m}^{(j)} : (\mathcal{A}_o^*, \underline{\mathrm{Cat}})^* \to (\underline{\mathrm{Cat}}, \mathcal{A}) .$$

<u>Theorem 3.1</u>. (preliminary structure-semantics adjointness).
<u>The functor</u> $\mathfrak{s}^{(j)}$ <u>is (right) adjoint to</u> $\mathfrak{m}^{(j)}$.

<u>Proof</u>. By the definition of pullbacks, a functor from $\mathcal{X}$ to $\mathcal{P}_V^j$ "is" a pair of functors from $\mathcal{X}$ making the diagram

$$\begin{array}{ccccc} \mathcal{X} & & \longrightarrow & & (\mathcal{C}, \mathfrak{S}) \\ \downarrow & & & & \downarrow \\ \mathcal{A} & \xrightarrow[Y]{} & (\mathcal{A}^*, \mathfrak{S}) & \xrightarrow[(j^*, \mathfrak{S})]{} & (\mathcal{A}_o^*, \mathfrak{S}) \end{array}$$

commute. Hence a morphism from $U : \mathcal{X} \to \mathcal{A}$ to $\mathfrak{m}^{(j)}(V)$ "is" a functor $F : \mathcal{X} \to (\mathcal{C}, \mathfrak{S})$ making diagram (3.2) commute, i.e., "is" an object of $M(j;U,V)$, as defined in Lemma 3.1.

It is even easier to see that the $(\mathcal{A}_o^*, \underset{\sim}{\text{Cat}})^*$-morphisms from $\mathfrak{s}^{(j)}(U)$ to V coincide with the objects of the category $S(j;U,V)$ of Lemma 3.1. Consequently, the desired natural equivalence

$$(\underset{\sim}{\text{Cat}}, \mathcal{A})(U, \mathfrak{m}^{(j)}(V)) \cong (\mathcal{A}_o^*, \underset{\sim}{\text{Cat}})^*(\mathfrak{s}^{(j)}(U), V)$$
$$(\cong (\mathcal{A}_o^*, \underset{\sim}{\text{Cat}})(V, \mathfrak{s}^{(j)}(U)))$$

is delivered by the isomorphisms (3.1) of Lemma 3.1.

Remark. In fact, $(\underset{\sim}{\text{Cat}}, \mathcal{A})$ and $(\mathcal{A}_o^*, \underset{\sim}{\text{Cat}})$ are hypercategories [8], both $\mathfrak{s}^{(j)}$ and $\mathfrak{m}^{(j)}$ are hyperfunctors, and the adjointness relation is a hyperadjointness. The same remark will apply to the adjointness of Theorem 4.1 ; however, we know of no use for the stronger information.

§ 4. Full images and the operational structure-semantics adjointness theorem.

It is time to introduce the full image cotriple in $(\mathcal{A}_o^*, \underset{\sim}{\text{Cat}})$. We recall that the full image of a functor $V : \mathcal{A}_o^* \to \mathcal{C}$ is the category $\mathcal{J}_V$ whose objects and maps are given by the formulas

$$|\mathcal{J}_V| = |\mathcal{A}_o| ,$$
$$\mathcal{J}_V(n,k) = \mathcal{C}(Vn, Vk) ,$$

and whose composition rule is that of $\mathcal{C}$. Then V admits a factorization $V = \underline{V} \circ \bar{V} : \mathcal{A}_o^* \to \mathcal{J}_V \to \mathcal{C}$, where $\bar{V}$ and $\underline{V}$ are the functors

$$\bar{V} : \mathcal{A}_o^* \to \mathcal{J}_V : \begin{cases} n \longmapsto n , \\ f \longmapsto Vf , \end{cases}$$

$$\underline{V} : \mathcal{J}_V \to \mathcal{C} : \begin{cases} n \longmapsto Vn , \\ g \longmapsto g . \end{cases}$$

Moreover, if $c : \mathcal{C} \to \mathcal{C}'$ is an $(\mathcal{A}_o^*, \underset{\sim}{\text{Cat}})$-morphism from V to V' (i.e., if $c \circ V = V'$), then

$$\mathcal{T}_c : \mathcal{T}_V \to \mathcal{T}_{V'} : \quad \begin{cases} n \longmapsto n , \\ g \longmapsto cg \end{cases}$$

(is the only functor that) makes the diagrams

$$\begin{array}{ccc} & \overset{\bar{V}}{\nearrow} & \mathcal{T}_V \\ \mathcal{A}_o^* & & \downarrow \\ & \underset{\overline{V'}}{\searrow} & \mathcal{T}_{V'} \end{array} \quad , \quad \begin{array}{ccc} \mathcal{T}_V & \xrightarrow{\underline{V}} & \mathcal{C} \\ \downarrow & & \downarrow c \\ \mathcal{T}_{V'} & \xrightarrow[\underline{V'}]{} & \mathcal{C}' \end{array}$$

commute. Thus $V \mapsto \bar{V}$, $c \mapsto \mathcal{T}_c$ is an endofunctor on $(\mathcal{A}_o^*, \underset{\sim}{Cat})$, and the maps $\underline{V} : \bar{V} \to V$ are $(\mathcal{A}_o^*, \underset{\sim}{Cat})$-natural in V . Since clearly $\bar{\bar{V}} = \bar{V}$, we are in the presence of an idempotent cotriple on $(\mathcal{A}_o^*, \underset{\sim}{Cat})$.

We use this cotriple first to define a <u>clone over</u> $\mathcal{A}_o$ as a functor $V \in |(\mathcal{A}_o^*, \underset{\sim}{Cat})|$ for which $\bar{V} = V$ -- the full subcategory of $(\mathcal{A}_o^*, \underset{\sim}{Cat})$ consisting of clones will be denoted $Cl(\mathcal{A}_o)$. Next, the formulas

$$\mathfrak{s}^j(U) = \overline{\mathfrak{s}^{(j)}(U)} : \mathcal{A}_o^* \longrightarrow \mathcal{T}_{\mathfrak{s}^{(j)}(U)} ,$$

$$\mathfrak{s}^j(x) = \mathcal{T}_{\mathfrak{s}^{(j)}(x)} ,$$

$$\mathfrak{m}^j = \mathfrak{m}^{(j)} \Big|_{Cl(\mathcal{A}_o)} ,$$

serve to define functors

$$\mathfrak{s}^j : (\underset{\sim}{Cat}, \mathcal{A}) \to (Cl(\mathcal{A}_o))^* ,$$

$$\mathfrak{m}^j : (Cl(\mathcal{A}_o))^* \to (\underset{\sim}{Cat}, \mathcal{A}) ,$$

called operational structure and operational semantics (rel. j), respectively; they will be said to assign to an $\mathcal{A}$-valued functor U (resp., a clone over $\mathcal{A}_o$) its structure clone (resp., its category of algebras in $\mathcal{A}$) (rel. j) .

Given $V = \bar{V} \in |Cl(\mathcal{A}_o)|$ and $U \in |(\underset{\sim}{Cat}, \mathcal{A})|$, Theorem 3.1 and the idempotence of the full image cotriple immediately yield

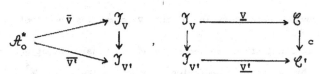

$$(\underset{\sim}{Cat},\mathcal{A})(U,\mathfrak{m}^j(V)) \cong (\mathcal{A}_o^*,\underset{\sim}{Cat})(V,\mathfrak{s}^{(j)}(U)) =$$
$$= (\mathcal{A}_o^*,\underset{\sim}{Cat})(\bar{V},\mathfrak{s}^{(j)}(U) \cong Cl(\mathcal{A}_o)(\bar{V},\overline{\mathfrak{s}^{(j)}(U)}) =$$
$$= Cl(\mathcal{A}_o)(V,\mathfrak{s}^j(U)) \ ,$$

identifications whose obvious naturality in U and V completes the proof of

Theorem 4.1. (operational structure semantics adjointness). Operational structure (rel. j), $\mathfrak{s}^j$, is (right) adjoint to operational semantics (rel. j), $\mathfrak{m}^j$.

§ 5. Remarks on § 4.

The first two remarks establish a generalization of Observation 1.1 to the (rel. j) case. They involve a fixed clone $V : \mathcal{A}_o^* \to \mathcal{C}$ and a fixed functor $j : \mathcal{A}_o \to \mathcal{A}$.

1. A one-one correspondence is set up between V-algebras in $\mathcal{A}$ (rel. j), i.e., objects $(A,\mathtt{A})$ of the pullback $\mathcal{P}_V^j$, and systems $(A,*)$ consisting of

i) an object A of $\mathcal{A}$,

ii) pairings $(\theta,a) \mapsto \theta * a : \mathcal{C}(n,k) \times A^{j(n)} \to A^{j(k)}$

satisfying the identities

(5.1) $\qquad (\theta'\circ\theta) * a = \theta' * (\theta * a) \qquad (\theta,\theta' \ \mathcal{C}\text{-morphisms})$,

(5.2) $\qquad V(f) * a = a \circ j(f) \qquad (f \text{ an } \mathcal{A}_o\text{-morphism})$,

by the equations

(5.3) $\qquad \{\mathtt{A}_{n,k}(\theta)\}(a) = \theta * a$,

(5.4) $\qquad \mathtt{A}(n) = \mathcal{A}(jn,A) = A^{j(n)}$.

Proof. If $(A,\mathtt{A}) \in |\mathcal{P}_V^j|$, formula (5.3) gives rise to a system ii) of pairings. That identities (5.1) and (5.2) are valid for the resulting $(A,*)$ is a consequence of the functoriality of $\mathtt{A}$ and the relation

(5.5) $\qquad \mathtt{A}\circ V = \mathcal{A}(j(-),A)$.

Converversely, if $(A,*)$ is a system i), ii) satisfying (5.1) and (5.2), the attempt to define a functor $\mathfrak{A}$ satisfying (5.5) by means of (5.3) and (5.4) is successful precisely because of ii), (5.1), and (5.2), while (5.5) then guarantees $(A,\mathfrak{A})$ is in $\mathcal{P}^j_V$. The biunivocity of these correspondences is clear.

2. With the functor $j : \mathcal{A}_o \to \mathcal{A}$ and the clone $V : \mathcal{A}^*_o \to \mathcal{C}$ still fixed, let $(A,\mathfrak{A})$ and $(B,\mathfrak{B})$ be objects of $\mathcal{P}^j_V$. Then, given $g \in \mathcal{A}(A,B)$, there is never more than one natural transformation $\varphi : \mathfrak{A} \to \mathfrak{B}$ with

(5.6) $$(g,\varphi) \in \mathcal{P}^j_V((A,\mathfrak{A}) \ , (B,\mathfrak{B})) \ ,$$

and there is one if and only if, in the notation of (5.3),

(5.7) $$g \circ (\theta * a) = \theta * (g\ a)$$

for all $a \in A^{jn}$, all $\theta \in \mathcal{C}(n,k)$, and all $n,k \in |\mathcal{A}_o|$. Conversely, given the natural transformation $\varphi : \mathfrak{A} \to \mathfrak{B}$, there is a $g \in \mathcal{A}(A,B)$ satisfying (5.6) if the composition

(5.8) $$\mathcal{A} \xrightarrow[Y]{} (\mathcal{A}^*,\mathfrak{S}) \xrightarrow[(j^*,\mathfrak{S})]{} (\mathcal{A}^*_o,\mathfrak{S})$$

is full, and there is at most one such g if (5.8) is faithful. Hence if j is <u>dense</u> (this means (5.8) is full and faithful -- cf. [ ] or [ ] -- Isbell [ ] uses the term <u>adequate</u>), the functor

(5.9) $$\mathcal{P}^j_V \longrightarrow (\mathcal{C},\mathfrak{S}) : \begin{cases} (A,\mathfrak{A}) \longmapsto \mathfrak{A} \ , \\ (g,\varphi) \longmapsto \varphi \end{cases}$$

arising in the pullback diagram (3.4) is full and faithful (indeed, the density of j is a necessary and sufficient condition for (5.9) to be full and faithful for <u>every</u> clone V on $\mathcal{A}_o$).

<u>Proof</u>. Given $g : A \to B$, the requirement that (5.6) hold forces the components of φ to be

$$\varphi_n = \mathcal{A}(j(n),g) : A^{j(n)} \to B^{j(n)} \ ,$$

and that takes care of uniqueness. That this system $\{\varphi_n\}_{n \in |\mathcal{A}_o|}$ is a natural transformation $\mathfrak{A} \to \mathfrak{B}$ iff g satisfies the identities (5.7) is elementary definition juggling. The converse assertions are evident; the next assertion follows from them; and the statement in parentheses is seen to be true by taking $V = \mathrm{id} : \mathcal{A}_0^* \to \mathcal{A}_0^*$ when j is not dense.

The next remark points out some dense functors $j : \mathcal{A}_0 \to \mathcal{A}$.

3. For any category $\mathcal{A}$, $\mathrm{id}_{\mathcal{A}} : \mathcal{A} \to \mathcal{A}$ is dense (this is just part of the Yoneda Lemma). Moreover, if I is any set and $\mathcal{S}_{\aleph}$ is the full subcategory of the category $\mathcal{S}$ of sets and functions consisting of the cardinals (or sets of cardinality) $< \aleph$ $(\aleph > 2)$, then the inclusion $\mathcal{S}_{\aleph} \to \mathcal{S}$ and the induced inclusion $(\mathcal{S}_{\aleph})^I \to \mathcal{S}^I$ are both dense.

The following remarks interpret the results of § 4 in the settings indicated in Remark 3, using Remarks 1 and 2 where necessary.

4. When $j : \mathcal{A}_0 \to \mathcal{A}$ is the inclusion in $\mathcal{S}$ of the full subcategory $\mathcal{S}_{\aleph_0}$ of finite cardinals, Theorem 4.1 is Lawvere's structure-semantics adjointness theorem [16].

5. If I is a set and $j : \mathcal{A}_0 \to \mathcal{A}$ is the full inclusion $(\mathcal{S}_{\aleph_0})^I \to \mathcal{S}^I$, Theorem 4.1 is Bénabou's structure-semantics adjointness theorem [5].

6. When $j = \mathrm{id}_{\mathcal{S}}$, Theorem 4.1 is the structure semantics adjointness theorem of [19, § 2].

7. When $j = \mathrm{id}_{\mathcal{A}}$, then, for any $U : \mathcal{X} \to \mathcal{A}$, $\mathfrak{s}^j(U) = \exp_U : \mathcal{A}^* \to \mathfrak{T}_U$, $\mathcal{P}^j_{\mathfrak{s}^j(U)} = U\text{-}\underline{\mathrm{Alg}}$, $\mathfrak{m}^j\mathfrak{s}^j(U) = |\ |_U$, Observation 1.1 is the content of Remarks 1 and 2, and $\Phi_U : \mathcal{X} \to U\text{-}\underline{\mathrm{Alg}}$ is just the functor corresponding, under the adjointness of Theorem 4.1, to

$$\mathrm{id}_{\mathfrak{s}^j(U)} \in \mathrm{Cl}(\mathcal{A})(\mathfrak{s}^j(U), \mathfrak{s}^j(U)) \cong (\underline{\mathrm{Cat}}, \mathcal{A})(U, \mathfrak{m}^j\mathfrak{s}^j(U)),$$

i.e., is the front adjunction for the operational structure-semantics adjointness.

8. When j is the inclusion $\mathcal{S}_{\aleph} \to \mathcal{S}$, Theorem 4.1 is the adjointness implicit in the first paragraphs of [19, § 6].

§ 6. Two constructions in algebras over a clone.

Proposition 6.1. ($\mathfrak{M}^j(V)$ creates (inverse) limits). Let $V : \mathcal{A}_o^* \to \mathcal{C}$ be a clone over $\mathcal{A}_o$, and let $j : \mathcal{A}_o \to \mathcal{A}$ be a functor. Given a functor $X : \Delta \to \mathcal{P}_V^j$, whose values at objects and morphisms of Δ are written $X_\delta = (A_\delta, \mathfrak{A}_\delta)$ and $X(i) = (g_i, \varphi_i)$, respectively, and given an object $A \in |\mathcal{A}|$ and $\mathcal{A}$-morphisms $p_\delta : A \to A_\delta$ $(\delta \in |\Delta|)$ making

$$A = \varprojlim \mathfrak{M}^j(V) \circ X : \Delta \to \mathcal{P}_V^j \to \mathcal{A},$$

there are an object Q of $\mathcal{P}_V^j$ and maps $q_\delta : Q \to X_\delta$ $(\delta \in |\Delta|)$, uniquely determined by the requirements

$$\mathfrak{M}^j(V)(q_\delta) = p_\delta \;; \tag{6.1}$$

moreover, via the projections q_δ, $Q = \varprojlim X$.

Proof. If $Q = (B, \mathfrak{B})$ and $q_\delta = (g_\delta, \varphi_\delta)$ satisfy (6.1), we must have $B = A$ and $g_\delta = p_\delta$. Remark 5.2 then identifies φ_δ. Thus it need only be seen that there is precisely one functor $\mathfrak{B} : \mathcal{C} \to \mathcal{S}$ such that, in the notation of (5.3),

$$p_\delta \circ (\theta * a) = \theta * (p_\delta \circ a) \qquad (\theta \in \mathcal{C}(n,k) \;,\; a \in A^{j(n)}) \;. \tag{6.2}$$

But that's an obvious consequence of the limit property of A. That $(A, \mathfrak{B})$ is then an object of $\mathcal{P}_V^j$ is, again using the limit property of A and Remark 5.1, an automatic verification, and (6.2), using Remark 5.2, bespeaks the fact that p_δ "is" a $\mathcal{P}_V^j$-morphism from $(A, \mathfrak{B})$ to $(A_\delta, \mathfrak{A}_\delta)$. Finally, given a compatible system of $\mathcal{P}_V^j$-morphisms $(K, \mathfrak{K}) \to (A_\delta, \mathfrak{A}_\delta)$, the $\mathcal{A}$-morphism components determine a unique $\mathcal{A}$-morphism $K \to A$, which, using (6.2) and Remark 5.2, it is not hard to see "is" a $\mathcal{P}_V^j$-morphism $(K, \mathfrak{K}) \to (A, \mathfrak{B})$. This completes the proof.

Proposition 6.2. ($\mathfrak{M}^j(V)$ creates $\mathfrak{M}^j(V)$-split coequalizers). Let $V : \mathcal{A}_o^* \to \mathcal{C}$ be a clone over $\mathcal{A}_o$, let $j : \mathcal{A}_o \to \mathcal{A}$ be a functor, let $(A, \mathfrak{A})$ and $(B, \mathfrak{B})$ be two V-algebras in $\mathcal{A}$, let $K \in |\mathcal{A}|$, and let

$$(A,\mathfrak{A}) \underset{(g,\varphi)}{\overset{(f,\psi)}{\rightrightarrows}} (B,\mathfrak{B})$$

and

$$A \xleftarrow{\;d_1\;} B \underset{d_o}{\overset{p}{\rightleftarrows}} K$$

be two $\mathcal{P}^j_V$*-morphisms and three* $\mathcal{A}$*-morphisms satisfying*

$$(6.3)\qquad \begin{cases} pf = pg \;, \\ pd_o = id_K \;, \\ d_op = gd_1 \;, \\ id_B = fd_1 \;. \end{cases}$$

Then there is a $\mathcal{P}^j_V$*-morphism* $(q,Q) : (B,\mathfrak{B}) \to (C,\mathfrak{K})$ *uniquely determined by the requirement that* $\mathfrak{M}^j(V)(q,Q) = p$; *moreover,* (q,Q) *is then a coequalizer of the pair* $((f,\psi),\ (g,\varphi))$.

Proof. Clearly $C = K$ and $q = p$, so Q will be forced; we must see there is a unique functor $\mathfrak{K} : \mathcal{C} \to \mathcal{S}$ making $(K,\mathfrak{K})$ a V-algebra and p a $\mathcal{P}^j_V$-morphism. Now, since $p : B \to K$ is a split epimorphism, each function $p^{j(n)} : B^{j(n)} \to K^{j(n)}$ is onto. This fact ensures the uniqueness of any functions $\mathfrak{K}(\theta)$ $(\theta \in \mathcal{C}(n,k))$ making the diagrams

$$\begin{array}{ccc} B^{j(n)} & \xrightarrow{\quad p^{j(n)}\quad} & C^{j(n)} \\ {\scriptstyle \mathfrak{B}(\theta)}\downarrow & & \downarrow{\scriptstyle \mathfrak{K}(\theta)} \\ B^{j(k)} & \xrightarrow[\quad p^{j(k)}\quad]{} & C^{j(k)} \end{array}$$

commute. Their existence is ensured, using the section d_o , by the formula

$$\mathfrak{K}(\theta) = p^{j(k)} \circ \mathfrak{B}(\theta) \circ d_o^{j(n)} \qquad (\text{i.e., } \theta * a = p\circ(\theta * (d_o \circ \dot{a}))\ )\ ,$$

as the following calculations relying on (6.3) show:

$$\begin{aligned}
\mathfrak{K}(\theta)\circ p^{j(n)} &= p^{j(k)}\circ\mathfrak{B}(\theta)\circ d_o{}^{j(n)}\circ p^{j(n)} = \\
&= p^{j(k)}\circ\mathfrak{B}(\theta)\circ g^{j(n)}\circ d_1{}^{j(n)} = p^{j(k)}\circ g^{j(k)}\circ\mathfrak{A}(\theta)\circ d_1{}^{j(n)} = \\
&= p^{j(k)}\circ f^{j(k)}\circ\mathfrak{A}(\theta)\circ d_1{}^{j(n)} = p^{j(k)}\circ\mathfrak{B}(\theta)\circ f^{j(n)}\circ d_1{}^{j(n)} = \\
&= p^{j(k)}\circ\mathfrak{B}(\theta) .
\end{aligned}$$

That the resulting $(K,\mathfrak{K})$ is in $\mathcal{P}^j_V$ easy to see, using only the fact that each $p^{j(n)}$ is surjective. Finally, to see that $p : (B,\mathfrak{B}) \to (K,\mathfrak{K})$ is the coequalizer of (f,ψ) and (g,φ), note that the $\mathcal{A}$-morphism component of any $\mathcal{P}^j_V$-morphism from $(B,\mathfrak{B})$ having equal compositions with f and g factors uniquely through K (via its composition with d_o) ; but this factorization is a $\mathcal{P}^j_V$-morphism from $(K,\mathfrak{K})$ (in the sense of Remark 5.2) by virtue simply of the surjectivity of each $p^{j(n)}$. This completes the description of the proof.

§ 7. Constructions involving triples.

We recall [9] that a triple $\mathbf{T}$ on a category $\mathcal{A}$ is a system $\mathbf{T} = (T,\eta,\mu)$ consisting of a functor

$$T : \mathcal{A} \to \mathcal{A}$$

and natural transformations

$$\eta : id_{\mathcal{A}} \to T , \qquad \mu : TT \to T$$

satisfying the relations

(7.1) $$\mu \circ T\eta = id_T ,$$

(7.2) $$\mu \circ \eta_T = id_T ,$$

(7.3) $$\mu \circ \mu_T = \mu \circ T\mu .$$

It is often possible to associate a triple on $\mathcal{A}$ to an $\mathcal{A}$-valued functor $U : \mathcal{X} \to \mathcal{A}$. For example, whenever U has a left adjoint $F : \mathcal{A} \to \mathcal{X}$ with front and back adjunctions $\eta : id_{\mathcal{A}} \to UF$, $\beta : FU \to id_{\mathcal{X}}$, it is well known (cf. [9, Prop. 2.1] or [11, Th. 4.2*]) that

(7.4) $$(UF, \eta, U\beta F)$$

is a triple on $\mathcal{A}$. More general situations in which a triple can be associated to U are discussed in § 8. In any event, it will turn out (in § 9) that, when $\mathbf{T}$ is a triple on $\mathcal{A}$ suitably associated to an $\mathcal{A}$-valued functor $U : \mathcal{X} \to \mathcal{A}$, the category U-Alg of § 1 is canonically isomorphic with the category $\mathcal{A}^{\mathbf{T}}$ (constructed in [9, Th. 2.2], for example) of $\mathbf{T}$-algebras and $\mathbf{T}$-homomorphisms. For the reader's convenience, the definition of $\mathcal{A}^{\mathbf{T}}$ will be reviewed. Since Kleisli's construction [18] of (what we shall call) the Kleisli category associated to a triple is needed in § 8, enters into one proof of the isomorphism theorem of § 9, and is relatively unfamiliar, we shall review it, here, too. Thereafter, we pave the way for § 10 with some trivial observations.

Given the triple $\mathbf{T} = (T, \eta, \mu)$ on the category $\mathcal{A}$, a $\mathbf{T}$-algebra in $\mathcal{A}$ is a pair (A, α), where

(7.5) $$\alpha : TA \to A$$

is an $\mathcal{A}$-morphism satisfying the relations

(7.6) $$\alpha \circ \eta_A = id_A ,$$

(7.7) $$\alpha \circ \mu_A = \alpha \circ T\alpha .$$

For example, equations (7.2) and (7.3) bespeak the fact that

$$F^{\mathbf{T}}(A) = (TA, \mu_A)$$

is a $\mathbf{T}$-algebra, whatever $A \in |\mathcal{A}|$.

The category $\mathcal{A}^{\mathbf{T}}$ of $\mathbf{T}$-algebras has as objects all $\mathbf{T}$-algebras in $\mathcal{A}$ and as morphisms from (A, α) to (B, β) all $\mathcal{A}$-morphisms $g : A \to B$ satisfying

(7.8) $$g \circ \alpha = \beta \circ Tg ;$$

the composition rule is that induced by composition of $\mathcal{A}$-morphisms. It follows that the passages

$$(A,\alpha) \longmapsto A\ ,$$
$$g \longmapsto g$$

define a functor $U^{\mathbf{T}} : \mathcal{A}^{\mathbf{T}} \to \mathcal{A}$, the underlying $\mathcal{A}$-object functor for $\mathbf{T}$-algebras.

On the other hand, it is easy to see that $Tf : TA \to TB$ is an $\mathcal{A}^{\mathbf{T}}$-morphism from $F^{\mathbf{T}}(A)$ to $F^{\mathbf{T}}(B)$ $(f \in \mathcal{A}(A,B))$, and it readily follows that the passages

$$A \longmapsto F^{\mathbf{T}}(A) = (TA,\mu_A)\ ,$$
$$f \longmapsto Tf$$

define a functor $F^{\mathbf{T}} : \mathcal{A} \to \mathcal{A}^{\mathbf{T}}$. Finally, it can be shown that $F^{\mathbf{T}}$ is left adjoint to $U^{\mathbf{T}}$ with front adjunction $id_{\mathcal{A}} \to U^{\mathbf{T}}F^{\mathbf{T}} = T$ given by η and back adjunction $\beta : F^{\mathbf{T}}U^{\mathbf{T}} \to id_{\mathcal{A}^{\mathbf{T}}}$ given by

$$\beta_{(A,\alpha)} = \alpha : (TA,\mu_A) = F^{\mathbf{T}}U^{\mathbf{T}}(A,\alpha) \to (A,\alpha)\ .$$

Consequently, the triple (7.4) arising from this adjunction is precisely the original triple $\mathbf{T} = (T,\eta,\mu)$ itself.

The Kleisli category associated to the triple $\mathbf{T} = (T,\eta,\mu)$ is the category $\mathcal{K}^{\mathbf{T}}$ whose objects and maps are given by

$$|\mathcal{K}^{\mathbf{T}}| = |\mathcal{A}|\ ,$$
$$\mathcal{K}^{\mathbf{T}}(k,n) = \mathcal{A}(k,Tn)\ ;$$

the composition rule sends the pair

$$(s,t) \in \mathcal{K}^{\mathbf{T}}(m,k) \times \mathcal{K}^{\mathbf{T}}(k,n) = \mathcal{A}(m,Tk) \times \mathcal{A}(k,Tn)$$

to the element

$$t \circ s = \mu_n \circ Tt \circ s \in \mathcal{A}(m,Tn) = \mathcal{K}^{\mathbf{T}}(m,n)\ ,$$

the composition symbols on the right denoting composition in $\mathcal{A}$.

Functors $f^{\mathbf{T}} : \mathcal{A} \to \mathcal{K}^{\mathbf{T}}$, $u^{\mathbf{T}} : \mathcal{K}^{\mathbf{T}} \to \mathcal{A}$ are defined by

$$f^{\mathbf{T}}(n) = n \ , \quad f^{\mathbf{T}}(f) = \eta_n \circ f \qquad (n \in |\mathcal{A}| \ , \quad f \in \mathcal{A}(k,n)) \ ,$$

$$u^{\mathbf{T}}(n) = Tn, \quad u^{\mathbf{T}}(t) = \mu_n \circ Tt \qquad (n \in |\mathcal{A}| \ , \quad t \in \mathcal{K}^{\mathbf{T}}(k,n)) \ ,$$

where, again, the composition symbols on the right denote the composition in $\mathcal{A}$. One observes that the equalities

$$\mathcal{K}^{\mathbf{T}}(f^{\mathbf{T}}k,n) = \mathcal{K}^{\mathbf{T}}(k,n) = \mathcal{A}(k,Tn) = \mathcal{A}(k,u^{\mathbf{T}}n)$$

are $\mathcal{A}$-natural in k and $\mathcal{K}^{\mathbf{T}}$-natural in n , hence bespeak the adjointness of $u^{\mathbf{T}}$ to $f^{\mathbf{T}}$. Moreover, the triple (7.4) arising from this adjointness turns out, once again, to be just $\mathbf{T}$.

Linking $\mathcal{A}^{\mathbf{T}}$ with $\mathcal{K}^{\mathbf{T}}$ is the observation that the passages

$$n \longmapsto (Tn,\mu_n) \qquad (n \in |\mathcal{A}|) \ ,$$

$$t \longmapsto \mu_n \circ Tt \qquad (t \in \mathcal{K}^{\mathbf{T}}(k,n) = \mathcal{A}(k,Tn))$$

set up a full and faithful functor $\mathcal{K}^{\mathbf{T}} \to \mathcal{A}^{\mathbf{T}}$ making the diagram

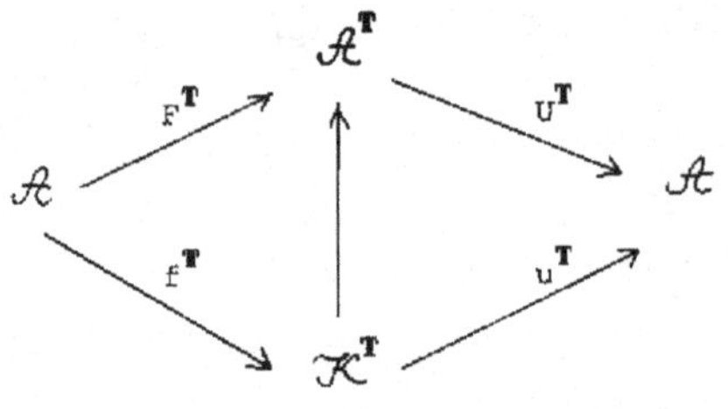

commute. This observation is based on the identifications

$$\mathcal{K}^{\mathbf{T}}(k,n) = \mathcal{A}(k,Tn) \cong \mathcal{A}^{\mathbf{T}}(F^{\mathbf{T}}k,F^{\mathbf{T}}n) \ ,$$

and results in an isomorphism (in $(\mathcal{A},\underset{\sim}{\text{Cat}})$) between $f^{\mathbf{T}}$ and the full image of $F^{\mathbf{T}}$.

§ 8. Codensity triples.

Given a functor $U : \mathfrak{X} \to \mathcal{A}$, there may be a triple $\mathbb{T}$ on $\mathcal{A}$ whose Kleisli category $\mathcal{K}^{\mathbb{T}}$ is isomorphic to $(\mathfrak{T}_U)^*$ in such a way -- say by an isomorphism $y : \mathfrak{T}_U \to (\mathcal{K}^{\mathbb{T}})^*$ -- that the triangle

$$\begin{array}{ccc} & & (\mathcal{K}^{\mathbb{T}})^* \\ & \nearrow^{(f^{\mathbb{T}})^*} & \uparrow y \;\cong \\ \mathcal{A}^* & & \\ & \searrow_{\exp_U} & \\ & & \mathfrak{T}_U \end{array}$$

commutes. In that event, the diagram

$$(8.0)\qquad \begin{array}{ccccc} U\text{-}\underset{\sim}{Alg} & \longrightarrow & (\mathfrak{T}_U, \mathfrak{S}) & \xrightarrow{(y^{-1},\mathfrak{S})} & ((\mathcal{K}^{\mathbb{T}})^*, \mathfrak{S}) \\ \Big\downarrow{\scriptstyle |\;|_U} & & & \searrow^{(\exp_U,\mathfrak{S})} & \Big\downarrow{\scriptstyle ((f^{\mathbb{T}})^*,\mathfrak{S})} \\ \mathcal{A} & & \xrightarrow[\quad Y \quad]{} & & (\mathcal{A}^*, \mathfrak{S}) \end{array}$$

commutes, and its vertices form a pullback diagram

$$(8.1)\qquad \begin{array}{ccc} U\text{-}\underset{\sim}{Alg} & \longrightarrow & ((\mathcal{K}^{\mathbb{T}})^*, \mathfrak{S}) \\ \Big\downarrow{\scriptstyle |\;|_U} & & \Big\downarrow{\scriptstyle ((f^{\mathbb{T}})^*,\mathfrak{S})} \\ \mathcal{A} & \xrightarrow[Y]{} & (\mathcal{A}^*, \mathfrak{S}) \end{array} \quad .$$

Since this pullback representation of U-$\underset{\sim}{Alg}$ is more convenient, for the purposes of § 9, than that (established in § 5) of Observation 1.1, the present section is devoted to the establishment of necessary and sufficient conditions for, and the interpretation of, the availability, given U, of such a triple and such an isomorphism.

In the ensuing discussion, we therefore fix an $\mathcal{A}$-valued functor $U : \mathfrak{X} \to \mathcal{A}$. We will need the comma categories $(n,U) = (\{pt.\},U^n)$, constructed (see [15] for related generalities) as follows for each $n \in |\mathcal{A}|$. The objects of (n,U) are all pairs (f,X) with $X \in |\mathfrak{X}|$ and $f \in U^n X = \mathcal{A}(n,UX)$. As morphisms from (f,X) to

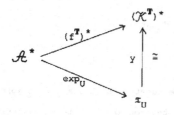

(f',X') are admitted all $\mathfrak{X}$-morphisms $\xi : X \to X'$ satisfying $U(\xi) \circ f = f'$. They are composed using the composition rule in $\mathfrak{X}$, so that the passages

$$(f,X) \longmapsto X ,$$
$$\xi \longmapsto \xi$$

constitute a functor from (n,U) to $\mathfrak{X}$, to be denoted

$$C_n : (n,U) \to \mathfrak{X}.$$

Now assume, <u>for this paragraph only</u>, that the functor U has a left adjoint. Then it is known (cf.,e.g.,[4] for details, including the converse) that U must preserve inverse limits and that the values of the left adjoint F serve as inverse limits for the functors C_n. (Indeed, the $(f,X)^{th}$ projection from Fn (to X) can be chosen to be the $\mathfrak{X}$-morphism $Fn \to X$ corresponding by the adjointness to $f : n \to UX$.) It follows that UFn serves as inverse limit of the composite

(8.2) $$(n,U) \xrightarrow[C_n]{} \mathfrak{X} \xrightarrow[U]{} \mathcal{A} .$$

<u>Definition</u> (cf. [1], [14], [26]). U admits a codensity triple if $\varprojlim UC_n$ exists for each $n \in |\mathcal{A}|$. A functor $T : |\mathcal{A}| \to |\mathcal{A}|$ will be said to be a codensity triple for U if each Tn $(n \in |\mathcal{A}|)$ is accompanied with a system of maps

(8.3) $$\{\langle f,X\rangle_n : Tn \to UX \mid (f,X) \in |(n,U)|\}$$

by virtue of which $Tn = \varprojlim UC_n$.

The reader who is disturbed by the fact that a codensity triple for U seems not to be a triple may use the maps $\langle f,X\rangle_n$ (which we shall often abbreviate to $\langle f\rangle_n$ or even $\langle f\rangle$) to define $\mathcal{A}$-morphisms $Tg : Tk \to Tn$ $(g \in \mathcal{A}(k,n))$, $\eta_n : n \to Tn$ $(n \in |\mathcal{A}|)$, and $\mu_n : TTn \to Tn$ $(n \in |\mathcal{A}|)$ by requiring their compositions with the projections $\langle f\rangle = \langle f\rangle_n$ to be

(8.4) $$\langle f\rangle \circ Tg = \langle f \circ g\rangle ,$$

(8.5) $\langle f\rangle \circ \eta_n = f$, and

(8.6) $\langle f\rangle \circ \mu_n = \langle\langle f\rangle\rangle \quad (= \langle\langle f\rangle_n\rangle_{Tn})$,

respectively. He may then verify that T becomes a functor, that η and μ are natural transformations, and that (T,η,μ) is thus a triple (the same triple as (7.4) if U has a left adjoint F and T is obtained by the prescription in the discussion preceding (8.2)). Finally, he can prove that $\mathfrak{X}_U$ <u>is isomorphic with the dual</u> $(\mathcal{K}^{\mathbf{T}})^*$ <u>of the Kleisli category</u> $\mathcal{K}^{\mathbf{T}}$ <u>of</u> $\mathbf{T}$ <u>in the manner described at the head of this section</u>. Since we are after somewhat more information, including a converse to the italicised statement above, we prefer what may seem a more roundabout approach.

A functor $U : \mathcal{X} \to \mathcal{A}$ and a functor $T : |\mathcal{A}| \to |\mathcal{A}|$ may be related in five apparently different ways, if certain additional information is specified; that T be a codensity triple for U is one of these ways. The five kinds of information we have in mind are:

I. maps $\langle f\rangle : Tn \to UX$ (one for each $f : n \to UX$ and $X \in |\mathcal{X}|$) , making T a codensity triple for U ;

II. functions $y \;(= y_{n,k}) : \mathfrak{X}_U(n,k) \to \mathcal{A}(k,Tn)$ making Tn represent the functor $\mathfrak{X}_U(n,\exp_U(-))$;

III. a left adjoint to $\exp_U : \mathcal{A}^* \to \mathfrak{X}_U$, with specified front and back adjunctions, such that T is the object function of the composition $\mathcal{A}^* \to \mathfrak{X}_U \to \mathcal{A}^*$;

IV. a triple $\mathbf{T}$ whose functor component has object function T , and an isomorphism $y : \mathfrak{X}_U \to (\mathcal{K}^{\mathbf{T}})^*$ satisfying $y \circ \exp_U = (f^{\mathbf{T}})^*$;

V. a triple $\mathbf{T}$ whose functor component has object function T and functions $y = y_{n,k} : \mathfrak{X}_U(n,k) \to \mathcal{A}(k,Tn)$ fulfilling the four conditions:

o) each $y_{n,k}$ is a one-one correspondence,

i) $y_{n,m}(\theta'\circ\theta) = \mu_n \circ T(y_{n,k}(\theta))\circ y_{k,m}(\theta')$,

ii) $U^{\eta_n} \circ y^{-1}_{n,Tn}(id_{Tn}) = U^{id_n}$,

iii) $y_{n,k}(U^f) = \eta_n \circ f$.

The theorem coming up asserts that if U and T are related in any one of these five ways, they are related in all of them. § 11 exploits the computational accessibility of the fifth way; the other ways are more satisfactory from a conceptual point of view.

Theorem 8.1. There are canonical one-one correspondences, given $T : |\mathcal{A}| \to |\mathcal{A}|$ *and* $U : \mathfrak{X} \to \mathcal{A}$, *among the five specified classes of information relating* U *and* T . *In particular, each codensity triple for* U *"is", in one and only one way, a triple* $\mathbf{T}$ *the dual* $(\mathcal{K}^{\mathbf{T}})^*$ *of whose Kleisli category is isomorphic to* $\mathfrak{T}_U$ *in a manner compatible with the injections* $(f^{\mathbf{T}})^*$ *and* exp_U *of* $\mathcal{A}^*$. *Moreover, this triple structure on* T *is the one described above in the formulas* (8.4), (8.5) *and* (8.6).

Proof. What is completely obvious is the one-one correspondence between information of type II and of type III : all that is being used is the fact (cf. [21, Prop. 8.3]) that left adjoints are defined pointwise. To go from information of type IV to that of type III, observe simply that $(y^*)^{-1} \circ f^{\mathbf{T}}$ serves as left adjoint to exp_U in the desired way; that this sets up a one-one correspondence between these kinds of information is due to the universal property (described in [2] and [22, Th. 1])of the Kleisli category. The major portion of the proof therefore consists in showing that informations of types I and II (resp., of types IV and V) are in one-one correspondence with each other.

For types I and II, we have to recourse to the

Lemma 8.1. Let $U : \mathfrak{X} \to \mathcal{A}$ *be a functor, let* n *and* k *be objects of* $\mathcal{A}$, *and let* $\bar{k} : (n,U) \to \mathcal{A}$ *be the constant functor with value* k . *Then there are canonical one-one correspondences,* $\mathcal{A}$*-natural in* k , *among* $n.t.(\bar{k}, UC_n)$, $n.t.(U^n, U^k)$, *and the class of all functors* $\theta : (n,U) \to (k,U)$ *satisfying* $C_k \circ \theta = C_n$. *Indeed, the information needed to specify a member of any of these classes is the same.*

Proof. An element θ of any of these classes involves a function assigning to each object $X \in |\mathfrak{X}|$ and each map $f : n \to UX$ a new map $\theta(f,X) = \theta_X(f) : k \to UX$, subject to side conditions. In the first instance, the side conditions are

$$U\xi \circ \theta(f,X) = \theta(U\xi \circ f,X) \qquad (f : n \to UX\ ,\quad \xi : X \to X')\ .$$

In the second instance, the side conditions are the commutativity of all the squares

$$\begin{array}{ccc} U^nX & \xrightarrow{\ \theta_X\ } & U^kX \\ {\scriptstyle U^n\xi}\downarrow & & \downarrow{\scriptstyle U^k\xi} \\ U^nX' & \xrightarrow[\ \theta_{X'}\ ]{} & U^kX' \end{array} \qquad\qquad (\xi : X \to X')\ .$$

In the third instance, the side conditions, in view of the requirement $C_k \circ \theta = C_n$ and the faithfulness of C_k and C_n , are the same as in the first instance. It now takes but a moment's reflection to see that the side conditions in the first two instances are also the same. The naturality in k will be left to the reader.

Continuing with the proof of Theorem 8.1, information of type I results in one-one correspondences $\mathcal{A}(k,Tn) \xrightarrow{\ \cong\ } \text{n.t.}(\bar{k},UC_n)$, natural in k , obtained by composing with the $\langle f \rangle$'s . Information of type II results in one-one correspondences, natural in k , $\text{n.t.}(U^n,U^k) \cong \mathcal{A}(k,Tn)$. The free passage, natural in k , allowed by Lemma 8.1, between $\text{n.t.}(\bar{k},UC_n)$ and $\text{n.t.}(U^n,U^k)$ thus takes ample care of the I $\leftrightarrow$ II relation.

[For later use, we remark that the resulting functions

$$y_{n,k}^{-1} : \mathcal{A}(k,Tn) \longrightarrow \text{n.t.}(U^n,U^k)$$

send $t : k \to Tn$ to the natural transformation $y_{n,k}^{-1}(t)$ given by

(8.7) $$\{y_{n,k}^{-1}(t)\}_X(a) = \langle id_{UX} \rangle \circ Ta \circ t \qquad (X \in |\mathfrak{X}|\ ,\quad a \in U^nX)\ .\ ]$$

At this point, the reader can easily verify for himself that, in the passage (via II and III) from I to IV, the codensity triple

$$T, \{\{\langle f \rangle \mid f \in |(n,U)|\} \mid n \in |\mathcal{A}|\}$$

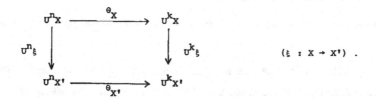

inherits the triple structure described in (8.4), (8.5) and (8.6); he need only use the fact that the triple $\mathbb{T}$ appearing in IV is the interpretation in $\mathcal{A}$ of the co-triple in $\mathcal{A}^*$ arising (by [9, Prop. 2.1*] or [11, Th. 4.2]) from the adjoint pair $\exp_U : \mathcal{A}^* \to \mathfrak{X}_U$, $\mathfrak{X}_U \to \mathcal{A}^*$ resulting in III from the reinterpretation of the co-density triple as information of type II.

The relation between information of types IV and V is taken care of by another lemma.

Lemma 8.2. *Let* $U : \mathfrak{X} \to \mathcal{A}$ *be a functor, and let* $\mathbb{T} = (T,\eta,\mu)$ *be a triple on* $\mathcal{A}$. *A one-one correspondence between the class of all systems of functions*

$$y_{n,k} : \text{n.t.}(U^n,U^k) \to \mathcal{A}(k,Tn) \qquad (n,k \in |\mathcal{A}|)$$

fulfilling the four requirements in V *above, and the class of all functors* $y : \mathfrak{X}_U \to (\mathcal{K}^{\mathbb{T}})^*$ *satisfying the two conditions*

iv) $y \circ \exp_U = (f^{\mathbb{T}})^*$,

v) y *is an isomorphism of categories,*

is induced by the passage from the functor y *to the system* $\{y_{n,k}\}$ *in which* $y_{n,k}$ *is the effect of the functor* y *on the* $\mathfrak{X}_U$*-morphisms from* n *to* k.

Proof. Given an isomorphism $y : \mathfrak{X}_U \to (\mathcal{K}^{\mathbb{T}})^*$ satisfying iv), and given n and k in $|\mathcal{A}|$, define $y_{n,k} : \text{n.t.}(U^n,U^k) \to \mathcal{A}(k,Tn)$ to be the composition of the sequence

$$\text{n.t.}(U^n,U^k) = \mathfrak{X}_U(n,k) \xrightarrow[y]{\cong} (\mathcal{K}^{\mathbb{T}})^*(n,k) = \mathcal{K}^{\mathbb{T}}(k,n) = \mathcal{A}(k,Tn) .$$

Condition V.o) follows from condition v). Condition V.i) follows immediately from the functoriality of y, once composition in $\mathfrak{X}_U$ and in $\mathcal{K}^{\mathbb{T}}$ are recalled. Condition V.iii) follows from iv). To establish V.ii), apply the isomorphism y to both sides. The right side is $y(U^{id_n}) = y(\exp_U(id_n)) = (f^{\mathbb{T}})^*(id_n)$, which, viewed as an $\mathcal{A}$-morphism, is just η_n. In view of the validity of V.i), V.iii), and a triple identity, the left side is

$$y(U^{\eta_n} \circ y^{-1}(id_{Tn})) = \mu_n \circ Tyy^{-1}id_{Tn} \circ y(U^{\eta_n}) =$$

$$= \mu_n \circ Tid_{Tn} \circ \eta_{Tn} \circ \eta_n = \mu_n \circ \eta_{Tn} \circ \eta_n = \eta_n \, .$$

Since this is what the right side of V.ii) is, after applying y, half the lemma is proved.

For the converse, take a system of functions as envisioned in the lemma, and attempt to define a functor $y : \mathfrak{T}_U \to (\mathcal{K}^{\mathbf{T}})^*$ by setting $y(n) = n$ and, for $\theta : U^n \to U^k$,

$$y(\theta) = y_{n,k}(\theta) \in \mathcal{A}(k,Tn) = \mathcal{K}^{\mathbf{T}}(k,n) = (\mathcal{K}^{\mathbf{T}})^*(n,k) \, .$$

This attempt is successful because V.i) and the definition of composition in $\mathcal{K}^{\mathbf{T}}$ show that y preserves composition, while V.iii) (with $f = id_n$) and the functoriality of $f^{\mathbf{T}}$ show that y preserves identity maps. Finally, V.iii) yields iv), using nothing but the definitions of $\exp_U$ and $f^{\mathbf{T}}$, and V.o) yields v), which completes the proof of the lemma, and hence of the theorem, too.

One last comment. If T is a codensity triple for U (made into a triple $\mathbf{T} = (T,\eta,\mu)$ by the procedure of (8.4), (8.5) and (8.6), say), define $\Phi_{U,T} : \mathcal{X} \to \mathcal{A}^{\mathbf{T}}$ by

(8.8)
$$\Phi_{U,T}(X) = (UX, \langle id_{UX} \rangle) \, ,$$
$$\Phi_{U,T}(\xi) = U\xi \, ,$$

where $\langle id_{UX} \rangle : TUX \to UX$ is the id_{UX} th projection. There is no trouble in checking that $\Phi_{U,T}$ is well defined, is a functor, and satisfies $U^{\mathbf{T}} \circ \Phi_{U,T} = U$. This functor will turn out to be the front adjunction for the structure-semantics adjointness in the context of triples. If $\mathbf{T}$ arises as the adjunction triple $(UF,\eta,U\beta F)$ resulting from a left adjoint F for U, with front and back adjunctions η, β, then the effect of $\Phi_{U,T}$ on objects is given equivalently by

$$\Phi_{U,T}(X) = (UX, U\beta_X) \, .$$

§ 9. The isomorphism theorem

In this and the following section, we shall write $\mathfrak{M} = \mathfrak{M}^{\mathrm{id}_{\mathcal{A}}}$ and $\mathcal{P}_V = \mathcal{P}_V^{\mathrm{id}_{\mathcal{A}}}$.

Theorem 9.1. *If* $\mathbf{T}$ *is a codensity triple for the* $\mathcal{A}$*-valued functor* $U : \mathcal{X} \to \mathcal{A}$, *there is a canonical isomorphism* $\Psi : U\text{-}\underline{\mathrm{Alg}} \to \mathcal{A}^{\mathbf{T}}$ *making the triangle*

$$\begin{array}{ccc} U\text{-}\underline{\mathrm{Alg}} & \xrightarrow{\ \Psi\ } & \mathcal{A}^{\mathbf{T}} \\ & {\scriptstyle |\ |_U} \searrow \quad \swarrow {\scriptstyle U^{\mathbf{T}}} & \\ & \mathcal{A} & \end{array}$$

commute.

Proof. Step 1. $|\ |_U : U\text{-}\underline{\mathrm{Alg}} \to \mathcal{A}$ is isomorphic to the $\mathcal{A}$-valued functor in the pullback diagram

$$\begin{array}{ccc} \mathcal{P}_{(f^{\mathbf{T}})^*} & \longrightarrow & ((\mathcal{K}^{\mathbf{T}})^*, \mathfrak{S}) \\ \downarrow {\scriptstyle \mathfrak{M}((f^{\mathbf{T}})^*)} & & \downarrow {\scriptstyle ((f^{\mathbf{T}})^*, \mathfrak{S})} \\ \mathcal{A} & \longrightarrow & (\mathcal{A}^*, \mathfrak{S}) \end{array} ,$$

because of the isomorphism $\mathfrak{X}_U \xrightarrow[Y]{\cong} (\mathcal{K}^{\mathbf{T}})^*$ provided in § 8.

Step 2. $\mathcal{P}_{(f^{\mathbf{T}})^*} \to \mathcal{A}$ has a left adjoint. Indeed, the commutativity of the diagram

$$\begin{array}{ccc} \mathcal{K}^{\mathbf{T}} & \xrightarrow{\ Y\ } & ((\mathcal{K}^{\mathbf{T}})^*, \mathfrak{S}) \\ \downarrow {\scriptstyle u^{\mathbf{T}}} & & \downarrow {\scriptstyle ((f^{\mathbf{T}})^*, \mathfrak{S})} \\ \mathcal{A} & \xrightarrow{\ Y\ } & (\mathcal{A}^*, \mathfrak{S}) \end{array}$$

provides a functor $\Gamma : \mathcal{K}^{\mathbf{T}} \to \mathcal{P}_{(f^{\mathbf{T}})^*}$.

Lemma 9.1. $\Gamma : \mathcal{K}^{\mathbf{T}} \to \mathcal{P}_{(f^{\mathbf{T}})^*}$ *is full and faithful*, $\Gamma \circ f^{\mathbf{T}} : \mathcal{A} \to \mathcal{P}_{(f^{\mathbf{T}})^*}$ *serves as left adjoint to* $\mathfrak{M}((f^{\mathbf{T}})^*)$, *and the resulting adjunction triple is* $\mathbf{T}$.

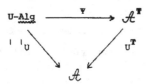

<u>Proof</u>. Since the diagram

$$\begin{array}{ccc} \mathcal{K}^{\mathbf{T}} & \xrightarrow{\quad Y \quad} & ((\mathcal{K}^{\mathbf{T}})^{*}, \mathcal{S}) \\ & {\scriptstyle \Gamma} \searrow \qquad \nearrow & \\ & \mathcal{P}_{(f^{\mathbf{T}})^{*}} & \end{array}$$

commutes, and both the Yoneda functor and

$$\text{(9.1)} \qquad \mathcal{P}_{(f^{\mathbf{T}})^{*}} \to ((\mathcal{K}^{\mathbf{T}})^{*}, \mathcal{S})$$

are full and faithful (see Remarks 5.2 and 5.3), Γ is full and faithful, too. For the adjointness statement, the Yoneda Lemma and the fullness and faithfulness of (9.1) deliver

$$\mathcal{P}_{(f^{\mathbf{T}})^{*}}(\Gamma \circ f^{\mathbf{T}}k\ ,\ (A,\lambda)) \cong \text{n.t.}(Yf^{\mathbf{T}}k, \lambda) \cong \lambda(f^{\mathbf{T}}(k)) =$$

$$= \mathcal{A}(k,A) = \mathcal{A}(k, \mathfrak{m}((f^{\mathbf{T}})^{*})(A,\lambda))\ ,$$

whose naturality in $k \in |\mathcal{A}|$ and $(A,\lambda) \in |\mathcal{P}_{(f^{\mathbf{T}})^{*}}|$ is left to the reader's verification. To compute the adjunction triple, note that $\mathfrak{m}((f^{\mathbf{T}})^{*}) \circ \Gamma \circ f^{\mathbf{T}} = u^{\mathbf{T}} \circ f^{\mathbf{T}} = T$, and that, when $(A,\lambda) = \Gamma \circ f^{\mathbf{T}}k$, the front adjunction, which is whatever $\mathcal{A}$-morphism $k \to Tk$ arises from the identity on $\Gamma f^{\mathbf{T}}k$, is the $\mathcal{A}$-morphism serving as the identity, in $\mathcal{K}^{\mathbf{T}}$, on k , namely η_k . It follows that, whatever n , $k \in |\mathcal{A}|$, the diagram

$$\begin{array}{ccccc} \mathcal{K}^{\mathbf{T}}(f^{\mathbf{T}}k, f^{\mathbf{T}}n) & & \xrightarrow[\quad \Gamma \quad]{\cong} & & \mathcal{P}_{(f^{\mathbf{T}})^{*}}(\Gamma f^{\mathbf{T}}k, \Gamma f^{\mathbf{T}}n) \\ & {\scriptstyle u^{\mathbf{T}}} \searrow & & \swarrow {\scriptstyle \mathfrak{m}((f^{\mathbf{T}})^{*})} & \\ {\scriptstyle =} \downarrow & & \mathcal{A}(Tk,Tn) & & \downarrow {\scriptstyle \cong} \\ & & \downarrow & & \\ \mathcal{A}(k, u^{\mathbf{T}}f^{\mathbf{T}}n) & = & \mathcal{A}(k,Tn) & = & \mathcal{A}(k,\ ((f^{\mathbf{T}})^{*})\ f^{\mathbf{T}}n) \end{array}$$

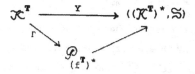

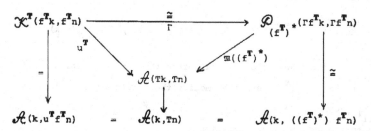

commutes, since $u^{\mathbf{T}} = \mathfrak{m}((f^{\mathbf{T}})^{*}) \circ \Gamma$ and the front adjunctions are the same. But then it follows that those back adjunctions that are obtained when $k = u^{\mathbf{T}}f^{\mathbf{T}}n$ (by reversing the vertical arrows and chasing the identity maps in $\mathcal{A}$ upwards) correspond to each other under Γ. This completes the proof of the lemma.

Step 3. Apply the most precise tripleableness theorem (e.g., [22, Th. 1.2.9]) -- it asserts that if the functor $U : \mathcal{X} \to \mathcal{A}$ has a left adjoint and creates U-split coequalizers, the canonical functor $\Phi_{U,T} : \mathcal{X} \to \mathcal{A}^{\mathbf{T}}$ (defined in (8.8), where $\mathbf{T}$ is the adjunction triple) is an isomorphism -- using Lemma 9.1 and Proposition 6.2, to the functor $\mathfrak{M}((f^{\mathbf{T}})^*) : \mathcal{P}_{(f^{\mathbf{T}})^*} \to \mathcal{A}$ to get an isomorphism $\mathfrak{M}((f^{\mathbf{T}})^*) \xrightarrow{\cong} U^{\mathbf{T}}$. Finally, combine this isomorphism with the isomorphism $|\ |_U \xrightarrow{\cong} \mathfrak{M}((f^{\mathbf{T}})^*)$ of Step 1, to obtain the desired isomorphism Ψ.

Because this proof is (relatively) short and conceptual, it is somewhat uninformative. We collect the missing information in

Theorem 9.2. If $\mathbf{T}$ is a codensity triple for $U : \mathcal{X} \to \mathcal{A}$, with associated isomorphism $y : \mathfrak{T}_U \to (\mathcal{K}^{\mathbf{T}})^*$, the isomorphism $\Psi : U\text{-}\underline{\text{Alg}} \to \mathcal{A}^{\mathbf{T}}$ provided by the proof of Theorem 9.1 has the following properties.

1. $\Psi((A,\lambda)) = (A,\alpha(\lambda))$, where $\alpha(\lambda) = y^{-1}(id_{TA}) *_{\lambda} id_A$.

2. $\Phi_{U,T} = \Psi \circ \Phi_U : \mathcal{X} \to U\text{-}\underline{\text{Alg}} \to \mathcal{A}^{\mathbf{T}}$.

3. $\Psi^{-1}((A,\alpha)) = (A,\lambda(\alpha))$, where $\theta *_{\lambda(\alpha)} a = \alpha \circ Ta \circ y(\theta)$;

furthermore, Ψ^{-1} is the functor portion $\Phi : \mathcal{A}^{\mathbf{T}} \to U\text{-}\underline{\text{Alg}}$ of the $(\underline{\text{Cat}},\mathcal{A})$-morphism $U^{\mathbf{T}} \to \mathfrak{M}(\exp_U)$ corresponding by structure-semantics adjointness to the isomorphism $\exp_U \to \exp_{U^{\mathbf{T}}}$ arising from the identifications

$$\mathfrak{T}_U(n,k) \cong \mathcal{A}(k,Tn) = (U^{\mathbf{T}})^k(F^{\mathbf{T}}(n)) \cong \text{n.t.}((U^{\mathbf{T}})^n, (U^{\mathbf{T}})^k) = \mathfrak{T}_{U^{\mathbf{T}}}(n,k).$$

Proof. For the first assertion, we calculate the back adjunction for the adjointness of $\mathfrak{M}((f^{\mathbf{T}})^*)$ to $\Gamma \circ f^{\mathbf{T}}$, and then modify the result appropriately by y. Given the $\mathcal{P}_{(f^{\mathbf{T}})^*}$-object (A,λ), we chase $id_A \in \mathcal{A}(A,A)$ through the adjunction identification to the natural transformation

$$(9.2) \qquad \mathcal{K}^{\mathbf{T}}(-,f^{\mathbf{T}}A) \to \lambda$$

sending $t \in \mathcal{K}^{T}(k, f^{T}A)$ to $\{\lambda(t)\}(a) \in \lambda(k) = A^{k}$. To find the $\mathcal{A}$-morphism component of the $\mathcal{P}_{(f^{T})^{*}}$-morphism having (9.2) as its natural transformation component, we must apply the Yoneda Lemma to its value under $((f^{T})^{*}, \mathfrak{S})$. The resulting functors are

$$\mathcal{K}^{T}(f^{T}(-), f^{T}A) = \mathcal{A}(-, TA)$$

and

$$\lambda \circ f^{T} = \mathcal{A}(-, A) \ ;$$

the natural transformation still has the same components; so the Yoneda Lemma produces $\lambda(id_{TA})(id_{A})$. Hence $\Phi_{U,T}((A,\lambda)) = (A, \lambda(id_{TA})(id_{A}))$, and the effect of Ψ is therefore as asserted.

For point 2, it suffices to observe that $\alpha(\lambda_{U}(X)) = \langle id_{UX} \rangle$ (see (1.5) and (8.8)). But, in fact, using (8.7),

$$\alpha(\lambda_{U}(X)) = y^{-1}(id_{TUX}) *_{\lambda_{U}(X)} id_{UX} = \{y^{-1}(id_{TUX})\}_{X}(id_{UX})$$

$$= \langle id_{UX} \rangle \circ T(id_{UX}) \circ id_{TUX} = \langle id_{UX} \rangle \ .$$

To settle point 3, it is enough to show, where $\Phi : \mathcal{A}^{T} \to U\text{-}\underset{\sim}{Alg}$ is obtained by the indicated adjointness, that $\Phi(A,\alpha) = (A, \lambda(\alpha))$ and that $\Psi \circ \Phi = id_{\mathcal{A}^{T}}$. Now Φ is the composition

(9.3) $$\mathcal{A}^{T} \xrightarrow{\Phi_{U^{T}}} U^{T}\text{-}\underset{\sim}{Alg} \xrightarrow{\cong} U\text{-}\underset{\sim}{Alg} \ ,$$

$\Phi_{U^{T}}((A,\alpha)) = (A, \lambda_{U^{T}}(A,\alpha))$, and, since $\alpha : (TA, \mu_{A}) \to (A,\alpha)$ is the back adjunction $F^{T}U^{T}(A,\alpha) \to (A,\alpha)$, it follows from (1.5) that

$$\theta' *_{\lambda_{U^{T}}(A,\alpha)} a = \alpha \circ Ta \bullet \theta' \ .$$

Applying the isomorphism, therefore, $\Phi(A,\alpha) = (A, \lambda(\alpha))$. That $\Psi \circ \Phi = id_{\mathcal{A}^{T}}$ now follows immediately from the computation

$$(9.4)\qquad \begin{aligned}\alpha(\lambda(\alpha)) &= y^{-1}\mathrm{id}_{TA} *_{\lambda(\alpha)} \mathrm{id}_A = \alpha \circ T\mathrm{id}_A \circ y(y^{-1}\mathrm{id}_{TA}) \\ &= \alpha \circ \mathrm{id}_{TA} \circ \mathrm{id}_{TA} = \alpha .\end{aligned}$$

Theorems 9.1 and 9.2 conspire jointly to prove Theorem 9.3 below, a more elementary, though less conceptual, proof of which appears in § 11. To set up Theorem 9.3, we place ourselves (at first) in a more general setting, letting $U : \mathcal{X} \to \mathcal{A}$ and $\mathbf{T} = (T,\eta,\mu)$ be an arbitrary $\mathcal{A}$-valued functor and a possibly unrelated triple on $\mathcal{A}$. Then, given functions

$$y_{n,k} : \mathrm{n.t.}(U^n,U^k) \to \mathcal{A}(k,Tn) \qquad (n,\ k \in |\mathcal{A}|)$$

and an $\mathcal{A}$-morphism $\alpha : TA \to A$, define a system $\lambda_y(\alpha) = \{(\lambda_y(\alpha))_{n,k} \mid n,k \in |\mathcal{A}|\}$ of functions

$$(\lambda_y(\alpha))_{n,k} : \mathrm{n.t.}(U^n,U^k) \to \mathcal{S}(A^n,A^k)$$

by setting

$$(9.5)\qquad \{(\lambda_y(\alpha))_{n,k}(\theta)\}(a)\ \ (= \theta * a) = \alpha \circ Ta \circ y_{n,k}(\theta)$$

whenever $\theta : U^n \to U^k$ and $a \in A^n$. Conversely, given functions

$$z_{n,k} : \mathcal{A}(k,Tn) \to \mathrm{n.t.}(U^n,U^k) \qquad (n,\ k \in |\mathcal{A}|)$$

and a U-algebra (A,λ), define an $\mathcal{A}$-morphism

$$\alpha_z(\lambda) : TA \to A$$

by posing

$$(9.6)\qquad \alpha_z(\lambda) = z_{A,TA}(\mathrm{id}_{TA}) *_\lambda \mathrm{id}_A = \{\lambda_{A,TA}(z_{A,TA}(\mathrm{id}_{TA}))\}(\mathrm{id}_A) .$$

Before stating Theorem 9.3, which closes the section, we use the formalism above to suggest another proof, not using the tripleableness argument we have employed, of the core of Theorems 9.1 and 9.2. Where $\mathbf{T}$ is a codensity triple for U, functions y as above are provided in § 8. Let $z_{n,k} = (y_{n,k})^{-1}$. The left adjoint $\Gamma \circ f^{\mathbf{T}}$ to

$| \ |_U$ provided by steps 1 and 2 in the proof of Theorem 9.1 permits construction of $\Psi = \Phi_{| \ |_U, T} : U\text{-}\underline{Alg} \to \mathcal{A}^{\mathbf{T}}$, and one can prove $\Psi(A,\lambda) = (A,\alpha_z(\lambda))$. Explicit analysis shows the back adjunction $\Gamma \circ f^{\mathbf{T}} \circ | \ |_U \to id_{U\text{-}\underline{Alg}}$ at (A,λ) is just $\alpha_z(\lambda)$, mapping $(TA, \lambda_y(\mu_A))$ to (A,λ). With this, one proves

$$(9.7) \qquad \lambda_y(\alpha_z(\lambda)) = \lambda ,$$

just as in Lemma 11.6. Since there is a functor $\Phi : \mathcal{A}^{\mathbf{T}} \to U\text{-}\underline{Alg}$ (defined as in the proof of assertion 3 of Theorem 9.2) sending (A,α) to $(A,\lambda_y(\alpha))$, it follows from (9.4), (9.7), and the fact that Φ and Ψ are compatible with the underlying $\mathcal{A}$-object functors that $\Phi = \Psi^{-1}$.

Theorem 9.3. If $\mathbf{T}$ *is a codensity triple for* $U : \mathcal{X} \to \mathcal{A}$, *if* $y : \mathfrak{T}_U \to (\mathcal{K}^{\mathbf{T}})^*$ *is the resulting isomorphism, and if* $z = y^{-1}$, *then*

$$(A,\alpha) \longmapsto (A,\lambda_y(\alpha)) , \qquad (A,\lambda) \longmapsto (A,\alpha_z(\lambda))$$

are the (bijective) object functions of a mutually inverse pair

$$\Phi : \mathcal{A}^{\mathbf{T}} \to U\text{-}\underline{Alg} , \qquad \Psi : U\text{-}\underline{Alg} \to \mathcal{A}^{\mathbf{T}}$$

of isomorphism making commutative the diagram

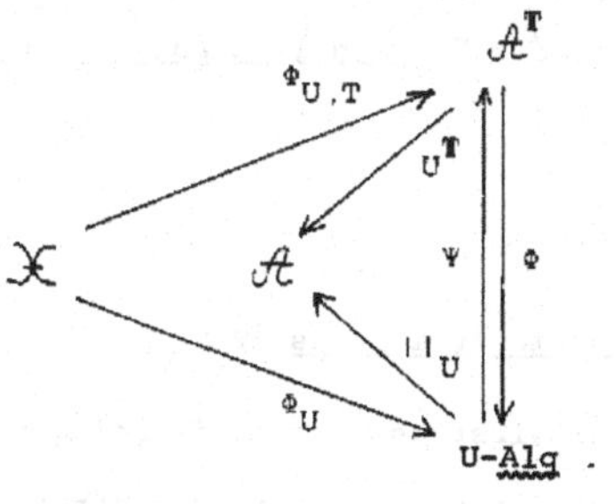

§ 10. Structure and semantics in the presence of a triple.

In this section, we use the isomorphism of § 9 to compare the structure-semantics adjointness of § 4, when $j = id_{\mathcal{A}}$, with that of Appelgate-Barr-Beck-Eilenberg-Huber-

Kleisli-Maranda-Moore-Tierney in the context of triples. For notational convenience, we shall write $\mathfrak{S} = \mathfrak{S}^{id_{\mathcal{A}}}$, and, as earlier, $\mathfrak{M} = \mathfrak{M}^{id_{\mathcal{A}}}$, $\mathcal{P}_V = \mathcal{P}_V^{id_{\mathcal{A}}}$. It will be useful to take terminological and notational account of the canonical isomorphism

$$(\mathcal{A}^*, \underset{\sim}{Cat}) = (\mathcal{A}, \underset{\sim}{Cat}) ,$$

obtained by reinterpreting $V : \mathcal{A}^* \to \mathcal{C}$ as $V^* : \mathcal{A} \to \mathcal{C}^*$, by speaking of V^* as a theory over $\mathcal{A}$ if V is a clone over $\mathcal{A}$; by referring to

$$\mathfrak{S}^*(U) = (\mathfrak{S}(U))^* = (\exp_U)^*$$

as the theory of $U : \mathcal{X} \to \mathcal{A}$; and by speaking of

$$\mathfrak{M}^*(\varphi) = \mathfrak{M}(\varphi^*) : \mathcal{P}_{\varphi^*} \to \mathcal{A}$$

as (the underlying $\mathcal{A}$-object functor on) the category of φ-algebras in $\mathcal{A}$, where φ is a theory over $\mathcal{A}$. It is clear that a clone has a left adjoint iff the corresponding theory has a right adjoint. By the same argument as was used in § 8, one can prove the first part of

Lemma 10.1. *To give a right adjoint* u, *with front and back adjunctions* η, β *for a theory* φ *on* $\mathcal{A}$ *is the same as to give an isomorphism* $\varphi \cong f^{\mathbf{T}}$, *where* $\mathbf{T}$ *is the triple* $(u\varphi, \eta, u\beta\varphi)$ *on* $\mathcal{A}$. *Moreover, given a triple* $\mathbf{T}$, $\mathfrak{B} = \mathcal{K}^{\mathbf{T}}$, $\varphi = f^{\mathbf{T}}$, $u = u^{\mathbf{T}}$ *provide the only theory* $\varphi : \mathcal{A} \to \mathfrak{B}$ *with left adjoint* u *satisfying* 1) *the adjunction equivalences*

$$\mathcal{A}(k, un) \xrightarrow{\ \cong\ } \mathfrak{B}(\varphi k, n)$$

are identity maps, and 2) *the adjunction triple is* $\mathbf{T}$.

Proof. We skip the proof of the first assertion, it being just like the proof of III → IV in Theorem 8.1. For the second assertion, it is clear that the objects of $\mathfrak{B}$ must be those of $\mathcal{A}$. Then $\mathfrak{B}$-morphisms $k \to n$ must be $\mathcal{A}$-morphisms $k \to Tn$, the identity in $\mathfrak{B}(n,n)$ must be $\eta_n \in \mathcal{A}(n, Tn)$, and, finally, for the identity functions to be natural, as required by 1), it is forced that the composition rule of $\mathfrak{B}$ is that of $\mathcal{K}^{\mathbf{T}}$.

We introduce the categories $\mathrm{Ad}(\mathrm{Cat},\mathcal{A})$ (resp. $\mathrm{Tr}(\mathrm{Cat},\mathcal{A})$) of $\mathcal{A}$-valued functors having specified left adjoints (resp. specified codensity triples), and $\mathrm{AdTheo}(\mathcal{A})$ (resp. $\mathrm{AdCl}(\mathcal{A})$) of theories (resp. clones) over $\mathcal{A}$ having specified right (resp. left) adjoints. These are so constructed, per definitionem, as to make the obvious forgetful functors to the similarly named categories, with the prefix Ad or Tr omitted, full and faithful. We shall also need the category $\mathrm{Trip}(\mathcal{A})$ whose objects are triples on $\mathcal{A}$: a triple morphism from $\mathbf{T} = (T,\eta,\mu)$ to $\mathbf{T}' = (T',\eta',\mu')$ will be any natural transformation $\tau : T \to T'$ for which

(10.1) $\qquad \tau \circ \eta = \eta'$, and

(10.2) $\qquad \tau \circ \mu = \mu' \circ \tau\tau$

(where $\tau\tau$ denotes either of the compositions

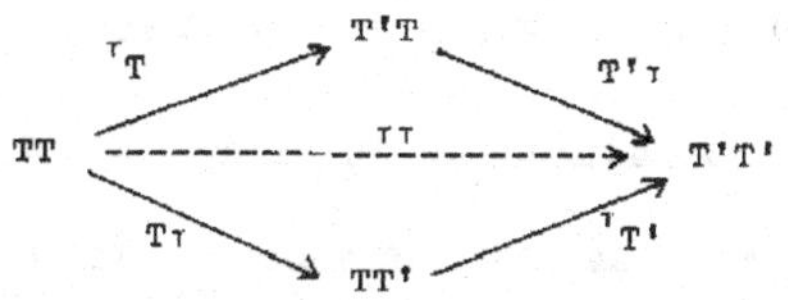

which are equal because τ is natural).

<u>Lemma 10.2</u>. <u>The attempts to define functors</u>

$$\mathrm{AdTheo}(\mathcal{A}) \xrightarrow{\ t\ } \mathrm{Trip}(\mathcal{A}) \ : \ \varphi,u,\eta,\beta \longmapsto (u\varphi,\eta,u\beta\varphi) \ ,$$

$$\mathrm{Trip}(\mathcal{A}) \xrightarrow{\ k\ } \mathrm{AdTheo}(\mathcal{A}) \ : \ \mathbf{T} \longmapsto \text{Kleisli cat. w/} f^{\mathbf{T}}, u^{\mathbf{T}} \ ,$$

<u>are successful and represent</u> $\mathrm{Trip}(\mathcal{A})$ <u>isomorphically as the full subcategory of</u> $\mathrm{AdTheo}(\mathcal{A})$, <u>equivalent to</u> $\mathrm{AdTheo}(\mathcal{A})$, <u>consisting of those adjointed theories for which condition</u> 1) <u>of Lemma</u> 10.1 <u>is valid</u>.

<u>Proof</u>. Elementary. For related information, see [2] or [23].

Theorem 8.1 shows that $\mathrm{Tr}(\mathrm{Cat},\mathcal{A})$ is the pullback of the pullback diagram

$$\begin{array}{ccc} \mathrm{Tr}(\mathbf{Cat},\mathcal{A}) & \xrightarrow{\ \mathfrak{S}_t\ } & (\mathrm{AdTheo}(\mathcal{A}))^* \\ \downarrow & & \downarrow \\ (\mathbf{Cat},\mathcal{A}) & \xrightarrow{\ \mathfrak{S}^*\ } & (\mathrm{Theo}(\mathcal{A}))^* \end{array}$$

and arguments like those for Lemma 9.1 provide a lifting $\mathfrak{M}_t$

$$\begin{array}{ccc} (\mathrm{AdTheo}(\mathcal{A}))^* & \xrightarrow{\ \mathfrak{M}_t\ } & \mathrm{Ad}(\mathbf{Cat},\mathcal{A}) \\ \downarrow & & \downarrow \\ (\mathrm{Theo}(\mathcal{A}))^* & \xrightarrow{\ \mathfrak{M}^*\ } & (\mathbf{Cat},\mathcal{A}) \end{array}$$

of $\mathfrak{M}^*$.

Theorem 10.1. *Let* $I : \mathrm{Ad}(\mathbf{Cat},\mathcal{A}) \to \mathrm{Tr}(\mathbf{Cat},\mathcal{A})$ *be the obvious functor (sending* $(U;F,\eta,\beta)$ *to* $(U;(UF,\eta,U\beta F)))$, *and let* $\mathfrak{M}' : (\mathrm{Trip}(\mathcal{A}))^* \to \mathrm{Ad}(\mathbf{Cat},\mathcal{A})$ *be the functor sending* $\mathbf{T}$ *to* $(U^{\mathbf{T}};F^{\mathbf{T}},\eta,\beta)$ *(where* $\beta_{(A,\alpha)} = \alpha$) . *Then:*

1) $\mathfrak{S}_t \circ I$ (*resp.* $\mathfrak{S}_t$) *is adjoint to* $\mathfrak{M}_t$ (*resp.* $I \circ \mathfrak{M}_t$) ,

2) $\mathfrak{M}' \circ t$ (*resp.* $\mathfrak{M}_t \circ k$) *is equivalent to* $\mathfrak{M}_t$ (*resp.* $\mathfrak{M}'$) ,

3) $t \circ \mathfrak{S}_t \circ I$ (*resp.* $t \circ \mathfrak{S}_t$) *is adjoint to* $\mathfrak{M}'$ (*resp.* $I \circ \mathfrak{M}'$) .

The proof, which is easy, uses Theorem 4.1, the above lemmas, and the isomorphisms produced in § 9 . $t \circ \mathfrak{S}_t \circ I$ and $\mathfrak{M}'$ are the most familiar structure and semantics functors in the context of triples and adjoint pairs; $t \circ \mathfrak{S}_t$ and $\mathfrak{M}'$ are those needed in the work of Appelgate and Tierney [1], [26].

Motivating the presentation of [10] is the realization that the Kleisli category arising from the adjunction triple of an adjoint pair U,F is isomorphic with the full image of F . This makes "free algebras" more amenable, and encourages yet another (equivalent) structure functor in the setting of adjointed theories and adjoint $\mathcal{A}$-valued functors.

A pleasant exercise (for private execution) is to tabulate all the isomorphisms and equivalences that have arisen in this work and will arise from them by composition with an $\mathfrak{S}$ or an $\mathfrak{M}$.

§ 11. Another proof of the isomorphism theorem.

This section is devoted to a straightforward computational proof of Theorem 9.3. The proof itself follows a sequence of lemmas; these lemmas depend only on the "information of type V" arising from the assumption that $\mathbf{T}$ is a condensity triple for U (see § 8). For convenience of reference, we recall the equations

(V.i) $\qquad y(\theta' \circ \theta) = \mu_n \circ T(y\theta) \circ y\theta'$,

(V.ii) $\qquad U^{\eta_n} \circ y^{-1}(id_{Tn}) = U^{id_n}$,

(V.iii) $\qquad y(U^f) = \eta_n \circ f$,

imposed on the one-one correspondences

$$y = y_{n,k} : n.t(U^n, U^k) \to \mathcal{A}(k, Tn) \qquad (n,k \in |\mathcal{A}|) ,$$

the diagrams

(7.6)
$$\begin{array}{ccccc} & & TA & & \\ & \nearrow^{\eta_A} & & \searrow^{\alpha} & \\ A & & \xrightarrow{\quad id_A \quad} & & A \end{array}$$

(7.7)
$$\begin{array}{ccc} TTA & \xrightarrow{\mu_A} & TA \\ {\scriptstyle T\alpha}\downarrow & & \downarrow{\scriptstyle \alpha} \\ TA & \xrightarrow[\alpha]{} & A \end{array}$$

whose commutativity betokens the assertion that $\alpha : TA \to A$ is a $\mathbf{T}$-algebra, and the diagram

(7.8)
$$\begin{array}{ccc} TA & \xrightarrow{Tg} & TB \\ {\scriptstyle \alpha}\downarrow & & \downarrow{\scriptstyle \beta} \\ A & \xrightarrow[g]{} & B \end{array}$$

on the basis of whose commutativity $g : A \to B$ is an $\mathcal{A}^{\mathbf{T}}$-morphism from (A,α) to (B,β) .

We begin to chip away at Theorem 9.3 by proving

Lemma 11.1. Suppose (A,α) is a $\mathbb{T}$-algebra. Then $(A,\lambda(\alpha))$ is a U-algebra (here $\lambda(\alpha) = \lambda_y(\alpha)$ is defined by (9.5)).

Proof. The definition of $\lambda(\alpha)$, (V.iii), naturality of η, and (7.6) allow us to verify ALG 1:

$$U^f * a = \alpha \circ Ta \circ y(U^f) = \alpha \circ Ta \circ \eta_n \circ f = \alpha \circ \eta_A \circ a \circ f = a \circ f .$$

Similarly, (9.5), (V.i), naturality of μ, (7.7), functoriality of T, and (twice more) (9.5) again, deliver ALG 2 :

$$\begin{aligned}(\theta'\circ\theta) * a &= \alpha \circ Ta \circ y(\theta'\circ\theta) = \alpha \circ Ta \circ \mu_n \circ T(y\theta) \circ y\theta' = \\ &= \alpha \circ \mu_A \circ TTa \circ T(y\theta) \circ y\theta' = \alpha \circ T\alpha \circ TTa \circ T(y\theta) \circ y\theta' = \\ &= \alpha \circ T(\alpha \circ Ta \circ y\theta) \circ y\theta' = \alpha \circ T(\theta * a) \circ y\theta' = \\ &= \theta' * (\theta * a) .\end{aligned}$$

As a start in going the other way, we offer

Lemma 11.2. Suppose (A,λ) is a U-algebra. Then

$$\alpha(\lambda) \circ \eta_A = id_A .$$

Proof. Using ALG 1, (9.6), (V.ii), and ALG 1 again, we see

$$\begin{aligned}\alpha(\lambda) \circ \eta_A &= U^{\eta_A} * \alpha(\lambda) = U^{\eta_A} * (y^{-1}(id_{TA}) * id_A) = \\ &=(U^{\eta_A} \circ y^{-1}(id_{TA})) * id_A = U^{id_A} * id_A = \\ &= id_A \circ id_A = id_A .\end{aligned}$$

To know that $\alpha(\lambda)$ is a $\mathbb{T}$-algebra, there remains the identity $\alpha(\lambda) \circ \mu_A = \alpha(\lambda) \circ T\alpha(\lambda)$. This identity, as well as the fact that each U-homomorphism $(A,\lambda) \to (B,\mathfrak{B})$ is also a $\mathbb{T}$-homomorphism $(A,\alpha(\lambda)) \to (B,\alpha(\mathfrak{B}))$, will result from the fact (Lemma 11.8) that each such U-homomorphism g makes diagram (7.8) (with $\alpha = \alpha(\lambda)$, $\beta = \alpha(\mathfrak{B})$) commute, and the fact (Lemma 11.5) that $\alpha(\lambda) : TA \to A$ is a

U-homomorphism. The next two lemmas pave the way for a proof of Lemma 11.5.

Lemma 11.3. *For any U-algebra* (A,λ) *and any* $\mathcal{A}$*-morphism* $f : k \to n$, *the diagram with solid arrows*

$$\begin{array}{ccccccc}
\mathcal{A}(n,TA) & \xrightarrow{\;y^{-1}\;} & \text{n.t.}(U^A,U^n) & \xrightarrow{\;\lambda_{A,n}\;} & \mathfrak{S}(A^A,A^n) & \xrightarrow{\;ev_{id_A}\;} & A^n \\
\downarrow{\scriptstyle \mathcal{A}(f,TA)} & & \vdots & & \vdots & & \downarrow{\scriptstyle A^f} \\
\mathcal{A}(k,TA) & \xrightarrow[\;y^{-1}\;]{} & \text{n.t.}(U^A,U^k) & \xrightarrow[\;\lambda_{A,k}\;]{} & \mathfrak{S}(A^A,A^k) & \xrightarrow[\;ev_{id_A}\;]{} & A^k
\end{array}$$

commutes. Moreover, starting with $n = TA$, *the effect of the top row on* $id_{TA} \in \mathcal{A}(n,TA)$ *is* $\alpha(\lambda) \in \mathcal{A}(TA,A) = A^n$. *Hence the diagram*

$$\begin{array}{ccccc}
\text{n.t.}(U^A,U^n) & \xrightarrow{\qquad\lambda_{A,n}\qquad} & & & \mathfrak{S}(A^A,A^n) \\
\downarrow{\scriptstyle y} & & & & \downarrow{\scriptstyle ev_{id_A}} \\
\mathcal{A}(n,TA) & \xrightarrow[\mathcal{A}(n,\alpha(\lambda))]{} & \mathcal{A}(n,A) & = & A^n
\end{array}$$

commutes, for each $n \in |\mathcal{A}|$.

Proof. For the dotted arrows use "composition with U^f" and "composition with A^f", respectively. That the right hand square then commutes follows from the naturality of ev_{id_A}. The central square commutes because

$$(\theta * a) \circ f = U^f * (\theta * a) = (U^f \circ \theta) * a ,$$

using ALG 1 and ALG 2. The left hand square commutes because (V.i), (V.iii), naturality of η , and one of the triple identities deliver the chain of equalities

$$y(U^f \circ \theta) = \mu_A \circ T(y\theta) \circ y(U^f) = \mu_A \circ T(y\theta) \circ \eta_n \circ f =$$

$$= \mu_A \circ \eta_{TA} \circ y\theta \circ f = y\theta \circ f \quad ;$$

setting $\theta = y^{-1}(t)$ and applying y^{-1} to both ends of this chain provides the the identity expressing commutativity of the left hand square. The assertion regarding $\alpha(\lambda)$ is just the definition of $\alpha(\lambda)$. The Yoneda Lemma then applies: the given natural transformation $\mathcal{A}(-,TA) \to \mathcal{A}(-,A)$ is of the form $\mathcal{A}(-,\alpha(\lambda))$. This proves the last assertion.

Lemma 11.4. *For any* U-*algebra* (A,λ) *and any natural transformation* $\theta : U^n \to U^k$, *the diagram*

$$\begin{array}{ccccccc}
\mathcal{A}(n,TA) & \xrightarrow{\;y^{-1}\;} & \text{n.t.}(U^A,U^n) & \xrightarrow{\;\lambda_{A,n}\;} & \mathfrak{S}(A^A,A^n) & \xrightarrow{\;ev_{id_A}\;} & A^n \\
\big\downarrow{\scriptstyle \lambda(\mu_A)_{n,k}(\theta)} & & \big\downarrow{\scriptstyle \text{compose with } \theta} & & & {\scriptstyle \lambda_{n,k}(\theta)} & \big\downarrow \\
\mathcal{A}(k,TA) & \xrightarrow[\;y^{-1}\;]{} & \text{n.t.}(U^A,U^k) & \xrightarrow[\;\lambda_{A,k}\;]{} & \mathfrak{S}(A^A,A^k) & \xrightarrow[\;ev_{id_A}\;]{} & A^k
\end{array}$$

Proof. The commutativity of the large right hand square is guaranteed by ALG 2. To deal with the small left hand square, note that (9.5) and (V.i) yield

$$\{\lambda(\mu_A)_{n,k}(\theta)\}(a) = \mu_A \circ Ta \circ y\theta = y(\theta \circ y^{-1}(a)) .$$

Apply y^{-1} to this equation to obtain the equation expressing the commutativity of the left hand square.

We can now prove that $\alpha(\lambda)$ is a U-homomorphism.

Lemma 11.5. *For every* U-*algebra* (A,λ), *the* $\mathcal{A}$-*morphism* $\alpha(\lambda) : TA \to A$ *is a* U-*algebra morphism from* $(TA,\lambda(\mu_A))$ *to* (A,λ).

Proof. Given $a : n \to TA$ and $\theta : U^n \to U^k$, the clockwise composition in the diagram of Lemma 11.4 sends a, according to the last assertion of Lemma 11.3, to $\theta * (\alpha(\lambda) \circ a)$. The counterclockwise composition, for the same reason, sends a to $\alpha(\lambda) \circ (\theta * a)$. Hence $\alpha(\lambda) \circ (\theta * a) = \theta * (\alpha(\lambda) \circ a)$, and the lemma is proved.

The only thing standing in the way of Lemma 11.8 is

Lemma 11.6. *Whenever* (A,λ) *is a* U-*algebra*, $a : n \to A$ *is an* $\mathcal{A}$-*morphism, and* $\theta : U^n \to U^k$ *is a natural transformation, then* $\{\lambda_{n,k}(\theta)\}(a) = \alpha(\lambda) \circ Ta \circ y\theta$.

Proof. Lemmas 11.2 and 11.5 allow us to write

$$\theta * a = \theta * (\alpha(\lambda) \circ \eta_A \circ a) = \alpha(\lambda) \circ (\theta * (\eta_A \circ a)) .$$

But, by (9.5), the definition of $\mathfrak{A}(\mu_A)$, we have

$$\theta * (\eta_A \circ a) = \mu_A \circ T(\eta_A \circ a) \circ y\theta \ .$$

Combining these equations, using the functoriality of T , and applying one of the triple identities, we obtain

$$\theta * a = \alpha(\mathfrak{A}) \circ \mu_A \circ T\eta_A \circ Ta \circ y\theta = \alpha(\mathfrak{A}) \circ Ta \circ y\theta \ ,$$

which proves the lemma.

Because it has to be proved sometime, we postpone the dénoument by means of

Lemma 11.7. *Whenever* (A,α) *is a* $\mathbf{T}$*-algebra and* $\mathfrak{A} = \mathfrak{A}(\alpha)$, *then* $\alpha = \alpha(\mathfrak{A})$.

Proof. Repeat the computation (9.4).

Lemma 11.8. *Let* $(A,\mathfrak{A})$ *and* $(B,\mathfrak{B})$ *be* U-*algebras*. *For each* U-*algebra morphism* $f : A \to B$ *between them*, *the diagram*

$$\begin{array}{ccc} TA & \xrightarrow{\ Tf\ } & TB \\ {\scriptstyle \alpha(\mathfrak{A})}\downarrow & & \downarrow{\scriptstyle \alpha(\mathfrak{B})} \\ A & \xrightarrow[\ f\ ]{} & B \end{array}$$

commutes.

Proof. Using (9.6), the hypothesis, and Lemma 11.6, we see

$$f \circ \alpha(\mathfrak{A}) = f \circ (y^{-1}(id_{TA}) * id_A) = y^{-1}(id_{TA}) * f =$$

$$= \alpha(\mathfrak{B}) \circ Tf \circ yy^{-1}(id_{TA}) = \alpha(\mathfrak{B}) \circ Tf \ .$$

It is time to reap our corollaries.

Corollary 1. *If* $(A,\mathfrak{A})$ *is a* U-*algebra*, $(A,\alpha(\mathfrak{A}))$ *is a* $\mathbf{T}$-*algebra*.

Proof. One of the necessary identities was proved as Lemma 11.2. The other is given by Lemma 11.8, applied (by virtue of Lemma 11.5) to $f = \alpha(\mathfrak{A}) : (TA,\mathfrak{A}(\mu_A)) \to (A,\mathfrak{A})$, and modified by taking into account Lemma 11.7 :

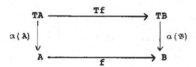

$$\alpha(\mathfrak{A}) \circ T\alpha(\mathfrak{A}) = \alpha(\mathfrak{A}) \circ \alpha(\mathfrak{A}(\mu_A)) = \alpha(\mathfrak{A}) \circ \mu_A .$$

Corollary 2. *If* $f : A \to B$ *is a* U-*algebra homomorphism from* $(A,\mathfrak{A})$ *to* $(B,\mathfrak{B})$, *it is also a* $\mathbf{T}$-*morphism from* $(A,\alpha(\mathfrak{A}))$ *to* $(B,\alpha(\mathfrak{B}))$.

Proof. This follows immediately from the diagram of Lemma 11.8 and from Corollary 1.

Corollary 3. *If* $f : A \to B$ *is a* $\mathbf{T}$-*algebra homomorphism from* (A,α) *to* (B,β) , *it is also a* U-*algebra map from* $(A,\mathfrak{A}(\alpha))$ *to* $(B,\mathfrak{A}(\beta))$.

Proof. Let $a : n \to A$, $\theta : U^n \to U^k$. By (9.5), the hypothesis, functoriality of T , and (9.5) again, we have

$$f \circ (\theta * a) = f \circ \alpha \circ Ta \circ y\theta = \beta \circ Tf \circ Ta \circ y\theta =$$
$$= \beta \circ T(f \circ a) \circ y\theta = \theta * (f \circ a) .$$

Proof of Theorem 9.3. Lemma 11.1, Corollary 1, and Lemmas 11.6 and 11.7 set up the desired isomorphism $|U\text{-}\underline{\text{Alg}}| \leftrightarrow |\mathcal{A}^{\mathbf{T}}|$. Corollaries 2 and 3, taken together with Lemmas 11.6 and 11.7, extend this to an isomorphism of categories $U\text{-}\underline{\text{Alg}} \leftrightarrow \mathcal{A}^{\mathbf{T}}$. It is clear from the constructions and from Corollaries 2 and 3 that the underlying $\mathcal{A}$-object functors are respected. The relation with the Φ's is settled by the proof of point 2) of Theorem 9.2.

Bibliography

[1] Appelgate, H. Acyclic models and resolvent functors. Dissertation, Columbia U., New York, 1965

[2] Barr, M. A note on a construction of Kleisli. Mimeographed, U. of Illinois, Urbana, 1965.

[3] Beck, J.M. Triples, algebras, and cohomology. Dissertation, Columbia U., New York, 1967.

[4] Bénabou, J. Critères de représentabilité des foncteurs. C.R. Acad. Sci. Paris **260** (1965), pp. 752-755.

[5] Bénabou, J. Structures algébriques dans les catégories. Thèse, Fac. Sci., U. de Paris, 1966.

[6] Birkhoff, G.D. On the structure of abstract algebras. Proc. Camb. Phil. Soc. **31** (1935), pp. 433-454.

[7] Cohn, P.M. Universal Algebra. Harper & Row, New York, 1965.

[8] Eilenberg, S., & Kelly, G.M. Closed categories. Proc. Conf. Categ. Alg. (La Jolla, 1965). Springer, Berlin, 1966, pp. 421-562.

[9] Eilenberg, S., & Moore, J.C. Adjoint functors and triples. Ill. J. Math. **9** (1965), pp. 381-398.

[10] Eilenberg, S., & Wright, J.B. Algebraic Categories and Automata Theory (to appear).

[11] Huber, P.J. Homotopy theory in general categories. Math. Annalen **144** (1961), pp. 361-385.

[12] Isbell, J. Adequate subcategories. Ill. J. Math. **4** (1960), pp. 541-552.

[13] Kleisli, H. Every standard construction is induced by a pair of adjoint functors. Proc. Amer. Math. Soc. **16** (1965), pp. 544-546.

[14] Kock, A. Continuous Yoneda representation of a small category. Mimeographed, E.T.H. Zürich, 1966.

[15] Lawvere, F.W. Functorial semantics of algebraic theories. Dissertation, Columbia U., New York, 1963. Summarized in Proc. Nat. Acad. Sci. **50** (1963), pp. 869-872.

[16] Lawvere, F.W. An elementary theory of the category of sets. Mimeographed, E.T.H. Zürich, 1965. Summarized in Proc. Nat. Acad. Sci. **52** (1964), pp. 1506-1511.

[17] Lawvere, F.W. The category of categories as a foundation for mathematics. Proc. Conf. Categ. Alg. (La Jolla, 1965), Springer, Berlin, 1966, pp. 1-20.

[18] Linton, F.E.J. Triples vs. theories (preliminary report). Notices Amer. Math. Soc. 13 (1966), p. 227.

[19] Linton, F.E.J. Some aspects of equational categories. Proc. Conf. Categ. Alg. (La Jolla, 1965), Springer, Berlin, 1966, pp. 84-94.

[20] Linton, F.E.J. Applied functorial semantics, I. To appear in Annali di Matematica.

[21] Mac Lane, S. Categorial algebra. Bull. Amer. Math. Soc. 71 (1965), pp. 40-106.

[22] Manes, E.G. A triple miscellany: some aspects of the theory of algebras over a triple. Dissertation, Wesleyan U., Middletown, Conn., 1967.

[23] Maranda, J.-M. On fundamental constructions and adjoint functors. Canad. Math. Bull. 9 (1966), pp. 581-591.

[24] Słomiński, J. The theory of abstract algebras with infinitary operations. Rozprawy Mat. 18, Warszawa, 1959.

[25] Sonner, J. On the formal definition of categories. Math. Z. 80 (1962), pp. 163-176.

[26] Tierney, M. Model-induced cotriples. This volume.

[27] Ulmer, F. Dense subcategories of functor categories. J. Alg. (to appear).

APPLIED FUNCTORIAL SEMANTICS, II

by

F. E. J. Linton *

Introduction.

In this note, we derive from Jon Beck's precise triple-ableness theorem (stated as Theorem 1 - for proof see [1]) a variant (appearing as Theorem 3) which resembles the characterization theorem for varietal categories (see [6 , Prop. 3] - in the light of [7], varietal categories are just categories tripleable over $\mathfrak{S}$). It turns out that this variant not only specializes to the theorem it resembles, but lies at the heart of a short proof of M. Bunge's theorem [2] (known also to P. Gabriel [5]) characterizing functor categories $\mathfrak{S}^{\mathfrak{C}} = (\mathfrak{C},\mathfrak{S})$ of all set-valued functors on a small category $\mathfrak{C}$.

* The research embodied here was supported by an N.A.S.-N.R.C. Postdoctoral Research Fellowship ; carried out at the Forschungsinstitut für Mathematik, E.T.H. Zürich, while the author was on leave from Wesleyan University, Middletown, Conn.; presented to the E.T.H. triples seminar; and improved, in §5, by gratefully received remarks of Jon Beck.

§ 1. The precise tripleableness theorem.

Our starting point is the assumption of familiarity with the precise tripleableness theorem [1] and its proof. This is summarized below as Theorem 1. The basic situation is a functor $U : \mathbb{C} \longrightarrow \mathbb{A}$ having a left adjoint $F : \mathbb{A} \longrightarrow \mathbb{C}$ with front and back adjunctions $\eta : id_{\mathbb{A}} \longrightarrow UF$, $\beta : FU \longrightarrow id_{\mathbb{C}}$. In this situation, one obtains a triple $\mathbb{T} = (UF, \eta, U\beta F)$ on $\mathbb{A}$ and a functor $\Phi : \mathbb{C} \longrightarrow \mathbb{A}^{\mathbb{T}}$ (satisfying $U^{\mathbb{T}} \circ \Phi = U$), defined by

$$\Phi X = (UX,\ U\beta_X)$$
$$\Phi \xi = U\xi \quad .$$

The concern of all tripleableness theorems is whether Φ is an equivalence.

We will have repeated occasion to consider so-called U-split coequalizer systems. These consist of a pair

(1.1) $$X \underset{g}{\overset{f}{\rightrightarrows}} Y$$

of $\mathbb{C}$-morphisms and three $\mathbb{A}$-morphisms

(1.2) $$UX \xleftarrow{\ d_1\ } UY \underset{d_o}{\overset{p}{\rightleftarrows}} Z$$

for which the four identities

(1.3) $$\begin{cases} pUF = pUg \\ pd_o = id_Z \\ d_o p = gd_1 \\ id_{UY} = fd_1 \end{cases}$$

are valid. An $id_{\mathbb{A}}$-split coequalizer system will be called simply a split coequalizer system.

Lemma 1. If

(1.4) $$A \underset{g}{\overset{f}{\rightrightarrows}} B \xrightarrow{\ p\ } C, \qquad d_1 : B \to A, \quad d_o : C \to B$$

is a split coequalizer system in $\mathbb{A}$, *then* $p = \text{coeq}(f,g)$. *Conversely, if* $B \xrightarrow{p} C$ *is a split epimorphism, with section* $d_o : C \longrightarrow B$, *and* $A \overset{f}{\underset{g}{\rightrightarrows}} B$ *is its kernel pair, defining* $d_1 : B \longrightarrow A$ *by the requirements* $fd_1 = id_B$, $gd_1 = d_o p$ *provides a split coequalizer diagram* (1.4).

Proof: Let $x : B \longrightarrow ?$ be any map. Then if $xf = xg$, $x = xfd_1 = xgd_1 = xd_o p$. Conversely, if $x = xd_o p$, then

$$xf = xd_o pf = xd_o pg = xg \ .$$

Consequently, $xf = xg$ iff x factors through p by xd_o . That settles the first statement. The second is even more trivial.

The class of all pairs of $\mathfrak{C}$-morphisms arising as (1.1) in a U-split coequalizer system (1.1), (1.2) will be denoted $\mathfrak{P}$. $\mathfrak{P}_F$ will denote those pairs in $\mathfrak{P}$ whose domain and codomain are values of F . Since we shall have to deal with yet other subclasses of $\mathfrak{P}$, we formulate the next three definitions in terms of an arbitrary class $\mathfrak{G}$ of pairs (1.1) of $\mathfrak{C}$-morphisms.

Definition. $\mathfrak{C}$ *has* $\mathfrak{G}$*-coequalizers* if each pair $(f,g) \in \mathfrak{G}$ has a coequalizer in $\mathfrak{C}$; U *reflects* $\mathfrak{G}$*-coequalizers* if, given a diagram

$$X \overset{f}{\underset{g}{\rightrightarrows}} Y \xrightarrow{p} Z \tag{1.5}$$

in $\mathfrak{C}$, with $(f,g) \in \mathfrak{G}$ and $Up = \text{coeq}(Uf,Ug)$, it follows that $p = \text{coeq}(f,g)$; U *preserves* $\mathfrak{G}$*-coequalizers* if, given a diagram (1.5) with $(f,g) \in \mathfrak{G}$ and $p = \text{coeq}(f,g)$, it follows that $Up = \text{coeq}\,(Uf,Ug)$.

Theorem 1. (Beck [1]). *If* U, F, $\mathbb{T}$ *and* $\Phi : \mathfrak{C} \longrightarrow \mathbb{A}^{\mathbb{T}}$ *are as in the basic situation above, then* Φ *is an equivalence if and only if* $\mathfrak{C}$ *has and* U *preserves and reflects* $\mathfrak{P}$*-coequalizers. More precisely, we have the following implications, some accompanied by their reasons.*

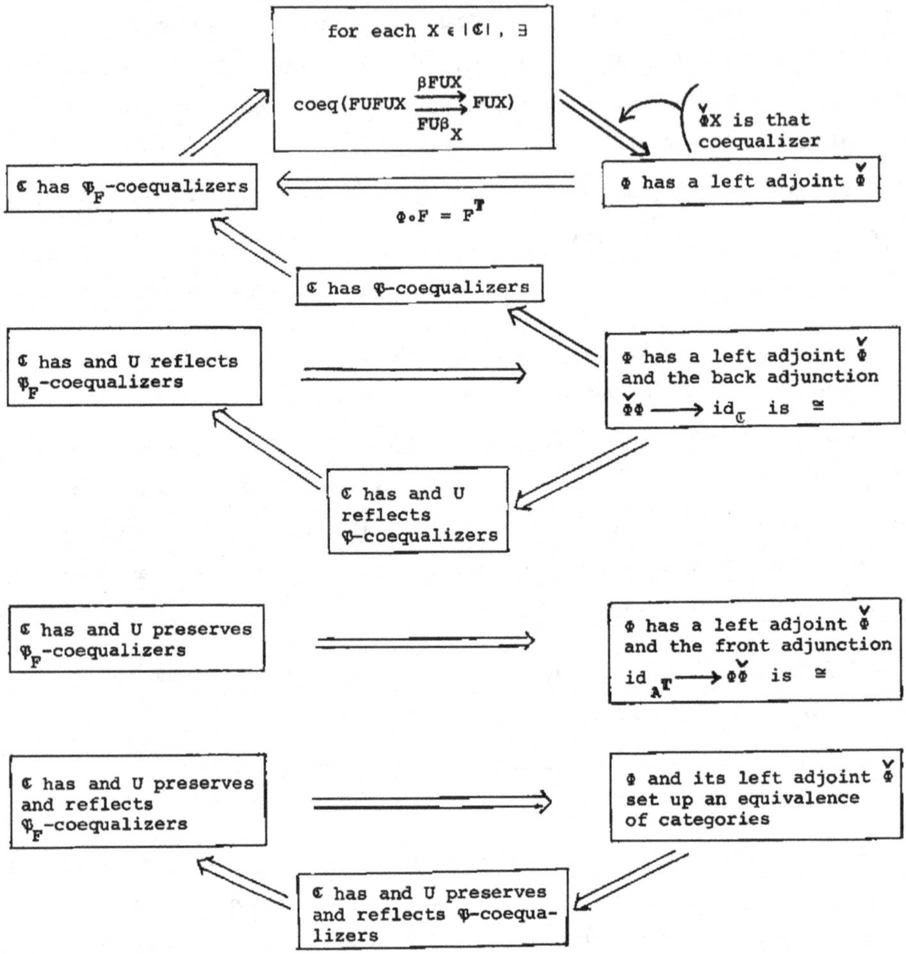

<u>Remark</u>. Φ will be an isomorphism if and only if it is an equivalence and U <u>creates isomorphisms</u>, in the sense: given X in $\mathbb{C}$ and an isomorphism

$$f : A \longrightarrow UX$$

in $\mathbb{A}$, there is one and only one $\mathbb{C}$-morphism $g : X' \longrightarrow X$ satisfying the single requirement

$$Ug = f$$

(which, of course, entails $UX' = A$), and that $\mathfrak{C}$-morphism is an isomorphism. For details on this fact, which will enter tangentially in §5, consult [8 , §0.8 and (1.2.9)].

§ 2. When $\mathfrak{A}$ has enough kernel pairs.

For the first variation on the theme of Theorem 1, we introduce the class $\mathfrak{P}_c$ of all pairs of $\mathfrak{C}$-morphisms

(2.1)
$$FE \overset{f}{\underset{g}{\rightrightarrows}} X$$

arising as follows:

i) there is a split epimorphism $p : UX \longrightarrow B$ in $\mathfrak{A}$;

ii) $E \overset{f_o}{\underset{g_o}{\rightrightarrows}} UX$ is its kernel pair ;

iii) $f = \beta_X \circ Ff_o$, $g = \beta_X \circ Fg_o$;

iv) $pUf = pUg$.

It follows that $p = \text{coeq}(Uf,Ug)$ and that

$$E \xrightarrow{\eta_E} UFE \overset{Uf}{\underset{Ug}{\rightrightarrows}} UX$$

is p's kernel pair (for iv) $\Longrightarrow \exists! \varepsilon : UFE \longrightarrow E$ with $f_o\varepsilon = Uf$, $g_o\varepsilon = Ug$, whence

$$f_o \circ \varepsilon \circ \eta_E = Uf \circ \eta_E = f_o$$

$$g_o \circ \varepsilon \circ \eta_E = Ug \circ \eta_E = g_o$$

whence $\varepsilon \circ \eta_E = id_E$; hence $qUf = qUg$ iff $qf_o = qg_o$, and $p = \text{coeq}(f_o,g_o) =$ $= \text{coeq}(Uf,Ug)$; the second assertion is obvious).

Conversely, if (2.1) is a pair of $\mathfrak{C}$-morphisms for which (Uf,Ug) has a coequalizer

p , if p is a split epimorphism, and if

$$(2.2) \qquad E \xrightarrow{\ \eta_E\ } UFE \overset{Uf}{\underset{Ug}{\rightrightarrows}} UX$$

is a kernel pair for p , then where $f_o = Uf\circ\eta_E$, $g_o = Ug\circ\eta_E$, (2.1) arises from p and f_o , g_o through steps i)...iv).

We use these remarks to prove

Lemma 2. $\mathfrak{P}_c \subseteq \mathfrak{P}$. *Moreover, if* (2.1) $\in \mathfrak{P}_c$ *and* $\xi : X \longrightarrow X'$, *then* $\xi f = \xi g$ *iff* $U(\xi f) = U(\xi g)$ $(f_o = Uf\circ\eta_E$, $g_o = Ug\circ\eta_E)$.

Proof. If (2.1) depicts a pair in $\mathfrak{P}_c$, (Uf, Ug) has as coequalizer a split epimorphism $p : UX \longrightarrow B$ in $\mathfrak{A}$, with section $d_o : B \longrightarrow UX$, whose kernel pair (2.2). By Lemma 1, there is a map $d_1 : UX \longrightarrow E$ making

$$E \overset{Uf\circ\eta_E}{\underset{Ug\circ\eta_E}{\rightrightarrows}} UX \overset{p}{\underset{d_o}{\rightleftarrows}} B \qquad (d_1 : UX \longrightarrow E)$$

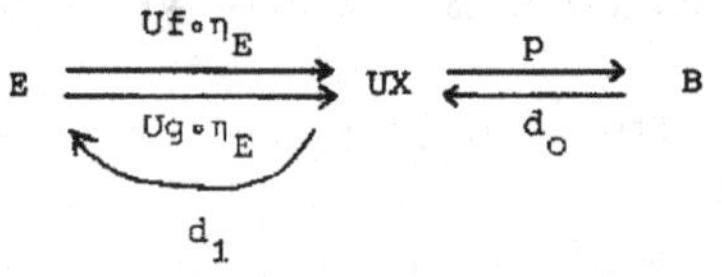

a split coequalizer diagram. Then so is

$$UFE \overset{Uf}{\underset{Ug}{\rightrightarrows}} UX \overset{p}{\underset{d_o}{\rightleftarrows}} B \ , \qquad (\eta_E d_1 : UX \longrightarrow UFE)$$

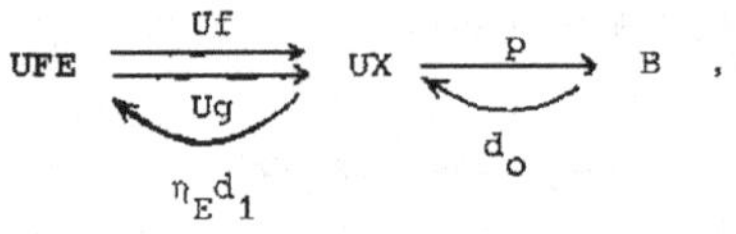

whence $\mathfrak{P}_c \subseteq \mathfrak{P}$. For the second assertion, the adjointness results in the equivalence of $\xi f = \xi g$ with $U\xi f_o = U\xi g_o$. But the relation coeq (Uf, Ug) = p = coeq (f_o, g_o) shows that $U\xi f_o = U\xi g_o$ iff $U\xi Uf = U\xi Ug$, which completes the proof.

Write $\mathfrak{P}_{Fc} = \mathfrak{P}_c \cap \mathfrak{P}_F$.

Lemma 3. *Assume* $\mathfrak{A}$ *has kernel pairs of split epimorphisms. Whenever* $FX \overset{f}{\underset{g}{\rightrightarrows}} FY$ *is a pair in* $\mathfrak{P}_F$, *there is a pair* $FE \overset{f'}{\underset{g'}{\rightrightarrows}} FY$ *in* $\mathfrak{P}_{Fc}$ *satisfying* $qf = qg$ iff $qf' = qg'$, *for every* $\mathfrak{C}$*-morphism* $q : FY \longrightarrow ?$.

<u>Proof</u>. Since $(f,g) \in \mathfrak{P}_F$, there are $\mathbb{A}$-morphisms

$$UFX \xleftarrow{\ d_1\ } UFY \underset{d_o}{\overset{p}{\rightleftarrows}} B$$

which, with (Uf, Ug), make a split coequalizer diagram in $\mathbb{A}$. Since $Uf = U^{\mathbb{T}}\Phi f$, $Ug = U^{\mathbb{T}}\Phi g$, we see that $\Phi f, \Phi g : \Phi FX \rightrightarrows \Phi FY$ and d_1 , p , d_o make a $U^{\mathbb{T}}$-split coequalizer system in $\mathbb{A}^{\mathbb{T}}$. Hence there is a $\mathbb{T}$-algebra structure $TB \longrightarrow B$ on B making $p : UFY \longrightarrow B$ a $\mathbb{T}$-homomorphism; letting (E,ε) be its kernel pair (possible because the kernel pair exists in $\mathbb{A}$ by hypothesis and lifts to $\mathbb{A}^{\mathbb{T}}$ by a property (cf. [4 , Prop. 5.1], [7 , §6], or [8 , (1.2.1.)]) of $U^{\mathbb{T}}$, we obtain an $\mathbb{A}$-object E , a pair of maps

$$E \underset{g_o'}{\overset{f_o'}{\rightrightarrows}} UFY$$

serving as a kernel pair of p , and a map $\varepsilon : UFE \longrightarrow E$ satisfying (at least)

(2.3) $$\varepsilon \circ \eta_E = id_E ,$$

and making the squares

(2.4) 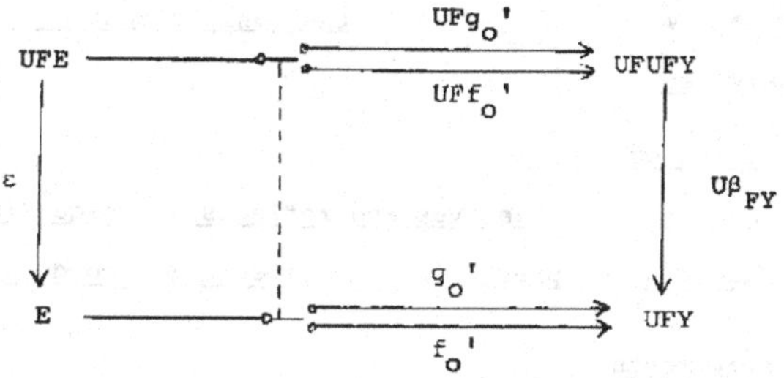

commute. Let $f' = \beta_{FY} \circ Ff_o'$, $g' = \beta_{FY} \circ Fg_o'$. Then $Uf' \circ \eta_E = f_o'$, $Ug' \circ \eta_E = g_o'$ (by adjointness) and, since $p = \mathrm{coeq}\,(f_o', g_o')$, (2.3) shows $p = \mathrm{coeq}\,(Uf', Ug')$, and

$$E \longrightarrow UFE \rightrightarrows UFY$$

is its kernel pair. Thus $(f', g') \in \mathfrak{P}_{Fc}$; finally, $\xi f = \xi g$ iff $U\xi f_o = U\xi g_o$ iff $U\xi$ factors through p iff $U\xi f_o' = U\xi g_o'$ iff $\xi f' = \xi g'$ (by adjointness, constructions and Lemma 2.)

Corollary. *Assume* $\mathfrak{A}$ *has kernel pairs of split epimorphisms. Then* $\mathfrak{C}$ *has* Φ_F*-coequalizers iff it has* Φ_{Fc}*-coequalizers,* U *preserves* Φ_F*-coequalizers iff it preserves* Φ_{Fc}*-coequalizers, and* U *reflects* Φ_F*-coequalizers iff it reflects* Φ_{Fc}*-coequalizers.*

Proof. The inclusion $\Phi_{Fc} \subset \Phi_F$ guarantees three of the implications. For the other three, we rely on Lemma 3 : given a pair $(f,g) \in \Phi_F$, let (f',g') be a pair in Φ_{Fc} having the property $\xi f = \xi g \Longleftrightarrow \xi f' = \xi g'$. Then any coequalizer for (f',g') must be a coequalizer for (f,g), and conversely. Hence if $\mathfrak{C}$ has Φ_{Fc}- coequalizers, (f',g'), and consequently (f,g), has a coequalizer. Similarly, if p is a coequalizer for (f,g) and U preserves Φ_{Fc}- coequalizers, Up is a coequalizer for (Uf', Ug'), hence a coequalizer for the kernel pair of the coequalizer of (Uf, Ug), hence a coequalizer of (Uf, Ug). Finally, if p is a map with Up a coequalizer of (Uf, Ug), Up is also a coequalizer for the kernel pair of p , hence p is a coequalizer of (f',g'), hence of (f,g).

From this corollary and Theorem 1 follows

Theorem 2. *Let* $\mathfrak{A}$ *be a category having kernel pairs of split epimorphisms, and let* U, F, T, $\Phi : \mathfrak{C} \longrightarrow \mathfrak{A}^{\mathbf{T}}$ *be as in the basic situation. Then the following statements are equivalent:*

1) Φ *is an equivalence*
2) $\mathfrak{C}$ *has and* U *preserves and reflects* Φ_C*-coequalizers*
3) $\mathfrak{C}$ *has and* U *preserves and reflects* Φ_{Fc}*-coequalizers*

Indeed, the statements

4) Φ *has a left adjoint*
5) $\mathfrak{C}$ *has* Φ_F*-coequalizers*
6) $\mathfrak{C}$ *has* Φ_{Fc}*-coequalizers*

are mutually equivalent, as are

7) Φ *has a left adjoint and* $\check{\Phi}\Phi \longrightarrow id_{\mathfrak{C}}$ *is an equivalence*

8) $\mathfrak{C}$ *has and* U *reflects* φ_c*-coequalizers*

9) $\mathfrak{C}$ *has and* U *reflects* φ_{Fc}*-coequalizers.*

Proof. Apply the Corollary to Lemma 3, and the inclusions

$$\begin{array}{ccccc} & & \varphi_F & & \\ & \nearrow & & \searrow & \\ \varphi_{Fc} & & & & \varphi \\ & \searrow & & \nearrow & \\ & & \varphi_c & & \end{array}$$

to Theorem 1, to prove $3n-2 \Longrightarrow 3n-1 \Longrightarrow 3n \Longrightarrow 3n-2$ $(n = 1,2,3)$.

§ 3. When $\mathfrak{A}$ is very like {sets} .

The second variation on Theorem 1 will eventually require more stringent restrictions on $\mathfrak{A}$. As in §2, we do the hypothesis juggling first, imposing the restrictions on $\mathfrak{A}$ as required. We stay in the basic situation of an $\mathfrak{A}$-valued functor $U : \mathfrak{C} \longrightarrow \mathfrak{A}$ having a left adjoint $F : \mathfrak{A} \longrightarrow \mathfrak{C}$. $\mathbf{T}$ is the resulting triple, and $\Phi : \mathfrak{C} \longrightarrow \mathfrak{A}^{\mathbf{T}}$ the semantical comparison functor for U , as before.

Lemma 4. Assume $\mathfrak{A}$ *has kernel pairs of split epimorphisms and that* U *reflects* φ_c*-coequalizers. Let* $p : X \longrightarrow Y$ *be a* $\mathfrak{C}$*-morphism with* Up *a split epimorphism. Then* p *is a coequalizer.*

Proof. Let $E \underset{g_o}{\overset{f_o}{\rightrightarrows}} UX$ be a kernel pair of Up . Then there is a map $\varepsilon : UFE \longrightarrow E$ making

$$(E,\varepsilon) \underset{g_o}{\overset{f_o}{\rightrightarrows}} \Phi X$$

a kernel pair of Φp . As in the proof of Lemma 3

$$FE \underset{g}{\overset{f}{\rightrightarrows}} X$$

(where $f = \beta_X \circ Ff_o$, $g = \beta_X \circ Fg_o$), is in φ_c , and so, since $Up = \text{coeq}\ (Uf, Ug)$, $p = \text{coeq}\ (f,g)$.

Lemma 5. Assume nothing about $\mathfrak{A}$, *but only that* Up *epi* $\rightarrow$ p *is a coequalizer. Then* U *is faithful, reflects monomorphisms, and reflects isomorphisms, and* $\beta_X : FUX \longrightarrow X$ *is coequalizer.*

Proof. For a functor U with left adjoint, the implication Up epi $\longrightarrow$ p epi guarantees (see [3 , Prop. II.1.5]) U to be faithful. A faithful functor obviously reflects monomorphisms. Finally, if U(p) is an isomorphism, p is a coequalizer and a monomorphism, hence an isomorphism.

Lemma 6. Assume $\mathfrak{C}$ *has kernel pairs,* $\mathfrak{A}$ *has coequalizers, and every epimorphism in* $\mathfrak{A}$ *splits. Suppose* U *is faithful and preserves* Φ_C*-coequalizers. Then, if the* $\mathfrak{C}$*-morphism* p *is a coequalizer,* Up *is (split) epi.*

Proof. Let $p : X \longrightarrow Y$ be a coequalizer, let $E \overset{f_o}{\underset{g_o}{\rightrightarrows}} X$ be its kernel pair. Then p is a coequalizer of (f_o, g_o). Now (Uf_o, Ug_o) is a kernel pair of Up (since U has a left adjoint) and hence fits in a split coequalizer diagram

Let $f,g : FUE \longrightarrow X$ correspond by adjointness to Uf_o , Ug_o . Then $(f,g) \in \Phi_C$ (roughly because $Ue \longrightarrow UFUE \longrightarrow UE = id_{UE}$) and, for any map $q : E \longrightarrow ?$, $qf_o = qg_o$ iff $UqUf_o = UqUg_o$ iff $qf = qg$ (using faithfulness of U and the adjointness naturality). So p is a coequalizer of $(f,g) \in \Phi_C$, and since U preserves Φ_C-coequalizers, Up is a coequalizer, too (of Uf,Ug), hence is (split) epi.

(Remark: need only suppose $\mathfrak{A}$ has coeq of kernel pairs, not of everything.)

Lemma 7. Assume $\mathfrak{C}$ *has kernel pairs and* Φ_C*-coequalizers,* $\mathfrak{A}$ *has coequalizers, and every epimorphism in* $\mathfrak{A}$ *splits. Suppose* U *is faithful, reflects isomorphisms, and preserves* Φ_C*-coequalizers. Then a pair of* $\mathfrak{C}$*-morphisms*

$$E \overset{f}{\underset{g}{\rightrightarrows}} X$$

is a kernel pair if (and, in view of U's left adjoint, only if)

$$UE \overset{Uf}{\underset{Ug}{\rightrightarrows}} UX$$

<u>is a kernel pair</u>.

<u>Proof</u>. Assume (Uf, Ug) is a kernel pair. Let $p : UX \longrightarrow Z$ be its coequalizer; then, since p is split epi and (Uf, Ug) is a kernel pair for p , we obtain a split coequalizer diagram

$$UE \rightrightarrows UX \xrightarrow{\ p\ } Z$$

Now $FUE \longrightarrow E \overset{f}{\underset{g}{\rightrightarrows}} X$ is therefore a $\mathfrak{P}_c$-pair (since $UE \longrightarrow UFUE \longrightarrow UE = id_{UE}$) , has a coequalizer $q : X \longrightarrow Y$ in $\mathfrak{C}$, which, because of the faithfulness of U , is a coequalizer for (f,g), too. Let $E' \overset{f'}{\underset{g'}{\rightrightarrows}} X$ be a kernel pair for q . We shall prove $E' \overset{f'}{\underset{g'}{\rightrightarrows}} X$ is isomorphic to $E \overset{f}{\underset{g}{\rightrightarrows}} X$ by using the hypothesis that U reflects isomorphisms. We have, in any case, a $\mathfrak{C}$-morphism $E \xrightarrow{\ e\ } E'$ with $f'e = f$, $g'e = g$, and the knowledge that (Uf', Ug') = ker pair (Uq), (Uf, Ug) = = ker pair (p). Since U preserves $\mathfrak{P}_c$-coequalizers, however, Uq = coeq (Uf, Ug). Thus p is isomorphic with Uq , whence $Ue : UE \longrightarrow UE'$ is an isomorphism, whence e is an isomorphism, and (f,g) is a kernel pair.

We can now prove one half of

<u>Theorem 3</u>. <u>Let $\mathfrak{A}$ be a category in which every epimorphism splits, and in which kernel pairs and difference cokernels are available. Let</u> $U : \mathfrak{C} \longrightarrow \mathfrak{A}$, F, $\mathbf{T}$, $\Phi : \mathfrak{C} \longrightarrow \mathfrak{A}^{\mathbf{T}}$ <u>be as in the basic situation. Then Φ is an equivalence of categories if and only if</u>

1) $\mathfrak{C}$ <u>has kernel pairs and $\mathfrak{P}_c$-coequalizers</u>

2) Up <u>epi</u> $\Longleftrightarrow$ p <u>is a coequalizer</u>

3) (f,g) <u>is a kernel pair if (and only if)</u> (Uf, Ug) <u>is a kernel pair</u>.

<u>Proof</u>. If Φ is an equivalence, Theorem 2 guarantees the $\mathfrak{P}_c$-coequalizers, and general principles guarantee the kernel pairs. Theorem 2 and Lemma 4 guarantee the implication Up epi $\Longrightarrow$ p a coequalizer. Lemma 5 applied to this implication,

Theorem 2, and Lemma 6 then provide the converse implication. Statement 3 follows from Lemma 7. The converse argument is outlined in statement 1 and the parenthetical remarks in statements 2 and 3 of the following theorem, whose proof, outlined below, is entirely contained in the three lemmas in §4.

Theorem 4. *With the situation as in Theorem 3, suppose throughout that* $\mathbb{C}$ *has* $\mathfrak{P}_C$*-coequalizers and kernel pairs. Then*

1) Φ *has a left adjoint.*

2) *If condition 2 of Theorem 3 holds, then* U *reflects* $\mathfrak{P}_C$*-coequalizers (whence the back adjunction* $\check{\Phi}\Phi \longrightarrow id_{\mathbb{C}}$ *is an equivalence) and any* $\mathbf{T}$*-algebra* (A,α) *admitting a jointly monomorphic family of maps to values of* Φ *is itself (isomorphic to) a value of* Φ *, namely* $\Phi\check{\Phi}(A,\alpha)$.

3) *If conditions 2 and 3 of Theorem 3 hold, then* U *preserves (and reflects)* $\mathfrak{P}_C$*-coequalizers (whence the front adjunction* $id_{A^{\mathbf{T}}} \longrightarrow \Phi\check{\Phi}$ *is an equivalence too, and* Φ *and* $\check{\Phi}$ *set up an equivalence of categories).*

Outline of proof. Theorem 2 proves 1). Lemma 8, Theorem 2, and Lemma 9 prove 2). 2), Lemma 9, Lemma 10, and Theorem 2 prove 3). Theorem 3 obviously follows. Lemmas 8, 9, 10 are proved in §4.

§ 4. Proof of Theorem 4.

Lemma 8. *With the situation as in Theorem 3, conditions 1 and 2 of Theorem 3 imply that* U *reflects* $\mathfrak{P}_C$*-coequalizers.*

Proof. Let $FE \overset{f}{\underset{g}{\rightrightarrows}} X$ be in $\mathfrak{P}_C$ and suppose $\xi : X \longrightarrow X'$ is a $\mathbb{C}$-morphism for which

$$(4.1) \qquad U\xi = \text{coeq}(Uf, Ug).$$

From condition 2 of Theorem 3, it follows that ξ is itself a coequalizer of something. Next, the equality $U\xi Uf = U\xi Ug$ (consequence of (4.1)), taken with the faithfulness of U (consequence of Lemma 5), shows

(4.2) $$\xi f = \xi g .$$

Condition 1 of Theorem 3 permits us to take a coequalizer $p : X \longrightarrow Z$ of the pair (f,g). Equation (4.2) then entails a unique $\mathbb{C}$-morphism $z : Z \longrightarrow X'$ satisfying

(4.3) $$z \circ p = \xi .$$

Since $pf = pg$, (4.1) affords a unique $\mathbb{A}$-morphism $z' : UX' \longrightarrow UZ$ satisfying

(4.4) $$z' \circ U\xi = Up .$$

Combining (4.3) and (4.4), we obtain the equations

(4.5) $$Up = z' \circ U\xi = z' \circ Uz \circ Up$$

(4.6) $$U\xi = Uz \circ Up = Uz \circ z' \circ U\xi$$

But Up is epi, since p is a coequalizer (using condition 2 of Theorem 3) and $U\xi$ is epi, being itself a coequalizer, so from (4.5) and (4.6) it follows that

$$z' \circ Uz = id , \quad Uz \circ z' = id ,$$

whence Uz is an isomorphism. Another appeal to Lemma 5 demonstrates that z is an isomorphism, from $p = \text{coeq}\,(f,g)$ to ξ, whence ξ is a coequalizer of (f,g), as needed to be shown.

__Lemma 9__. __With the situation as in Theorem 3, conditions 1 and 2 of Theorem 3 imply any object $X \in \mathbb{A}^{\mathbb{T}}$ admitting a jointly monomorphic family of maps to values of Φ is itself (isomorphic to) a value of Φ, namely $\Phi\check{\Phi}X$.__

__Proof__. Condition 1 of Theorem 3 and Theorem 2 guarantee a left adjoint $\check{\Phi}$ for Φ. Now, given a family of $\mathbb{A}^{\mathbb{T}}$-morphisms

$$f_i : X \longrightarrow \Phi Y_i \qquad (Y_i \in |\mathbb{C}| , X = (A,\alpha) , i \in I)$$

for which the implication $f_i \circ a = f_i \circ b \,\forall_i \Longrightarrow a = b$ holds for all $\mathbb{A}^{\mathbb{T}}$-morphisms a,b with codomain X, form the maps

$$\hat{f}_i : \check{\Phi}X \longrightarrow Y_i$$

resulting by adjointness, and, applying Φ to them, consider the diagrams

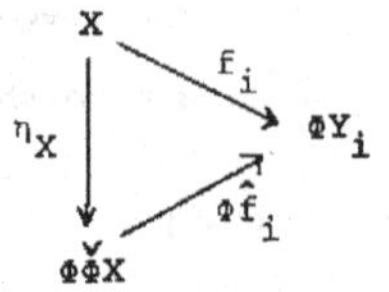

where η_X is the front adjunction for the adjointness of Φ to $\check{\Phi}$. If $\eta \circ a = \eta \circ b$, then $f_i \circ a = \Phi \hat{f}_i \circ \eta \circ a = \Phi \hat{f}_i \circ \eta \circ b = f_i \circ b$, whence η is a monomorphism. It is a matter of indifference whether this statement is understood in $\mathbf{A}$ or in $\mathbf{A}^{\mathbf{T}}$, for, being faithful and having a left adjoint, $U^{\mathbf{T}}$ preserves and reflects monomorphisms. To show η is an isomorphism, as required, it thus suffices to prove $U^{\mathbf{T}}\eta$ is (split) epi, since $U^{\mathbf{T}}$ certainly reflects isomorphisms.

To do this, we must recall the construction of $\check{\Phi}X$. $\check{\Phi}X$ is the coequalizer, via some projection $p : FA \longrightarrow \check{\Phi}X$, of the $\mathcal{G}_C$-pair $FE \rightrightarrows FA$ arising by adjointness from the kernel pair of $\alpha : UFA \longrightarrow A$. Now the coequalizer of $\Phi FE \rightrightarrows \Phi FA$ (which is $F^{\mathbf{T}}E \rightrightarrows F^{\mathbf{T}}A$) is just $\alpha : F^{\mathbf{T}}A \longrightarrow X = (A,\alpha)$ itself, hence there is a unique map $X = (A,\alpha) \longrightarrow \Phi\check{\Phi}X$ making the diagram

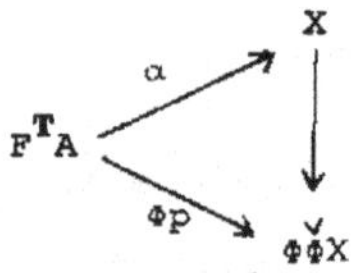

commute: that map is η_X. Since p is a coequalizer, $Up = U^{\mathbf{T}}\Phi p$ is epi ; hence $U^{\mathbf{T}}\eta$ is (split) epi. This completes the proof.

Lemma 10. *With the situation as in Theorem 3,* U *preserves* $\mathcal{G}_C$*-coequalizers if*

i) $\mathbf{C}$ *has* $\mathcal{G}_C$*-coequalizers (all that's really needed is a left adjoint* $\check{\Phi}$ *for* Φ *)*

ii) Up *is epi if* p *is a coequalizer*

iii) (f, g) *is a kernel pair if* (Uf, Ug) *is a kernel pair, and*

iv) <u>the conclusion of Lemma 9 holds</u>.

<u>Proof</u>. Given a pair $FE \overset{f}{\underset{g}{\rightrightarrows}} X$ in φ_C, and a map $p : X \longrightarrow Z$, coequalizer of f, g, we must show

$$Up = \operatorname{coeq}(Uf, Ug) .$$

Since $E \xrightarrow[\eta_E]{} UFE \overset{Uf}{\underset{Ug}{\rightrightarrows}} X$ is the kernel pair of (Uf, Ug)'s coequalizer in $\mathfrak{A}$, there is a unique $\mathfrak{A}$-morphism $\varepsilon : UFE \longrightarrow E$ making the diagrams

$$\begin{array}{ccc} UFE & \xrightarrow{Uf} & UX \\ {\scriptstyle\varepsilon}\downarrow & & \uparrow{\scriptstyle Uf} \\ E & \xrightarrow[\eta_E]{} & UFE \end{array} \qquad\qquad \begin{array}{ccc} UFE & \xrightarrow{Ug} & UX \\ {\scriptstyle\varepsilon}\downarrow & & \uparrow{\scriptstyle Ug} \\ E & \xrightarrow[\eta_E]{} & UFE \end{array}$$

commute. It is left as an easy exercise to prove that (E,ε) is then a $\mathbf{T}$-algebra and that

$$(4.7) \qquad (E,\varepsilon) \overset{Uf\circ\eta_E}{\underset{Ug\circ\eta_E}{\rightrightarrows}} \Phi X$$

is a jointly monomorphic pair of $\mathbf{T}$-homomorphisms. By iv), $(E,\varepsilon) \cong \Phi\check{\Phi}(E,\varepsilon)$; and there are maps

$$(4.8) \qquad \check{\Phi}(E,\varepsilon) \overset{\hat{f}}{\underset{\hat{g}}{\rightrightarrows}} X$$

corresponding, by the adjointness of Φ to $\check{\Phi}$, to (4.7). Using the adjointness relations and the definition of Φ, a $\mathfrak{C}$-morphism $q : X \longrightarrow Z$ satisfies $qf = qg$ iff $Uq(Uf\circ\eta_E) = Uq(Ug\circ\eta_E)$ iff $\Phi q\circ(Uf\circ\eta_E) = \Phi q\circ(Ug\circ\eta_E)$ iff $q\circ\hat{f} = q\circ\hat{g}$. Consequently, $p = \operatorname{coeq}(f, g) = \operatorname{coeq}(\hat{f}, \hat{g})$.

Next, since

$$(4.9) \qquad (U\hat{f}, U\hat{g}) = (U^{\mathbf{T}}\Phi\hat{f}, U^{\mathbf{T}}\Phi\hat{g}) \cong (Uf\circ\eta_E, Ug\circ\eta_E) ,$$

and the latter is a kernel pair (since $f,g \in \varphi_C$), $(U\hat{f}, U\hat{g})$ is a kernel pair, too, whence, by iii), (4.8) is a kernel pair. Since p is its coequalizer, (4.8) is a kernel pair for p . It follows, since U has a left adjoint, that $(U\hat{f}, U\hat{g})$ is a kernel pair for Up . Then (4.9) shows (4.7) is a kernel pair for Up , too. On the other hand, Up is (split) epi, by ii), since p is a coequalizer. Consequently, Up is the coequalizer of its kernel pair, namely of (4.7). Finally, since (4.7) has the same coequalizer as (Uf, Ug), Up = coeq (Uf, Ug), as had to be shown.

Schematically, the proof of Theorem 4 and the rest of Theorem 3 follows the following pattern: if $\mathfrak{A}$ has kernel pairs and coequalizers and every $\mathfrak{A}$-epimorphism splits, and if $\mathfrak{C}$ has kernel pairs and $\mathfrak{P}_C$-coequalizers, then Φ has a left adjoint $\check{\Phi}$ (by Theorem 2) and:

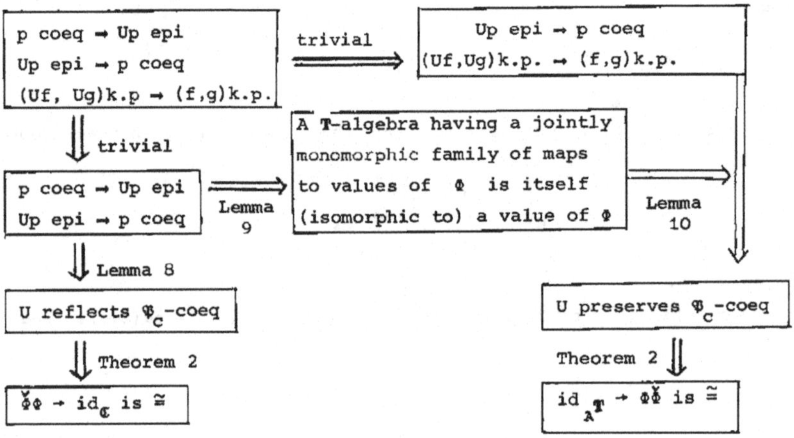

§ 5. Applications.

For the first application of Theorem 3, we take $\mathfrak{A} = \mathfrak{S} = \{sets\}$. Then, modulo the easily supplied information that any category tripleable over sets has all small limits and colimits, this instance of Theorem 3 is just the characterization theorem of [6] for varietal categories, since, by [7], varietal and tripleable over $\mathfrak{S}$ mean the same thing. For the second application of Theorem 3, which will also be the last to be presented here, we prove the theorem of Bunge-Gabriel.

<u>Theorem 5</u>. (<u>M</u>. <u>Bunge</u> [2], <u>P</u>. <u>Gabriel</u> [5]). <u>A category</u> $\mathfrak{B}$ <u>is equivalent to the functor category</u> $\mathfrak{S}^{\mathfrak{C}} = (\mathfrak{C},\mathfrak{S})$ <u>of all set valued functors on a small category</u> $\mathfrak{C}$ <u>if and only if</u>

1. $\mathfrak{B}$ <u>has kernel pairs and coequalizers</u>
2. <u>there is a set</u> X <u>and a function</u> $\psi : X \longrightarrow |\mathfrak{B}|$:
   a. $\mathfrak{B}$ <u>contains all small coproducts of images of</u> ψ
   b. $p : B \longrightarrow B'$ <u>is a coequalizer if and only if</u> $\mathfrak{B}(\psi x, p) : \mathfrak{B}(\psi x, B) \longrightarrow \mathfrak{B}(\psi x, B')$ <u>is onto</u>, $\forall x \in X$
   c. $E \underset{g}{\overset{f}{\rightrightarrows}} B$ <u>is a kernel pair if and only if</u> $\mathfrak{B}(\psi x,E) \underset{\mathfrak{B}(\psi x,g)}{\overset{\mathfrak{B}(\psi x,f)}{\rightrightarrows}} \mathfrak{B}(\psi x,B)$ <u>is a kernel pair</u>, $\forall x \in X$
   d. $x \in X \Longrightarrow \mathfrak{B}(\psi x, -)$ <u>preserves coproducts</u>.

<u>Indeed, if</u> $\mathfrak{B} \approx \mathfrak{S}^{\mathfrak{C}}$, $\psi : X \longrightarrow |\mathfrak{B}|$ <u>may be taken as the object function of</u> $\mathfrak{C}^* \longrightarrow \mathfrak{S}^{\mathfrak{C}} \xrightarrow{\approx} \mathfrak{B}$, while if $\psi : X \longrightarrow |\mathfrak{B}|$ <u>is given</u>, $\mathfrak{C}$ <u>may be taken as the full image of</u> $X \longrightarrow |\mathfrak{B}| \longrightarrow \mathfrak{B}^*$.

<u>Proof</u>. We dispense with the easy part of the proof first. To begin with, suppose $\mathfrak{B} = \mathfrak{S}^{\mathfrak{C}}$. Then surely condition 1. is valid. To check condition 2. where $X = |\mathfrak{C}|$ and $\psi(x) = \mathfrak{C}(x,=)$, note that

$$\mathfrak{S}^{\mathfrak{C}}(\psi(x), -) = ev_x : \mathfrak{S}^{\mathfrak{C}} \longrightarrow \mathfrak{S}$$

and so conditions 2b, 2c, 2d are automatic. So far as condition 2a is concerned, $\mathfrak{S}^{\mathfrak{C}}$ has all small coproducts. These, however, are all properties preserved under equivalence of categories, and that finishes the "only if" part of the proof.

For the converse, view the set X as a discrete category and let

$$X \xrightarrow{\bar{\varphi}} \mathfrak{C} \xrightarrow{\underline{\varphi}} \mathfrak{B}^*$$

be the full image factorization of the composition

$$\varphi : X \xrightarrow{\psi} |\mathfrak{B}| \xrightarrow{\text{incl.}} \mathfrak{B}^*$$

of the function ψ given by condition 2. with the inclusion of (the discrete category) $|\mathfrak{B}|$ as the class of objects of $\mathfrak{B}^*$. Here is an outline of the argument that $\mathcal{S}^{\underline{\varphi}} \circ Y : \mathfrak{B} \longrightarrow \mathcal{S}^{(\mathfrak{B}^*)} \longrightarrow \mathcal{S}^{\mathfrak{C}}$ is an equivalence.

<u>Step 1.</u> The $\mathcal{S}^X$- valued functors

$$U = \mathcal{S}^{\varphi} \circ Y : \mathfrak{B} \longrightarrow \mathcal{S}^{(\mathfrak{B}^*)} \longrightarrow \mathcal{S}^X$$

$$U_1 = \mathcal{S}^{\varphi} : \mathcal{S}^{(\mathfrak{B}^*)} \longrightarrow \mathcal{S}^X$$

$$U_2 = \mathcal{S}^{\bar{\varphi}} : \mathcal{S}^{\mathfrak{C}} \longrightarrow \mathcal{S}^X$$

all have left adjoints.

<u>Step 2.</u> If $\mathbf{T}, \mathbf{T}_1, \mathbf{T}_2$ are the triples on $\mathcal{S}^X$ and

$$\Phi : \mathfrak{B} \longrightarrow (\mathcal{S}^X)^{\mathbf{T}}$$

$$\Phi_1 : \mathfrak{B} \longrightarrow (\mathcal{S}^X)^{\mathbf{T}_1}$$

$$\Phi_2 : \mathfrak{B} \longrightarrow (\mathcal{S}^X)^{\mathbf{T}_2}$$

are the semantical comparison functors arising from U, U_1, and U_2, respectively, then Φ and Φ_2 are equivalences (in fact, Φ_2 is an isomorphism!)

<u>Step 3.</u> The commutativity of the lower triangles in the diagram

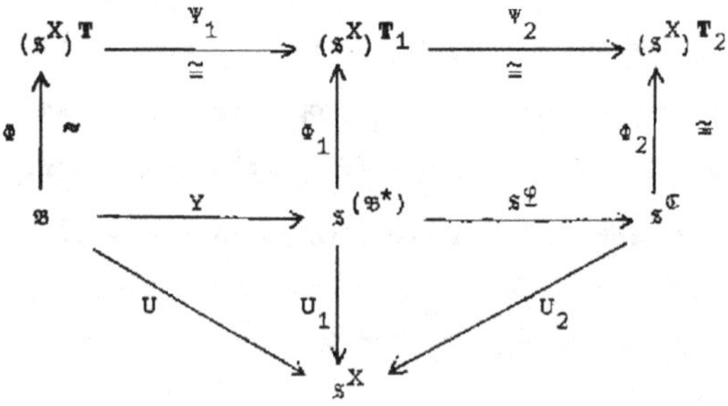

gives triple maps $\mathbf{T}_1 \xrightarrow{\tau_1} \mathbf{T}$, $\mathbf{T}_2 \xrightarrow{\tau_2} \mathbf{T}_1$, whose semantical interpretations Ψ_1, Ψ_2 on the categories of algebras make the upper squares commute; both τ_1 and

τ_2 will be shown to be isomorphisms.

It follows that Ψ_1 and Ψ_2 are isomorphisms and that

$$\mathfrak{S}^{\underline{\varphi}} \circ Y = \Phi_2^{-1} \circ \Psi_2 \circ \Psi_1 \circ \Phi : \mathfrak{B} \longrightarrow \mathfrak{S}^{\mathfrak{C}}$$

is an equivalence (indeed, an isomorphism if (and only if) $\Phi : \mathfrak{B} \longrightarrow (\mathfrak{S}^X)^T$ is an isomorphism, condition which can be expressed as an additional requirement on ψ :

2.e. Given $b \in \mathfrak{B}$, sets A_x $(x \in X)$ and one-one correspondences $f_x : A_x \xrightarrow{\cong} \mathfrak{B}(x,b)$, there is a $\mathfrak{B}$-morphism $f : b' \longrightarrow b$ uniquely determined by the single requirement

$$\mathfrak{B}(x, f) = f_x ;$$

moreover, this f is an isomorphism.

The details regarding this refinement will be omitted, being easy, and of little interest. See the remark following Theorem 1.)

<u>Step 1</u>. Using condition 2a, we produce a left adjoint F to $U = \mathfrak{S}^{\varphi} \circ Y : \mathfrak{B} \longrightarrow \mathfrak{S}^X$

$$F(G) = \bigoplus_{x \in X} Gx \cdot \psi x \qquad (G \in \mathfrak{S}^X) .$$

The identifications

$$\begin{aligned} \mathfrak{B}(F(G), b) &\cong \mathfrak{B}(\bigoplus_{x \in X} Gx \cdot \psi x, b) \\ &\cong \prod_{x \in X} \mathfrak{B}(Gx \cdot \psi x, b) \\ &\cong \prod_{x \in X} (\mathfrak{B}(\psi x, b))^{Gx} \\ &= \prod_{x \in X} \mathfrak{S}(Gx, U(b)(x)) \\ &= \mathfrak{S}^X (G, U(b)) \end{aligned}$$

show that this works. In the same way, the fact that $\mathfrak{S}^{(\mathfrak{B}^*)}$ and $\mathfrak{S}^{\mathfrak{C}}$ have all small coproducts allows us to define

$$F_1(G) = \bigoplus_{x\in X} (Gx \cdot Y(\psi x)$$

$$F_2(G) = \bigoplus_{x\in X} Gx \cdot (\mathcal{S}^{\Psi}Y(\psi x))$$

and to prove, by much the same calculations, that F_1 and F_2 serve as left adjoints to U_1 and U_2.

Step 2. Condition 1), 2), and 3) of Theorem 3 are provided precisely by conditions 1, 2b, and 2c of Theorem 5. Since $\mathfrak{A} = \mathcal{S}^X$ obviously has the properties envisioned of it in Theorem 3, the functor

$$\Phi : \mathfrak{B} \longrightarrow (\mathcal{S}^X)^{\mathbf{T}}$$

is an equivalence. In the same way, it is obvious that $U_2 : \mathcal{S}^{\mathfrak{C}} \longrightarrow \mathcal{S}^X$ fulfills the hypotheses of Theorems 1 and 3 (whichever the reader prefers to think of), and so $\Phi_2 : \mathcal{S}^{\mathfrak{C}} \longrightarrow (\mathcal{S}^X)^{\mathbf{T}_2}$ is an equivalence (in fact, an isomorphism, since U_2 creates isomorphisms).

Step 3. To see that $\tau_2 : \mathbf{T}_2 \longrightarrow \mathbf{T}_1$ is an isomorphism, refer back to step 1, and observe that $F_2 = \mathcal{S}^{\Psi}F_1$. Indeed,

$$\mathcal{S}^{\Psi}F_1(G) = \mathcal{S}^{\Psi} \bigoplus_{x\in X} Gx \cdot Y(\psi x) = \bigoplus_{x\in X} Gx \cdot \mathcal{S}^{\Psi}Y(\psi x) = F_2(G) ,$$

since $\mathcal{S}^{\Psi}$ preserves coproducts. Thus

$$U_2F_2 = U_2\mathcal{S}^{\Psi}F_1 = U_1F_1 ,$$

and this identity is the triple map τ_2.

To see that $\tau_1 : \mathbf{T}_1 \longrightarrow \mathbf{T}$ is an isomorphism, we need to invoke condition 2d, which bespeaks the fact that $U = \mathcal{S}^{\varphi} \circ Y$ preserves coproducts. Now

$$\begin{aligned} U_1F_1(G) &= U_1 \bigoplus_{x\in X} Gx \cdot Y(\psi x) \\ &= \mathcal{S}^{\varphi}(\bigoplus_{x\in X} Gx \cdot Y(\psi x)) \\ &= \bigoplus_{x\in X} Gx \cdot \mathcal{S}^{\varphi}Y(\psi x) \end{aligned}$$

and

$$UF(G) = U(\bigoplus_{x\in X} Gx \cdot \psi x)$$

$$= \bigoplus_{x\in X} Gx \cdot U\psi x = \bigoplus_{x\in X} Gx \cdot \mathcal{S}^{\varphi}Y(\psi x)$$

and it is clear that $\tau_1 : U_1F_1 \longrightarrow UF$ is this isomorphism.

This completes the proof.

Bibliography

[1] Beck, J. : Untitled manuscript, Cornell, 1966.

]2] Bunge, M. : Characterization of diagrammatic categories. Dissertation, University of Pennsylvania, 1966.

[3] Eilenberg, S., and Moore, J.C. : Foundation of relative homological algebra. Memoirs Amer. Math. Soc. 55 (1965).

[4] Eilenberg, S., and Moore, J.C. : Adjoint functors and triples. Ill. J. Math. 9 (1965), pp. 381-398.

[5] Gabriel, P. : Handwritten draft of §2 of Chevalley and Gabriel, Catégories et Foncteurs. to appear.

[6] Linton, F.E.J. : Some aspects of equational categories. Proc. Conf. Categ. Alg. (La Jolla, 1965), Springer, Berlin, 1966, pp. 84-94.

[7] Linton, F.E.J. : Outline of functorial semantics. This volume.

[8] Manes, E.G. : A triple miscellany: some aspects of the theory of algebras over a triple. Dissertation, Wesleyan University, Middletown, Conn., 1967.

COEQUALIZERS IN CATEGORIES OF ALGEBRAS *

by

F. E. J. Linton

Wesleyan University and E.T.H. Zürich

Introduction

It is well known [2,§6] that (inverse) limits in a category of algebras over $\mathcal{A}$ - in particular, in the category $\mathcal{A}^{\mathbf{T}}$ of algebras over a triple $\mathbf{T} = (T,\eta,\mu)$ on $\mathcal{A}$ - can be calculated in $\mathcal{A}$. Despite the fact that such a statement is, in general, false for colimits (direct limits), a number of colimit constructions can be carried out in $\mathcal{A}^{\mathbf{T}}$ provided they can be carried out in $\mathcal{A}$ and $\mathcal{A}^{\mathbf{T}}$ has enough coequalizers.

The coequalizers $\mathcal{A}^{\mathbf{T}}$ should have, at a minimum, are, as we shall see in §1, those of reflexive pairs: a pair

$$X \overset{f}{\underset{g}{\rightrightarrows}} Y$$

of maps f, g in a category $\mathcal{X}$ is reflexive if there is an $\mathcal{X}$-morphism

$$\Delta : Y \longrightarrow X$$

satisfying the identities

$$f\circ\Delta = id_Y = g\circ\Delta .$$

(This terminology arises from the fact that, when $\mathcal{X} = \mathcal{S}$ = {sets} , (f,g) is reflexive if and only if the image of the induced function

$$X \xrightarrow[f,g]{} Y \times Y$$

contains the diagonal of $Y \times Y$.)

---

* Research supported by an N.A.S.-N.R.C. Postdoctoral Research Grant while the author was on leave from the first named institution.

In §2 we give two criteria for $\mathcal{A}^{\mathbf{T}}$ to have coequalizers of reflexive pairs, neither of them necessary, of course. In §1, it will turn out, so long as $\mathcal{A}^{\mathbf{T}}$ has such coequalizers, that each functor

$$\mathcal{A}^{\tau} : \mathcal{A}^{\mathbf{T}} \longrightarrow \mathcal{A}^{\mathbf{S}} ,$$

induced by a map of triples $\tau : \mathbf{S} \to \mathbf{T}$, has a left adjoint, that $\mathcal{A}^{\mathbf{T}}$ has coproducts if $\mathcal{A}$ does, indeed, has all small colimits if $\mathcal{A}$ has coproducts, and that $\mathcal{A}^{\mathbf{T}}$ has tensor products if $\mathcal{A}$ does. These are, of course, known facts when $\mathcal{A} = \mathcal{S} =$ {sets}; however, at the time of this writing, it is unknown, for example, whether the category of contramodules over an associative coalgebra, presented (in [1]) as $\mathcal{A}^{\mathbf{T}}$ with $\mathcal{A} =$ {ab. groups}, has coequalizers of reflexive pairs.

§ 1. Constructions using coequalizers of reflexive pairs.

We begin with a lemma that will have repeated use. It concerns the following definition, which clarifies what would otherwise be a recurrent conceptual obscurity in the proofs of this section.

Let $U : \mathcal{X} \to \mathcal{A}$ be a functor, let $X \in |\mathcal{X}|$, and let $(f,g) = \{(f_i,g_i)/i \in I\}$ be a family of $\mathcal{A}$-morphisms

$$A_i \overset{f_i}{\underset{g_i}{\rightrightarrows}} UX \quad (i \in I). \tag{1.1}$$

An $\mathcal{X}$-morphism $p : X \to P$ is a <u>coequalizer</u> (<u>rel</u>. U) of the family of pairs (1.1) if

1) $\forall i \in I$, $Up \circ f_i = Up \circ g_i$, and
2) if $q : X \to Y$ satisfies $Uq \circ f_i = Uq \circ g_i$ $(\forall i \in I)$,

then $\exists!\, x : P \to Y$ with $q = x \circ p$.

If $U = \mathrm{id}_{\mathcal{X}} : \mathcal{X} \to \mathcal{X}$, a coequalizer (rel. U) of the family (1.1) will be called simply a <u>coequalizer</u> of (1.1).

<u>Lemma 1</u>. <u>If</u> U <u>has a left adjoint</u> $F : \mathcal{A} \to \mathcal{X}$ <u>and</u> $\bar{f}_i, \bar{g}_i : FA_i \to X$ <u>are the</u> $\mathcal{X}$-<u>morphisms corresponding to</u> f_i, g_i <u>by adjointness, then</u> $p : X \to P$ <u>is a coequalizer</u> (rel. U) of (f,g) <u>if and only if it is a coequalizer of</u> $(\bar{f},\bar{g})$. <u>If</u> U <u>is faithful</u>

and $\bar{f}_i, \bar{g}_i : X_i \to X$ *are* $\mathcal{X}$*-morphisms with* $U\bar{f}_i = f_i$, $U\bar{g}_i = g_i$, *then* $p : X \to P$ *is a coequalizer* (rel U) *of* (f,g) *if and only if it is a coequalizer of* $(\bar{f},\bar{g})$.

Proof. In the first case, the naturality of the adjunction isomorphisms yields

$$q \cdot \bar{f}_i = q \cdot \bar{g}_i \Leftrightarrow Uq \cdot f_i = Uq \cdot g_i$$

for every $\mathcal{X}$-morphism q defined on X . In the second case, that relation follows from the faithfulness of U . Clearly, that relation is all the proof required.

Proposition 1. *Let* $\mathbb{S} = (S,\eta',\mu')$ *and* $\mathbb{T} = (T,\eta,\mu)$ *be triples on* $\mathcal{A}$, *suppose the natural transformation* $\tau : S \to T$ *is a map of triples from* $\mathbb{S}$ *to* $\mathbb{T}$, *and let* (A,α) *be an* $\mathbb{S}$*-algebra*, (B,β) *a* $\mathbb{T}$*-algebra*, $p : TA \to B$ *a* $\mathbb{T}$*-homomorphism from* (TA,μ_A) *to* (B,β) , *and* $\iota = p \circ \eta_A : A \to B$. *Then the following statements are equivalent.*

1) p *is a coequalizer of the pair*

$$(TSA,\mu_{SA}) \xrightarrow{T(\tau_A)} (TTA,\mu_{TA}) \xrightarrow{\mu_A} (TA,\mu_A) \quad ; \qquad (TSA,\mu_{SA}) \xrightarrow{T(\alpha)} (TA,\mu_A)$$

2) p *is a coequalizer* (rel $U^{\mathbb{T}}$) *of the pair*

$$SA \xrightarrow{\tau_A} TA \quad ; \qquad SA \xrightarrow{\alpha} A \xrightarrow{\eta_A} TA$$

3) ι *is an* $\mathbb{S}$*-homomorphism* $(A,\alpha) \to (B,\beta \cdot \tau_B)$ *making the composition*

$$\begin{array}{c} \mathcal{A}^{\mathbf{T}}((B,\beta),X) \\ \downarrow \\ \mathcal{A}^{\mathbf{S}}(\mathcal{A}^{T}(B,\beta),\mathcal{A}^{T}X) \\ \downarrow = \\ \mathcal{A}^{\mathbf{S}}((B,\beta\cdot\tau_B),\mathcal{A}^{T}X) \\ \downarrow \iota \\ \mathcal{A}^{\mathbf{S}}((A,\alpha),\mathcal{A}^{T}X) \end{array}$$

<u>a one-one correspondence</u>, $\forall X \in |\mathcal{A}^{\mathbf{T}}|$.

<u>Proof</u>. The equivalence of statements 1) and 2) follows from Lemma 1, since $\mu_A \cdot T(\tau_A)$ is the $\mathcal{A}^{\mathbf{T}}$-morphism corresponding to τ_A by adjointness and $\eta_A \cdot \alpha = T(\alpha) \cdot \eta_{SA}$ is the $\mathcal{A}$-morphism corresponding to $T(\alpha)$ by adjointness.

Next, if $g : TA \to X$ is a $\mathbf{T}$-homomorphism from (TA,μ_A) to a $\mathbf{T}$-algebra (X,ξ), having equal compositions with τ_A and $\eta_A \cdot \alpha$, we show that $g \cdot \eta_A : A \to X$ is an $\mathbf{S}$-morphism from (A,α) to $\mathcal{A}^{T}(X,\xi) = (X,\xi\cdot\tau_X)$, i.e., that

$$g \cdot \eta_A \cdot \alpha = \xi \cdot \tau_X \cdot S(g \cdot \eta_A) \ .$$

Clearly this requires only the proof of

$$g \cdot \tau_A = \xi \cdot \tau_X \cdot Sg \cdot S\eta_A \ ,$$

for which, consider the diagram

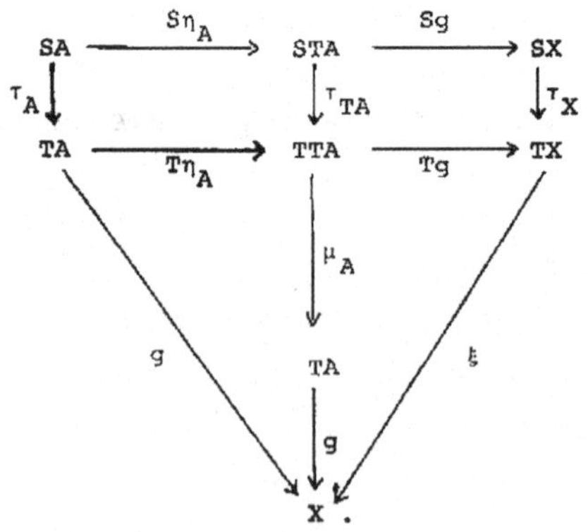

The upper squares commute because τ is natural, the left hand triangle, because $\mu_A \cdot T\eta_A = id_{TA}$, the right hand triangle, because g is a $\mathbf{T}$-homomorphism.

Finally, given an $\mathbf{S}$-homomorphism $f : A \to X$ from (A,α) to $\mathcal{A}^{\tau}(X,\xi) = (X,\xi \cdot \tau_X)$, it turns out that $\xi \cdot Tf$ is a $\mathbf{T}$-homomorphism $(TA,\mu_A) \to (X,\xi)$ having equal compositions with τ_A and $\eta_A \cdot \alpha$. For, the diagram

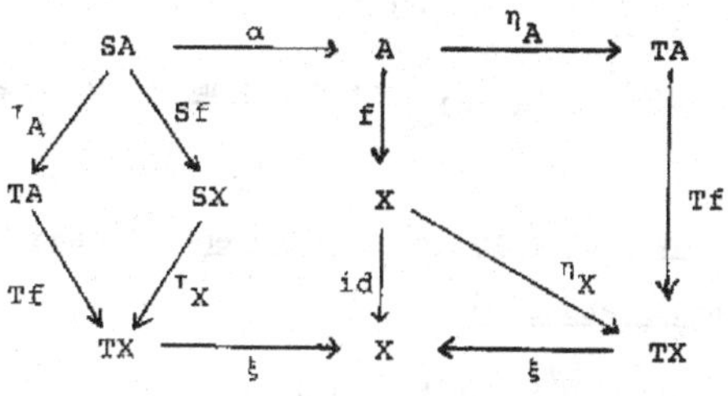

commutes, since τ is natural, f is an $\mathbf{S}$-homomorphism, η is natural, and $\xi \cdot \eta_X = id_X$.

These arguments form the core of a proof of Proposition 1.

Corollary 1. *If $\mathcal{A}^{\mathbf{T}}$ has coequalizers of reflexive pairs, then each functor $\mathcal{A}^{\tau} : \mathcal{A}^{\mathbf{T}} \to \mathcal{A}^{\mathbf{S}}$, induced by a triple map $\tau : \mathbf{S} \to \mathbf{T}$, has a left adjoint $\hat{\tau}$.*

Proof. For each $(A,\alpha) \in |\mathcal{A}^{\mathbf{S}}|$, the pair

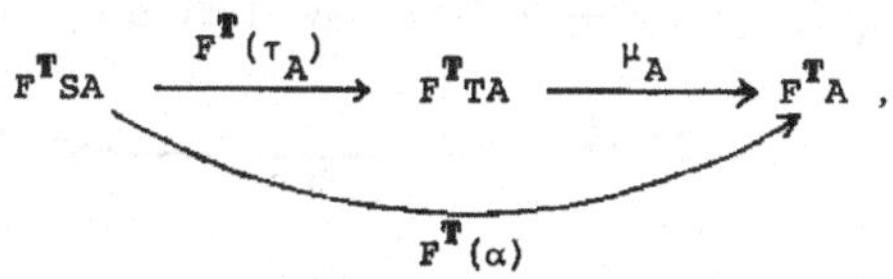

whose coequalizer, if any, is (by Proposition 1) the value $\hat{\tau}(A,\alpha)$ of $\hat{\tau}$ at (A,α), is reflexive by virtue of

$$\Delta = F^{\mathbf{T}}(\eta'_A) .$$

Proposition 2. *Let $(A_i,\alpha_i)(i \in I)$ be a family of $\mathbf{T}$-algebras, and assume the coproduct $\bigoplus_{i\in I} A_i$ exists in $\mathcal{A}$, say with injections $j_i : A_i \to \oplus A_i$. Let $p : T(\oplus A_i) \to P$ be a $\mathbf{T}$-homomorphism. Then the following statements are equivalent.*

1) p <u>is a coequalizer</u> (rel $U^{\mathbf{T}}$) <u>of the family of pairs</u>

$$\begin{array}{ccccc} TA_i & & \xrightarrow{\quad T(j_i) \quad} & & T(\oplus A_i) \\ & {\scriptstyle \alpha_i}\searrow & & \nearrow{\scriptstyle \eta_{\oplus A_i}} & \\ & A_i & \xrightarrow[j_i]{} & \oplus A_i & \end{array} \qquad (i\varepsilon I)$$

2) <u>each map</u> $h_i = p \cdot \eta_{\oplus A_i} \cdot j_i : A_i \to P$

<u>is a</u> $\mathbf{T}$<u>-homomorphism and the family</u> $(h_i)_{i\varepsilon I}$ <u>serves to make</u> P <u>the coproduct in</u> $\mathcal{A}^{\mathbf{T}}$ <u>of</u> $(A_i)_{i\varepsilon I}$.

<u>Moreover, if</u> $\oplus\, TA_i$ <u>is available in</u> $\mathcal{A}$, <u>statements</u> 1) <u>and</u> 2) <u>are equivalent to each of the following statements about</u> p :

3) p <u>is a coequalizer</u> (rel $U^{\mathbf{T}}$) <u>of the pair</u>

$$\begin{array}{ccc} \oplus\, TA_i & \xrightarrow{\quad (\ldots T(j_i) \ldots) \quad} & T(\oplus A_i) \\ & {\scriptstyle \oplus \alpha_i}\searrow \quad \nearrow{\scriptstyle \eta_{\oplus A_i}} & \\ & \oplus A_i & \end{array}$$

4) p <u>is a coequalizer of the pair</u>

$$\begin{array}{c} (T(\oplus\, TA_i),\mu) \xrightarrow{\quad T(\ldots T(j_i) \ldots) \quad} (TT(\oplus A_i),\mu) \xrightarrow{\ \mu\ } (T(\oplus A_i),\mu) \\ (T(\oplus\, TA_i),\mu) \xrightarrow{\quad T(\oplus \alpha_i) \quad} (T(\oplus A_i),\mu) \end{array}$$

<u>Proof</u>. The equivalence of statements 1) and 3) is obvious. The equivalence of 3) with 4) is due to Lemma 1, since the top (bottom) maps correspond to each other by adjointness.

Next, let $g : T(\oplus A_i) \to X$ be a $\mathbf{T}$-homomorphism from $F^{\mathbf{T}}(\oplus A_i)$ to (X,ξ), having equal compositions with both components of all the pairs in 1). Then $g \cdot \eta_{\oplus A_i} \cdot j_i : A_i \to X$ is a $\mathbf{T}$-homomorphism $(A_i,\alpha_i) \to (X,\xi)$, for all i , as is shown by the commutativity of the diagrams

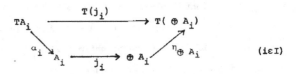

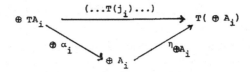

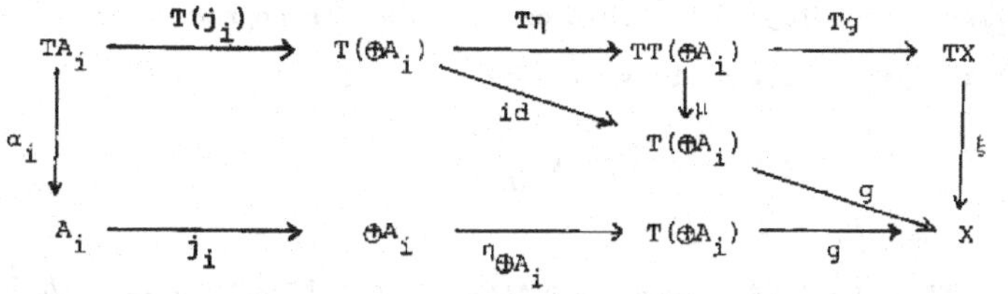

(i ε I)

Finally, if $f_i : A_i \to X$ is a family of $\mathbf{T}$-homomorphisms $(A_i,\alpha_i) \to (X,\xi)$, then the map

$$g = \xi \cdot T(\ldots f_i \ldots) : T(\oplus A_i) \to TX \xrightarrow{\xi} X$$

is a $\mathbf{T}$-homomorphism (the only one) having $g \cdot \eta_{\oplus A_i} \cdot j_i = f_i$, and, as the diagram below shows, has equal compositions with both members of all the pairs in 1).

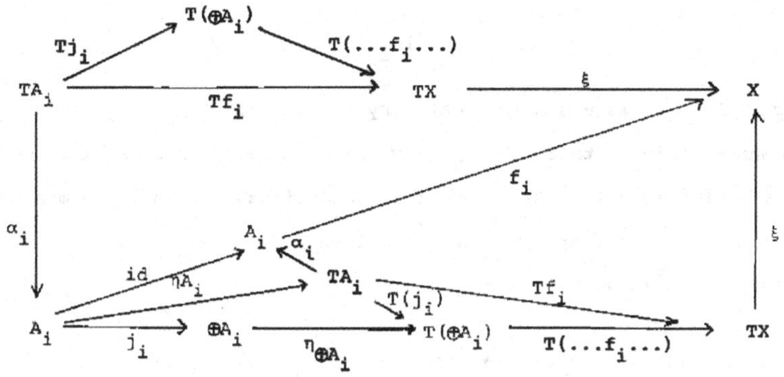

This essentially concludes the proof of Proposition 2.

__Corollary 2__. __If $\mathcal{A}^{\mathbf{T}}$ has coequalizers of reflexive pairs, and if $\mathcal{A}$ has all small coproducts, then $\mathcal{A}^{\mathbf{T}}$ has all small colimits (direct limits)__.

__Proof__. In the first place, $\mathcal{A}^{\mathbf{T}}$ has small coproducts, because, given $(A_i,\alpha_i) \in |\mathcal{A}^{\mathbf{T}}| (i \in I)$, the pair

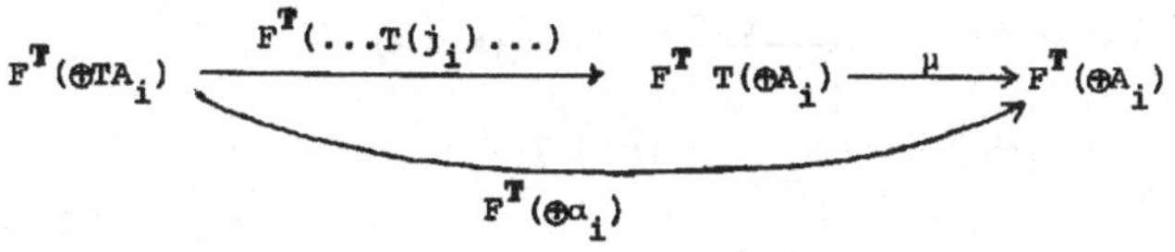
F^T(⊕TA_i)
F^T(...T(j_i)...)
F^T T(⊕A_i)
μ
F^T(⊕A_i)
F^T(⊕α_i)

,

whose coequalizer, according to Proposition 2, serves as coproduct in $\mathcal{A}^{\mathbf{T}}$ of the family $\{(A_i,\alpha_i) \ / \ i \in I\}$, is reflexive by virtue of

$$\Delta = F^{\mathbf{T}}(\oplus \ \eta_{A_i}) \ .$$

But then, having coproducts and coequalizers of reflexive pairs, $\mathcal{A}^{\mathbf{T}}$ has all small colimits. Indeed, the pair

$$\bigoplus_{\delta \in |\mathcal{D}^{2}|} D_{\mathrm{dom}\,\delta} \ \underset{(\ldots j_{\mathrm{dom}\,\delta} \ldots)}{\overset{(\ldots j_{\mathrm{cod}\,\delta} \cdot D_{\delta} \ldots)_{\delta \in |\mathcal{D}^{2}|}}{\rightrightarrows}} \ \bigoplus_{i \in |\mathcal{D}|} D_i \ ,$$

whose coequalizer is well known to serve as colimit of the functor $D : \mathcal{D} \longrightarrow ?$, is reflexive by virtue of

$$\Delta = (\ldots j_{\mathrm{id}_i} \ldots)_{i \in |\mathcal{D}|} \ .$$

Remark. If $\mathcal{A}$ is a monoidal category [O] and $\mathbf{T} = (T,\eta,\mu)$ is a suitable triple (meaning at least that $T : \mathcal{A} \to \mathcal{A}$ is a monoidal functor [O], so that there are maps $\tilde{T} : TA \otimes TB \to T(A \otimes B)$ subject to conditions, and η is a monoidal natural transformation, as should probably be μ), then, given $\mathbf{T}$-algebras (A,α), (B,β), a coequalizer (rel $U^{\mathbf{T}}$) of the pair

$$T(TA \otimes TB) \xrightarrow{T(\tilde{T})} TT(A \otimes B) \xrightarrow{\mu} T(A \otimes B) \ , \qquad T(TA \otimes TB) \xrightarrow{T(\alpha \otimes \beta)} T(A \otimes B)$$

which is reflexive by virtue of

$$\Delta = T(\eta_A \otimes \eta_B) \ ,$$

serves equally well as a coequalizer (rel $U^{\mathbf{T}}$) of the pair

$$TA \otimes TB \xrightarrow{\tilde{T}} T(A \otimes B) \ , \qquad TA \otimes TB \xrightarrow{\alpha \otimes \beta} A \otimes B \xrightarrow{\eta} T(A \otimes B)$$

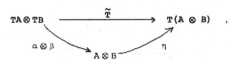

and, if $\mathcal{A}$ is closed monoidal [o] , can be interpreted (in terms of "bilinear maps") as a tensor product, in $\mathcal{A}^{\mathbf{T}}$, of (A,α) and (B,β). Such phenomena hope to be treated in detail elsewhere.

§2. Criteria for the existence of such coequalizers.

In view of §1, it behooves us to find workable sufficient conditions, on $\mathcal{A}$, on $\mathbf{T}$, or on both, that $\mathcal{A}^{\mathbf{T}}$ have coequalizers of reflexive pairs. The first such condition, though rather special, depends on knowing when coequalizers in $\mathcal{A}^{\mathbf{T}}$ can be calculated in $\mathcal{A}$.

Proposition 3. Let $\mathbf{T} = (T,\eta,\mu)$ *be a triple in* $\mathcal{A}$, *and let*

$$(A,\alpha) \underset{g}{\overset{f}{\rightrightarrows}} (B,\beta)$$

be a pair of $\mathcal{A}^{\mathbf{T}}$*-morphisms. Assume*

1) *there is an* $\mathcal{A}$*-morphism* $p : B \to C$ *which is a coequalizer (in* $\mathcal{A}$ *) of* $A \underset{g}{\overset{f}{\rightrightarrows}} B$;

2) Tp *is a coequalizer of* (Tf,Tg) ;

3) TTp *is epic*.

Then: there is a map $\gamma : TC \to C$, *uniquely determined by the single requirement that*

$$\begin{array}{ccc} TB & \xrightarrow{Tp} & TC \\ {\scriptstyle\beta}\downarrow & & \downarrow{\scriptstyle\gamma} \\ B & \xrightarrow[p]{} & C \end{array}$$

commute; (C,γ) *is a* $\mathbf{T}$*-algebra; and* $p : (B,\beta) \to (C,\gamma)$ *is a coequalizer in* $\mathcal{A}^{\mathbf{T}}$ *of* (f,g).

Proof. The equations $p{\circ}\beta{\circ}Tf = p{\circ}f{\circ}\alpha = p{\circ}g{\circ}\alpha = p{\circ}\beta{\circ}Tg$, occurring because of assumption 1, force, because of assumption 2, a unique $\gamma : TC \to C$ with $\gamma Tp = p\beta$.

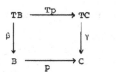

Then $\gamma \circ \eta_C \circ p = \gamma \circ Tp \circ \eta_B = p \circ \beta \circ \eta_B = p$, but p is epic (by 1) and so $\gamma \circ \eta_C = id_C$. Similarly, using assumption 3, the equation $\gamma \circ T\gamma = \gamma \circ \mu_C$ follows from the calculation

$$\begin{aligned} \gamma \circ T\gamma \circ TTp &= \gamma \circ Tp \circ T\beta = p \circ \beta \circ T\beta \\ &= p \circ \beta \circ \mu_B = \gamma \circ Tp \circ \mu_B \\ &= \gamma \circ \mu_C \circ TTp \ . \end{aligned}$$

Finally, if $q : B \to X$ is an $\mathcal{A}^{\mathbf{T}}$-morphism from (B,β) to (X,ξ) factoring through C , the factorization must be an $\mathcal{A}^{\mathbf{T}}$-morphism, because the diagram

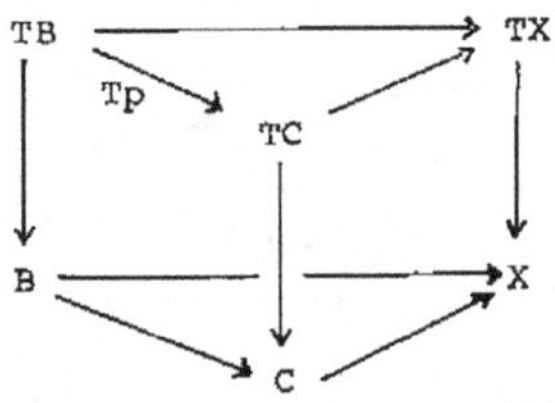

commutes everywhere else, and Tp is epic, by assumption 2.

__Corollary 3__. __If__ $\mathcal{A}$ __has and__ T __preserves coequalizers of reflexive pairs, then__ $\mathcal{A}^{\mathbf{T}}$ __has coequalizers of reflexive pairs, and Corollaries 1 and 2 apply.__

__Examples__. If $\mathbf{T}$ is an adjoint triple, then T preserves all coequalizers (and all colimits, in fact). Samples of such triples:

a) $\mathbf{T} = (- \otimes \Lambda \ , \ id \otimes u, \ id \otimes m)$, where the ground category $\mathcal{A}$ is {k-modules} (k comm.) and Λ is an associative k-algebra with unit $u : k \to \Lambda$ and multiplication $m : \Lambda \otimes \Lambda \to \Lambda (\otimes = \otimes_k)$. $\mathcal{A}^{\mathbf{T}}$ = Λ-modules.

b) $\mathbf{T} = (- \times \mathbf{2}, \ id \times (\mathbf{1} \xrightarrow{0} \mathbf{2}), \ id \times (\mathbf{2} \times \mathbf{2} \xrightarrow{\max} \mathbf{2}))$, where the ground category $\mathcal{A}$ is $\mathcal{C}at$ and $\mathbf{2}$ is the p.o. set $0 \to 1$. $\mathcal{C}at^{\mathbf{T}}$ = {categories with idempotent triples} . Where $\mathbf{S}$ is constructed like $\mathbf{T}$, replacing $\mathbf{2}$ by the category Δ given by

$$|\Delta| = \{0,1,2,\ldots,n,\ldots\}$$

$$\Delta(n,k) = \text{order preserving maps } \{0 \ldots n-1\} \to \{0 \ldots k-1\} \ ,$$

with the obvious composition, $0 : \mathbb{1} \to \Delta$ the inclusion of the object 0, $m : \Delta \times \Delta \to \Delta$ the functor given by

$$\begin{array}{ccc} n,\ n' & \longmapsto & n + n' \\ n \xrightarrow{f} k,\ n' \xrightarrow{f'} k' & \longmapsto & \begin{array}{ccc} 0 & \xrightarrow{\quad f \quad} & 0 \\ \vdots & & \vdots \\ n-1 & & k-1 \\ n & & k \\ \vdots & \xrightarrow{k+f'(-\ -n)} & \vdots \\ n+n'-1 & & k+k'-1 \end{array} \end{array} ,$$

$\mathcal{C}at^{\mathbb{S}}$ is {categories equipped with a triple} . Define $\tau : \mathbb{S} \to \mathbb{T}$ by crossing with the only functor $\Delta \to \mathbb{2}$ sending $n \neq 0$ to 1 , and 0 to 0 . Then $\mathcal{C}at^{\tau} : \mathcal{C}at^{\mathbb{T}} \to \mathcal{C}at^{\mathbb{S}}$ is the functor interpreting an idempotent triple as a triple on the same category. These constructions and observations are all due to Lawvere. Since $\mathcal{C}at$ has coequalizers and $\mathbb{T}$ is an adjoint triple, $\mathcal{C}at^{\tau}$ has a left adjoint, by Corollary 1 ; roughly speaking, it assigns to a triple in a category, a best idempotent triple on an as closely related other category as possible.

2. Let $\mathcal{A}$ be an additive category, let $m : G \times G \to G$ be an $\mathcal{A}$-morphism satisfying $m(m\times G) = m(G\times m)$, $m(id,0) = id = m(0,id)$. Define a triple $\mathbb{T} = (-\times G,\ (id,0),\ (id\times m))$ on $\mathcal{A}$. Then T preserves all coequalizers because $A \times G = A \oplus G$.

3. Any functor preserves split coequalizer systems [3]. In particular every triple does, and so Proposition 3 guarantees that coequalizers of $U^{\mathbb{T}}$-split pairs of $\mathcal{A}^{\mathbb{T}}$-morphisms can be computed in $\mathcal{A}$, as was stated in greater generality in [2 ,§6].

The other criterion involves images. We treat images axiomatically, in a manner suggestive of (and perhaps equivalent to) bicategories. Recall that $\mathbb{1}$, $\mathbb{2}$ and $\mathbb{3}$ are the categories depicted as the partially ordered sets

$$\begin{aligned} \mathbb{1} &= \{0\} , \\ \mathbb{2} &= \{0 \to 1\} , \\ \mathbb{3} &= \left\{ 0 \xrightarrow{a} 1 \xrightarrow{b} 2,\quad 0 \xrightarrow{c=b\circ a} 2 \right\} . \end{aligned}$$

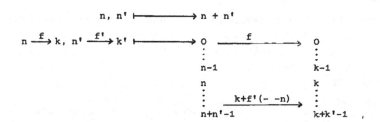

We will need the functors $\mathbf{2} \xrightarrow{c} \mathbf{3}$ and $\mathbf{1} \xrightarrow{1} \mathbf{3}$ (whose values serve as their names). These induce functors $\mathcal{A}^c : \mathcal{A}^{\mathbf{3}} \longrightarrow \mathcal{A}^{\mathbf{2}}$ and $\mathcal{A}^1 : \mathcal{A}^{\mathbf{3}} \longrightarrow \mathcal{A}^{\mathbf{1}} \cong \mathcal{A}$, for any category $\mathcal{A}$.

By an <u>image factorization functor</u> for the category $\mathcal{A}$, we mean a functor

$$\mathcal{I} : \mathcal{A}^{\mathbf{2}} \longrightarrow \mathcal{A}^{\mathbf{3}},$$

having the property

1) $\mathcal{A}^{\mathbf{2}} \xrightarrow{\mathcal{I}} \mathcal{A}^{\mathbf{3}} \xrightarrow{\mathcal{A}^c} \mathcal{A}^{\mathbf{2}}$ = identity on $\mathcal{A}^{\mathbf{2}}$, and three more properties which we state using the notations

$$\mathcal{A}^1(\mathcal{I}(f)) = I_f$$
$$\mathcal{I}f = \circ \xrightarrow{f_a} I_f \xrightarrow{f_b} \circ \ :$$

2) $f \in |\mathcal{A}^{\mathbf{2}}| \Longrightarrow f_a$ is an epimorphism,

3) $f \in |\mathcal{A}^{\mathbf{2}}| \Longrightarrow f_b$ is a monomorphism,

4) $f \in |\mathcal{A}^{\mathbf{2}}| \Longrightarrow (f_b)_a$ and $(f_a)_b$ are isomorphisms.

A functor T <u>preserves $\mathcal{I}$-images</u> if there is a natural equivalence, whose composition with $\mathcal{A}^C$ is the identity, between $T^{\mathbf{3}} \circ \mathcal{I}$ and $\mathcal{I} \circ T^{\mathbf{2}}$. This entails, for each $f \in \mathcal{A}(A,B)$, an isomorphism $\iota_f : T(I_f) \to I_{Tf}$ making the triangle

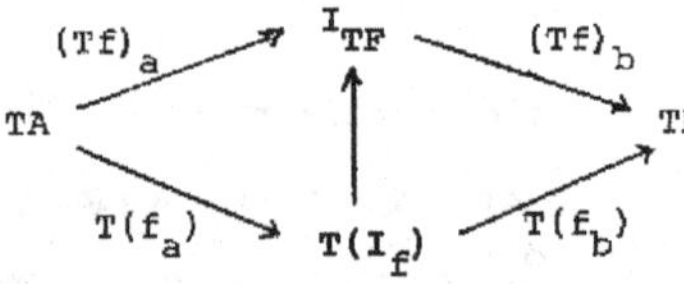

commute.

<u>A triple</u> $\mathbf{T} = (T,\eta,\mu)$ on $\mathcal{A}$ preserves $\mathcal{I}$-images if the functor T does.

<u>Lemma 2</u>. <u>If $\mathcal{I} : \mathcal{A}^{\mathbf{2}} \to \mathcal{A}^{\mathbf{3}}$ is an image factorization functor for $\mathcal{A}$ and $\mathbf{T}$ is a triple that preserves $\mathcal{I}$-images, then there is one and only one image factorization functor $\mathcal{I}^{\mathbf{T}}$ for $\mathcal{A}^{\mathbf{T}}$ with the property</u>

$$(U^{\mathbf{T}})^{\mathbf{3}} \circ \mathcal{I}^{\mathbf{T}} = \mathcal{I} .$$

Proof. Given $f : A \to B$, an $\mathcal{A}^{\mathbf{T}}$-morphism from (A,α) to (B,β), the commutativity of the square

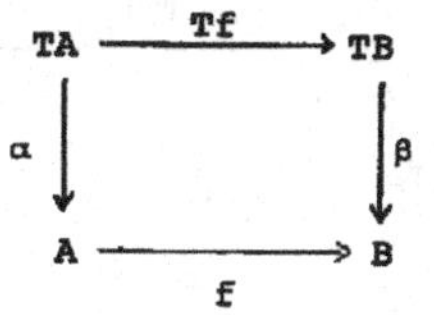

yields a commutative diagram

$$
\begin{array}{ccccc}
TA & \xrightarrow{(Tf)_a} & I_{Tf} & \xrightarrow{(Tf)_b} & TB \\
\downarrow{\scriptstyle\alpha} & \mathcal{A}(\alpha,\beta) & \downarrow & & \downarrow{\scriptstyle\beta} \\
A & \xrightarrow{f_a} & I_f & \xrightarrow{f_b} & B
\end{array}
$$

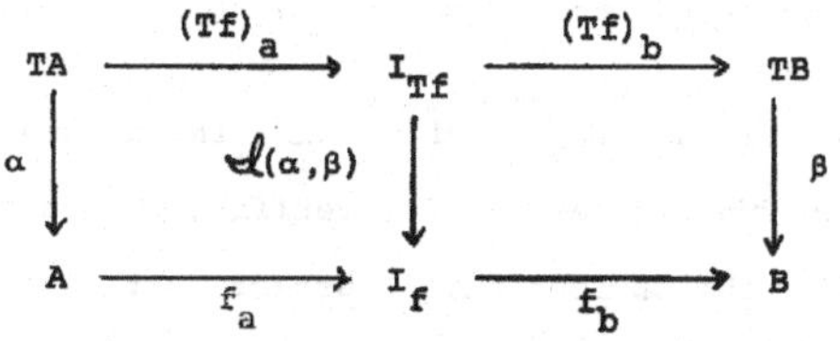

Combining this with the commutative diagram arising from the definition of "T preserves $\mathcal{A}$-images", we obtain a map $\gamma = \mathcal{A}(\alpha,\beta) \circ \iota_f : T(I_f) \to I_f$ making the diagram

$$
\begin{array}{ccccc}
TA & \xrightarrow{T(f_a)} & T(I_f) & \xrightarrow{T(f_b)} & TB \\
\downarrow{\scriptstyle\alpha} & & \downarrow{\scriptstyle\gamma} & & \downarrow{\scriptstyle\beta} \\
A & \xrightarrow{f_a} & I_f & \xrightarrow{f_b} & B
\end{array}
$$

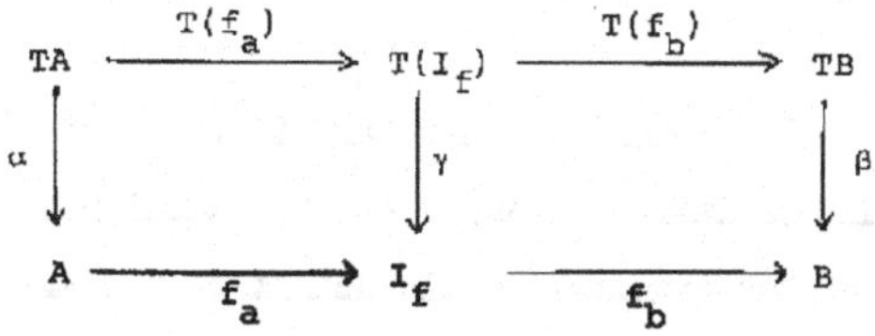

commute. There is only one such map γ because $T(f_a)$ is epic and f_b is monic. We show now that (I_f,γ) is a **T**-algebra. Write simply $I = I_f$. $\gamma \circ \eta_I = id_I$ follows from the commutativity of

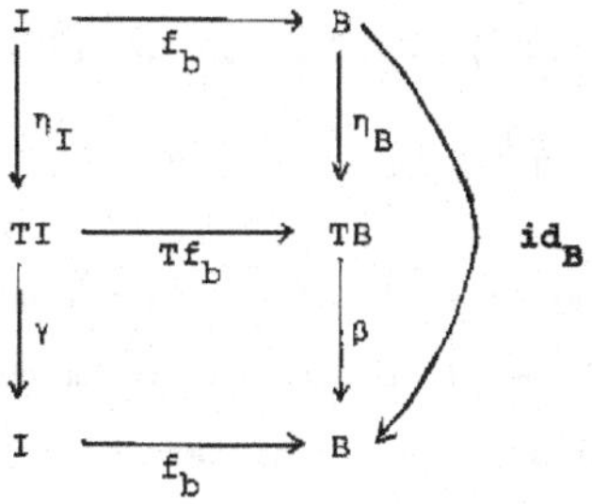

and the fact that f_b is monic. $\gamma \circ T\gamma = \gamma \circ \mu_I$ follows from the commutativity of

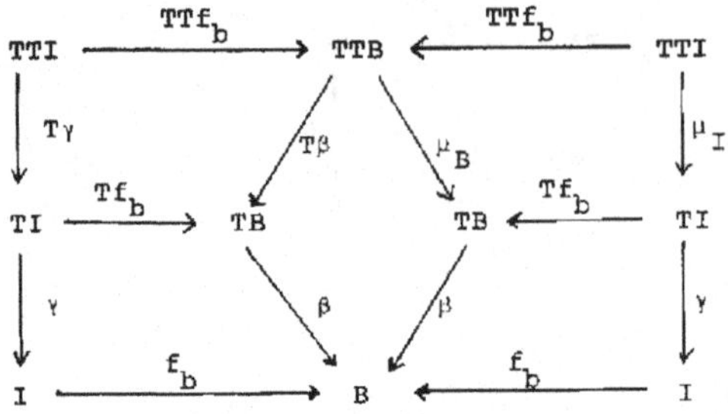

and the fact that f_b is monic. Since $(U^{\mathbf{T}})^{\mathbf{3}} \circ \mathcal{I}^{\mathbf{T}} = \mathcal{I}$, the axioms $\mathcal{I}^{\mathbf{T}}$ must satisfy, to be an image factorization functor, are easily verified. The uniqueness assertion is taken care of, essentially, by the obvious uniqueness of $\gamma : TI \to I$, subject to the commutativity relations expressed in the diagram

$$\begin{array}{ccccc} TA & \longrightarrow & TI & \longrightarrow & TB \\ \downarrow & & \downarrow & & \downarrow \\ A & \longrightarrow & I & \longrightarrow & B \end{array} .$$

Proposition 4. *Let* $\mathcal{I}$ *be an image factorization functor for* $\mathcal{A}$, *and let* $\mathbf{T}$ *be a triple on* $\mathcal{A}$ *preserving* $\mathcal{I}$*-images. Assume* $\mathcal{A}$ *has small products and is co-well-powered* (*or even just that the isomorphism classes of each class*

$$\mathcal{I}\text{-epi }(A) = \{f/f : A \to B,\ f_b : I_f \xrightarrow{\cong} B,\ B \in |\mathcal{A}|\}$$

constitute a set (*isomorphisms that are the identity on* A, *of course*)). *Then* $\mathcal{A}^{\mathbf{T}}$ *has all coequalizers.*

Proof. Given a pair $(E,\varepsilon) \overset{f}{\underset{g}{\rightrightarrows}} (A,\alpha)$ of $\mathcal{A}^{\mathbf{T}}$-morphisms, let

$$\mathcal{E}_{f,g} = \{h/h \in |\mathcal{A}^{\mathbf{2}}|,\ h : (A,\alpha) \to (X,\xi),\ hf = hg,\ h = h_a\} .$$

Observe that an isomorphism class of $\mathcal{E}_{f,g}$ in the sense of $\mathcal{A}^{\mathbf{T}}$ or in the sense of $\mathcal{A}$ is the same thing, because $\mathbf{T}$ preserves $\mathcal{I}$-images, and the maps h, Th are epic. Pick representatives of the isomorphism classes of $\mathcal{E}_{f,g}$, say

$$h_i : (A,\alpha) \longrightarrow (X_i,\xi_i) \qquad (i \in I)$$

and form the induced map (an $\mathcal{A}^{\mathbf{T}}$-morphism by [2, §6]

$$k = \langle \ldots h_i \ldots \rangle : (A,\alpha) \longrightarrow (\Pi_i X_i,\ \Pi_i \xi_i) \ .$$

Then k_a is a coequalizer of (f,g) in $\mathcal{A}^{\mathbf{T}}$. Indeed, given $h : (A,\alpha) \to (Z,\zeta)$ with $hf = hg$, we have $h_a \in \mathcal{E}_{f,g}$ and so $\exists i_o \in I$ with $h_a \cong h_{i_o}$. Then the composition

$$(I^{\mathbf{T}})_k \xrightarrow{\ k_b\ } (\Pi_i X_i, \Pi_i \xi_i) \xrightarrow{\ pr_{i_o}\ } (X_{i_o}, \xi_{i_o}) \xrightarrow{\ \cong\ } (I^{\mathbf{T}})_h \xrightarrow{\ h_b\ } (Z,\zeta)$$

makes the triangle

$$\begin{array}{ccc} (A,\alpha) & \xrightarrow{\ k_a\ } & (I^{\mathbf{T}})_k \\ & \underset{h}{\searrow} & \downarrow \\ & & (Z,\zeta) \end{array}$$

commute, and since k_a is epic, it is the only such map. That $k_a \in \mathcal{E}_{f,g}$ is obvious, and completes the proof of the existence of coequalizers.

Remark. Much the same arguments prove, under the same hypotheses, that $\mathcal{A}^{\mathbf{T}}$ has coequalizers of families of pairs of maps.

Example. $\mathcal{A} = \mathcal{S}$, $\mathcal{U}$ = usual epic-monic factorization. Then any triple preserves $\mathcal{U}$-images (proof below), and consequently $\mathcal{S}^{\mathbf{T}}$ has coequalizers, and, by virtue of Corollary 2, all colimits. That $\mathcal{A}^{\mathbf{T}}$ has all limits if $\mathcal{A}$ has is well known, and this then takes care of the completeness properties of varietal categories.

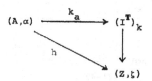

To see that every triple in $\mathcal{S}$ preserves images, it suffices to see that every triple in $\mathcal{S}$ preserves monomorphisms since the usual epic-monic factorization is determined to within isomorphism by the requirement that it be an epic-monic factorization, and every functor preserves epimorphisms, since they split. So let $\mathbf{T} = (T,\eta,\mu)$ be a triple in $\mathcal{S}$. The only monomorphisms f T has a chance of not preserving are those that are not split, i.e., those with empty domain. Now if $T(\emptyset)$ is $\emptyset$, Tf is surely monic. But if $T\emptyset$ has at least one element, Tf, which may be thought of as a $\mathbf{T}$-morphism from $F^{\mathbf{T}}(\emptyset)$ to $F^{\mathbf{T}}(n)$, admits a retraction, namely the extension to a $\mathbf{T}$-homomorphism of any function

$$n \to U^{T}F^{T}\emptyset = T\emptyset \ .$$

That the composition on $F^{T}(\emptyset)$ is the identity is due to the fact that $F^{T}(\emptyset)$ is a left zero (is initial, is a copoint) in $\mathcal{S}^{T}$.

Bibliography

[0] Eilenberg, S., and Kelly, G.M. : Closed categories. Proc. Conf. Categ. Alg. (La Jolla, 1965). Springer, Berlin, 1966, pp. 421-562. esp. pp. 471-476.

[1] Eilenberg, S., and Moore, J.C. : Adjoint functors and triples. Ill. J. Math. 9 (1965), pp. 381-398, esp. p. 398.

[2] Linton, F.E.J. : An outline of functorial semantics. This volume.

[3] Linton, F.E.J. : Applied functorial semantics, II. This volume.

A TRIPLE THEORETIC CONSTRUCTION OF COMPACT ALGEBRAS

by Ernest Manes

Let $\mathbb{T}$ be a triple in the category of sets. Using the Yoneda Lemma, it is possible to reinterpret $\mathbb{T}$-algebras in the classical way as sets with (not necessarily finitary) operations; (the "equations" are built into $\mathbb{T}$ and need not be mentioned). The objects in the <u>category of compact $\mathbb{T}$-algebras</u> are defined to be sets provided with $\mathbb{T}$-algebra structure and compact T2 topology in such a way that $\mathbb{T}$-operations are continuous, whereas the morphisms are defined to be continuous $\mathbb{T}$-homomorphisms. The end result of this paper is the proof that "compact $\mathbb{T}$-algebras" is itself the category of algebras over a triple in the category of sets; that is, a compact $\mathbb{T}$-algebra -a compact T2 space in particular- is an example of a set with algebraic structure. When $\mathbb{T}$-algebras = G-sets, compact $\mathbb{T}$-algebras = compact topological dynamics with discrete phase group G . The general case of compact topological dynamics, when G is a (not necessarily compact) topological group, is also algebraic, indeed is a Birkhoff subcategory (= variety) of the discrete case. (For more on the interplay between compact topological dynamics and universal algebra see [TM,§ 2.4, 2.5]). This motivates our general study of Birkhoff subcategories in § 3. Otherwise, the paper pursues a suitably geodesic course to our main result 7.1, so long as "suitable" means "intended to convince the reader with little background in triple theoretic methods".

Virtually all of this paper appears in the author's thesis [TM]. Many of the ideas were developed in conversations with Jon Beck and F.E.J. Linton to whom the author is grateful.

§ 1. Preliminaries.

We assume that the reader is conversant with elementary category theory at the level of, say, the first five chapters of [M]. Most of the main prerequisites are listed in this section.

1.1 Miscellaneous preliminaries. If f,g are morphisms in a category we compose first on the left so that fg (which we may also write "$f \cdot g$") $= \xrightarrow{f}\xrightarrow{g}$. We use "$=_{df}$" for "is defined to be" and "$=_{dn}$" for "is denoted to be". We write "$\overset{f}{\rightarrowtail}$" (resp. "$\overset{f}{\twoheadrightarrow}$") to assert that the morphism f is mono (resp., epi); "mono" and "epi" are defined below in 1.2. A function f is bijective $=_{df}$ f is 1-to-1 and onto. If f is a function and if x is an element of the domain of f, we write "xf" or "$\langle x,f \rangle$" for the element of Y that f assigns to x. "End of proof" $=_{dn}$ [] .

Let $\mathcal{K}$ be a category. $|\mathcal{K}|$ or obj $\mathcal{K}$ $=_{dn}$ the class of $\mathcal{K}$-objects. For $X \in$ obj $\mathcal{K}$, 1_X or $X \xrightarrow{1} X$ $=_{dn}$ the identity morphism of X. $\mathcal{S}$ $=_{df}$ the category of sets and functions. $\mathcal{K}$ is legitimate $=_{df}$ for all $X,Y \in$ obj $\mathcal{K}$ the class $(X,Y)\mathcal{K}$ of $\mathcal{K}$-morphisms for X to Y is a set. A class $\mathcal{F}$ of $\mathcal{K}$-morphisms has a representative set $=_{df}$ there exists a set $\mathcal{R}$ of $\mathcal{K}$-morphisms such that for every $X \xrightarrow{f} Y \in \mathcal{F}$ there exists $A \xrightarrow{r} B \in \mathcal{R}$ and $\mathcal{K}$-isomorphisms $X \xrightarrow{a} A$, $Y \xrightarrow{\beta} B$ such that $f\beta = ar$.

1.2. Monos and epis.

1.2.1. Definition. Let $A \xrightarrow{f} B \in \mathcal{K}$. f is split epi if there exists $B \xrightarrow{\tilde{f}} A \in \mathcal{K}$ with $\tilde{f}f = 1_B$. f is a coequalizer if there exist $g,h \in \mathcal{K}$ with $f = \text{coeq}(g,h)$. Define $\text{reg}(f) =_{df} [A \xrightarrow{g} Y \in \mathcal{K}$: for every $(a,b) : X \to A$, $af = bf$ implies $ag = bg]$. f is a regular epi if for every $g \in \text{reg}(f)$ there exists a unique $\tilde{g} \in \mathcal{K}$ with $f\tilde{g} = g$. f is epi if for every $(a,b) : B \to X$ in $\mathcal{K}$, $fa = fb$ implies $a = b$. Dually, we have split mono, equalizer, regular mono, mono.

1.2.2 Proposition. Let $A \xrightarrow{f} B \in \mathcal{K}$. Then f split epi implies f coequalizer implies f regular epi implies f epi.

Proof. If $\tilde{f}f = 1_B$, $f = \text{coeq}(1_A, f\tilde{f})$. If $f = \text{coeq}(a,b)$ then for every $g \in \text{reg}(f)$

we have $ag = bg$ so that the coequalizer property induces unique $\tilde{g}$ with $f\tilde{g} = g$. Finally, suppose f is regular epi and that $fa = fb$. Defining $g =_{df} fa$, $g \in reg(f)$ so there exists unique $\tilde{g}$ with $f\tilde{g} = g$, and $a = \tilde{g} = b$. []

<u>1.2.3 Proposition</u>. In $\mathcal{S}$ the following notions are equivalent: split epi, coequalizer, regular epi, epi, onto function.

<u>Proof</u>. To see that epis are onto, consider functions to a two-element set. The axiom of choice implies that onto functions are split epi (and conversely, by the way.) []

<u>1.2.4 Proposition</u> Let $A \xrightarrow{f} B \xrightarrow{g} C \in \mathcal{K}$. Then f (split) epi and g (split) epi implies fg (split) epi. fg (split) epi implies g (split) epi. []

<u>1.2.5 Proposition</u> Let $A \xrightarrow{f} B \in \mathcal{K}$. Then f iso iff f regular epi and mono.

<u>Proof</u>. [iso] implies [split epi and mono] implies [regular epi and mono]. Conversely, if f is regular epi and mono, $1_A \in reg(f)$ and so induces $\tilde{f}$ with $f\tilde{f} = 1_A$. As $f\tilde{f}f = f$ and f is epi, $\tilde{f}f = 1_B$. []

<u>1.2.6 Definition</u> Let $A \xrightarrow{f} B \in \mathcal{K}$. A <u>regular coimage factorization</u> of f is a factorization $f = A \xrightarrow{p} Q \xrightarrow{i} B$ with p regular epi and i mono. $\mathcal{K}$ <u>has regular coimage factorizations</u> if every $\mathcal{K}$-morphism admits a regular coimage factorization.

<u>1.2.7 Proposition</u>. Regular coimage factorizations are unique within isomorphism.

<u>Proof.</u> Suppose p,p' are regular epis and i,i' are monos with $pi = p'i'$. p' is in $reg(p)$ as i' is mono, so h is uniquely induced with $ph = p'$. $hi' = i$ because p is epi. h^{-1} is induced similarly. []

<u>1.2.8 Proposition</u>. Assume that $\mathcal{K}$ has regular coimage factorizations. Let $A \xrightarrow{f} B \xrightarrow{g} C \in \mathcal{K}$. Then f,g regular epi implies fg regular epi. fg regular epi implies g regular epi. The hypothesis on $\mathcal{K}$ is necessary in both cases.

<u>Proof.</u> Suppose fg is regular epi. Consider the diagram

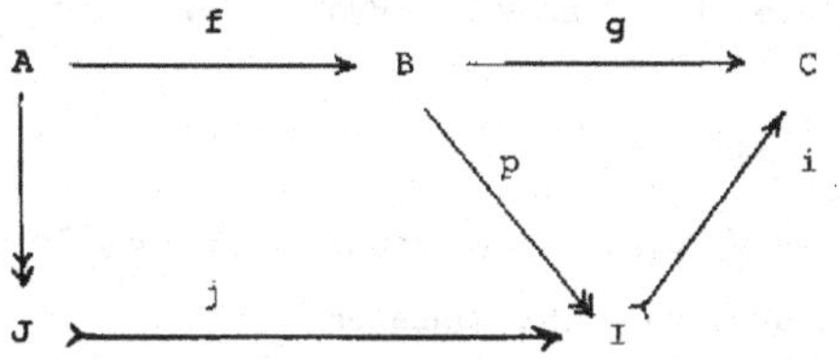

where pi is the regular coimage factorization of g and then J is the regular coimage factorization of fp . By 1.2.7, ji is an isomorphism. But then i is mono and split epi, hence an isomorphism, by 1.2.5; and then g is regular epi because p is.

Now suppose that f,g are regular epi and let fg = pi be a regular coimage factorization of fg. As i is mono, p is in reg(f) inducing $\tilde{p}$ such that $f\tilde{p} = p$. As just proved above, i is regular epi (noting that $\tilde{p}i = g$ because f is epi). As i is also mono, i is iso and hence fg is regular epi because p is.

The third assertion is left to the reader with the hint to look at some simple finite categories. []

<u>1.3 Limits</u>. If D is a $\mathcal{K}$-valued functor, the inverse limit of D (determined only within isomorphism if it exists at all) is denoted "$\varprojlim D$" , or more precisely "$\varprojlim D \to D$" ; similarly, we use "$\varinjlim D \to D$" for direct limits. The ith projection of a product $=_{dn} \Pi X_i \xrightarrow{pr_i} X_i$. The coequalizer of $(f,g) : X \to Y =_{dn}$ coeq(f,g) . If $(A_a \overset{i_a}{\rightarrowtail} X : a \in I)$ is a family of monomorphisms, their inverse limit $=_{dn} \cap A_a \overset{i}{\rightarrowtail} X$ (i is, in fact, a monomorphism). The class of monomorphisms into X is partially ordered by $A \overset{i}{\rightarrowtail} X < B \overset{j}{\rightarrowtail} X =_{df}$ there exists $A \xrightarrow{k} B$ such that kj = i (in which case k is unique and is a monomorphism); $\cap A_a = \inf A_a$ with respect to this ordering.

<u>1.4 Godement's cinq règles; see [GA].</u> Suppose that W, X, Y, Z are functors and that a is a natural transformation from X to Y . Natural transformations $WX \xrightarrow{Wa} WY$ and $XZ \xrightarrow{aZ} YZ$ are induced by defining $K(Wa) =_{df} (KW)a$ and $K(aZ) =_{df} (Ka)Z$ for every object K . The five rules concerning these operations are as follows.

$(WX)a = W(Xa) : WXY \to WXZ$; $a(YZ) = (aY)Z : WYZ \to XYZ$; $WaZ =_{df}$

$(Wa)Z = W(aZ) : WXZ \to WYZ$; $V(a.b)Z = VaZ.VbZ : VWZ \to VYZ$; $ab =_{df}$

$aY.Xb = Wb.aZ : WY \to XZ$.

<u>1.5 The Yoneda Lemma</u>. Let $\mathcal{K} \xrightarrow{H} \mathcal{S}$ be a set-valued functor, and let X be a $\mathcal{K}$-object such that $(X,-)\mathcal{K}$ is set-valued. Then the passages

$$((X,-)\mathcal{K},H)\text{n.t.} \longrightarrow XH, \qquad XH \longrightarrow ((X,-)\mathcal{K},H)\text{n.t.}$$

$$a \longmapsto \langle 1_X, Xa\rangle \qquad x \mapsto (X,-)\mathcal{K} \xrightarrow{\ a\ } H$$

$$(X,Y)\mathcal{K} \xrightarrow{\ Ya\ } YH$$

$$X \xrightarrow{f} Y \longmapsto \langle x, fH\rangle$$

(where "n.t." means natural transformations) are mutually inverse. In particular, $((X,-)\mathcal{K},H)$n.t. is a set. For a proof see [M,pp. 97-99].

1.6 <u>Adjoint functors.</u> Let $\mathcal{L} \overset{i}{\to} \mathcal{K}$ be a (not necessarily full) subcategory of $\mathcal{K}$, and let X be a $\mathcal{K}$-object. A reflection of X in $\mathcal{L} =_{df}$ a $\mathcal{K}$-morphism $X \xrightarrow{X\eta} X_{\mathcal{L}}$ such that $X_{\mathcal{L}} \in \text{obj}\,\mathcal{L}$ and such that whenever $X \xrightarrow{f} L \in \mathcal{K}$ with $L \in \text{obj}\,\mathcal{L}$ then there exists a unique $X_{\mathcal{L}} \xrightarrow{\tilde f} L \in \mathcal{L}$ such that $X\eta.\tilde f = f$. If every $\mathcal{K}$-object has a reflection in $\mathcal{L}$ then $\mathcal{L}$ is a <u>reflective subcategory of</u> $\mathcal{K}$ and there is a <u>reflector functor</u> $\mathcal{K} \overset{R}{\to} \mathcal{L}$ defined so as to make $1 \overset{\eta}{\to} Ri$ natural. R is determined within natural equivalence. $\mathcal{L}$ is full iff R may be chosen with $iR = 1_{\mathcal{L}}$; (however, the definition of reflectors requires a suitable axiom of choice).

A <u>left adjointness</u> consists of functors $\mathcal{K} \overset{F}{\to} \mathcal{A}$, $\mathcal{A} \overset{U}{\to} \mathcal{K}$ and natural transformations $UF \overset{\varepsilon}{\to} 1$, $1 \overset{\eta}{\to} FU$ (called <u>adjunctions</u>) subject to the <u>adjointness axioms</u> $F \overset{\eta F}{\to} FUF \overset{F\varepsilon}{\to} F = 1_F$, $U \overset{U\eta}{\to} UFU \overset{\varepsilon U}{\to} U = 1_U$. We denote this by "$F \dashv U$", read "F is left adjoint to U" and let η,ε be understood. U has a left adjoint $=_{df}$ there exists $F \dashv U$. If $\mathcal{A}$ and $\mathcal{K}$ are legitimate, then a left adjointness may be expressed in terms of a natural equivalence $((-)F,-)\mathcal{A} \overset{\alpha}{\to} (-,(-)U)\mathcal{K}$ where $\langle f, (X,A)\alpha\rangle = X\eta.fU$, $\langle f, (X,A)\alpha^{-1}\rangle = gF.A\varepsilon$ and conversely $X\eta = \langle 1_{XF}, (X,XF)\alpha\rangle$, $A\varepsilon = \langle 1_{AU}, (AU,A)\alpha^{-1}\rangle$.

If $F \dashv U$ and $\tilde F \dashv U$ then F and $\tilde F$ are naturally equivalent. A subcategory is reflective iff its inclusion functor has a left adjoint. Notice that a subcategory inclusion i is a full reflective subcategory iff there exists $R \dashv i$ with $iR \overset{\varepsilon}{\to} 1$ a natural equivalence.

Finally, we state the adjoint functor theorem first proved by Freyd. Let $\mathcal{A} \xrightarrow{U} \mathcal{K}$ be a functor. U satisfies the solution set condition $=_{df}$ for every $K \in \text{obj}\,\mathcal{K}$ there exists a set, R_K, of $\mathcal{A}$-objects such that whenever $A \in \text{obj}\,\mathcal{A}$ and $K \xrightarrow{f} AU \in \mathcal{K}$, there exist $R \in R_K$, $K \xrightarrow{a} RU \in \mathcal{K}$, $R \xrightarrow{b} A \in \mathcal{A}$ with $f = a.bU$; (such a set is called a solution set for K). Let $\mathcal{A}, \mathcal{K}$ be legitimate and assume further that $\mathcal{A}$ has $\underleftarrow{\lim}$'s. The adjoint functor theorem says: there exists $F \dashv U$ iff U preserves $\underleftarrow{\lim}$'s and satisfies the solution set condition.

1.7 Regular categories. The category $\mathcal{K}$ is regular if it satisfies the following four axioms.

REG 1. $\mathcal{K}$ has regular coimage factorizations.

REG 2. $\mathcal{K}$ has $\underleftarrow{\lim}$'s.

REG 3. $\mathcal{K}$ is legitimate.

REG 4. For every X in obj $\mathcal{K}$ the class of regular epimorphisms with domain X has a representative set.

1.8 Contractible pairs (Jon Beck).

1.8.1 Definition. Let $(f,g) : X \to Y$ be $\mathcal{K}$-morphisms. (f,g) is contractible $=_{df}$ there exists $Y \xrightarrow{d} X$ such that $df = 1_Y$ and $fdg = gdg$.

1.8.2 Proposition. A coequalizer of a contractible pair is a split epi.

Proof. If (f,g) is contractible and $q = \text{coeq}(f,g)$ then as $fdg = gdg$ there exists $Q \xrightarrow{h} Y$ with $qh = dg$. As q is epi and $qhq = dgq = dfq = q$, $hq = 1_Q$. []

For more on the theory of contractible pairs see [TM, §0.7].

1.9 Creation of constructions. Let $\mathcal{A} \xrightarrow{U} \mathcal{K}$ be a functor. U creates $\underleftarrow{\lim}$'s $=_{df}$ for each functor $\Delta \xleftarrow{H} \mathcal{A}$ and for each model $HU \xrightarrow{\alpha} X$ for $\underleftarrow{\lim}\,HU$ there exists a unique natural transformation $H \xleftarrow{\tilde{\alpha}} A$ with domain H such that $\tilde{\alpha}U = \alpha$; and moreover $\tilde{\alpha} = \underleftarrow{\lim}\,H$.

U creates regular coimage factorizations $=_{df}$ for each $\mathcal{A}$-morphism $A \xrightarrow{\tilde{f}} B$ and for each regular coimage factorization $AU \xrightarrow{p} I \xrightarrow{i} BU$ of $\tilde{f}U$ there exists unique

$\tilde{p}, \tilde{i} \in \mathcal{A}$ with $\tilde{p}U = p$ and $\tilde{i}U = i$; and moreover, $\tilde{p}.\tilde{i}$ is a regular coimage factorization of $\tilde{f}$.

U creates coequalizers of U-contractible pairs $=_{df}$ for each pair of $\mathcal{A}$-morphisms $(\tilde{f},\tilde{g}) : A \to B$ such that $(\tilde{f}U,\tilde{g}U)$ is contractible and for each model $BU \xrightarrow{q} Q$ of $\mathrm{coeq}(\tilde{f}U,\tilde{g}U)$ in $\mathcal{K}$, there exists unique $B \xrightarrow{\tilde{q}} Q$ with domain B such that $\tilde{q}U = q$; and moreover, $\tilde{q} = \mathrm{coeq}(\tilde{f},\tilde{g})$.

§ 2. Algebras over a triple.

In this section we study just enough about the category of algebras over a triple to suit our later needs. See [TM, chapter 1] for more results in a similar vein.

2.1 Definition. Let $\mathcal{K}$ be a category. $\mathbb{T} = (T,\eta,\mu)$ is a triple in $\mathcal{K}$ with unit η and multiplication μ if $\mathcal{K} \xrightarrow{T} \mathcal{K}$ is a functor and if $1 \xrightarrow{\eta} T$, $TT \xrightarrow{\mu} T$ are natural transformations subject to the three axioms:

$\mathbb{T}$-unitary axioms

$$\begin{array}{ccccc} T & \xrightarrow{T\eta} & TT & \xleftarrow{\eta T} & T \\ & {\scriptstyle 1}\searrow & \downarrow{\scriptstyle \mu} & \swarrow{\scriptstyle 1} & \\ & & T & & \end{array}$$

$\mathbb{T}$-associativity axiom

$$\begin{array}{ccc} TTT & \xrightarrow{\mu T} & TT \\ {\scriptstyle T\mu}\downarrow & & \downarrow{\scriptstyle \mu} \\ TT & \xrightarrow{\mu} & T \end{array}$$

Let $\mathbb{T} = (T,\eta,\mu)$ be a triple in $\mathcal{K}$. A $\mathbb{T}$-algebra $=_{df}$ a pair (X,ξ) with $X \in |\mathcal{K}|$ and $XT \xrightarrow{\xi} X$ a $\mathcal{K}$-morphism subject to the two axioms

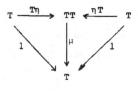

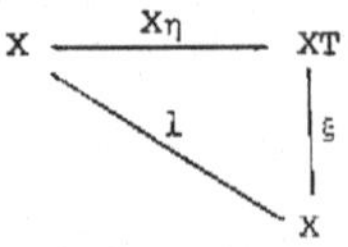

ξ-unitary axiom

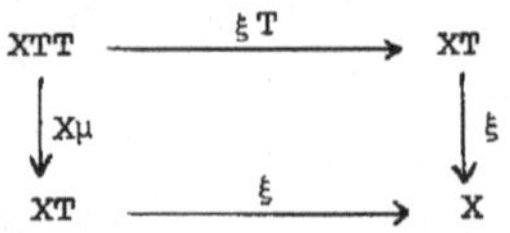

ξ-associativity axiom

X is the <u>underlying $\mathcal{K}$-object of</u> (X,ξ) and ξ is the <u>structure map of</u> (X,ξ) . If (X,ξ) and (Y,θ) are $\mathbb{T}$-algebras, a $\mathbb{T}$-<u>homomorphism</u>, $(X,\xi) \xrightarrow{f} (Y,\theta)$ <u>from</u> (X,ξ) to (Y,θ) is a $\mathcal{K}$-morphism $X \xrightarrow{f} Y$ subject to the

$\mathbb{T}$-homomorphism axiom

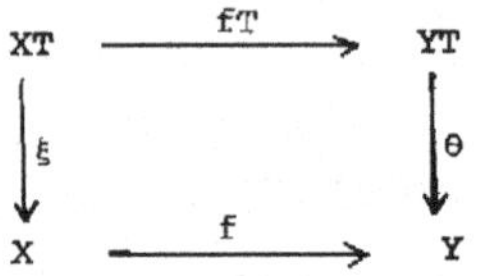

$\mathcal{K}^{\mathbb{T}} =_{dn}$ the resulting category of $\mathbb{T}$-algebras. $U^{\mathbb{T}} =_{dn}$ the faithful underlying $\mathcal{K}$-object functor.

A functor $\mathcal{A} \xrightarrow{U} \mathcal{K}$ is <u>tripleable</u> if there exists a triple $\mathbb{T}$ in $\mathcal{K}$ and an isomorphism of categories $\mathcal{A} \xrightarrow{\Phi} \mathcal{K}^{\mathbb{T}}$ such that $\Phi U^{\mathbb{T}} = U$.

<u>2.2 Heuristics in $\mathcal{K}^{\mathbb{T}}$.</u> Categories of algebras in the classical sense are tripleable (see the paper of Linton elsewhere in this volume; also [TM, 1.1.7]). We observe now that there are always free $\mathbb{T}$-algebras for $\mathbb{T} = (T,\eta,\mu)$ a triple in $\mathcal{K}$. If X is a $\mathcal{K}$-object, then $(XT,X\mu)$ is a $\mathbb{T}$-algebra (as is immediate from the triple axioms). Observe that if (X,ξ) is a $\mathbb{T}$-algebra, then $(XT,X\mu) \xrightarrow{\xi} (X,\xi)$ is a $\mathbb{T}$-homomorphism by the ξ-associativity axiom. Since μ is natural, for each $\mathcal{K}$-morphism $X \xrightarrow{f} Y$, $(XT,X\mu) \xrightarrow{fT} (YT,Y\mu)$ is a $\mathbb{T}$-homomorphism. We wish to think of $(XT,X\mu)$ as the "free $\mathbb{T}$-algebra on X generators" with $X \xrightarrow{X\eta} XT$ as "inclusion of the generators". Indeed, if (Y,θ) is a $\mathbb{T}$-algebra and if $X \xrightarrow{f} Y$ is a $\mathcal{K}$-morphism then it is easy to check that there exists a unique $\mathbb{T}$-homomorphism $(XT,X\mu) \xrightarrow{\tilde{f}} (Y,\theta)$ such that $X\eta.\tilde{f} = f$, namely $\tilde{f} =_{df} fT.\theta$. Note that a $\mathbb{T}$-algebra is characterized by the unique extension of the identity map on generators to $(XT,X\mu)$.

2.3 Example: the triple associated with a monoid. Let G be a monoid. "cartesian product with the underlying set of G" is a functor

$$\mathcal{S} \xrightarrow{- \times G} \mathcal{S}$$

Define $X\eta =_{df} X \xrightarrow{(1,e)} X \times G$ and $X\mu =_{df} X \times G \times G \xrightarrow{1 \times m} X \times G$ where e is the monoid unit and m is the monoid multiplication. Then $\mathcal{G} = (- \times G, \eta, \mu)$ is a triple in $\mathcal{S}$. $\mathcal{G}$-algebras are right G-sets. $\mathcal{G}$ is called the triple associated with G.

2.4 Proposition. $U^{\mathbb{T}}$ creates $\varprojlim$'s.

Proof. Suppose $\Delta \xrightarrow{D} \mathcal{K}^{\mathbb{T}}$ is a functor and $L \xrightarrow{\Gamma_i} X_i$ is a model for $\varprojlim DU^{\mathbb{T}}$. For every $i \xrightarrow{\delta} j \in \Delta$ we have

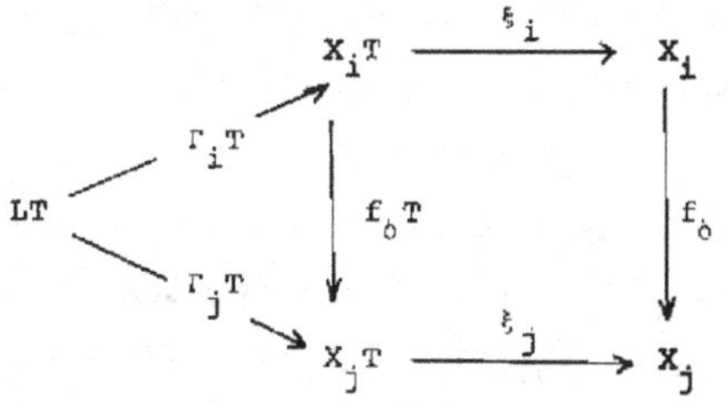

which induces a unique $\mathcal{K}$-morphism ξ such that $\Gamma_i T.\xi_i = \xi.\Gamma_i$ for all i. It is routine to check that $(L,\xi) \xrightarrow{\Gamma_i} (X_i,\xi_i)$ is the created $\varprojlim$ of D. []

2.5 Subalgebras. Let (X,ξ) be a $\mathbb{T}$-algebra and let $A \overset{i}{\rightarrowtail} X$ be a $\mathcal{K}$-monomorphism. Say that i (or by abuse of language, A) is a subalgebra of (X,ξ) if there exists a $\mathcal{K}$-morphism $AT \xrightarrow{\xi_0} A$ such that $\xi_0.i = iT.\xi$. It is easy to check that, indeed, (A,ξ_0) is a $\mathbb{T}$-algebra. To denote that (A,ξ_0) is a subalgebra of (X,ξ) we write "$(A,\xi_0) < (X,\xi)$".

2.6 Proposition. Let T preserve regular coimage factorizations. Then U creates regular coimage factorizations.

Proof. Let $(X,\xi) \xrightarrow{f} (Y,\xi)$ be a $\mathbb{T}$-homomorphism, and suppose f has regular coimage factorization $f = X \overset{p}{\twoheadrightarrow} I \overset{i}{\rightarrowtail} Y$ in $\mathcal{K}$. By hypothesis, $fT = XT \overset{pT}{\twoheadrightarrow} IT \overset{iT}{\rightarrowtail} YT$ is a regular coimage factorization in $\mathcal{K}$. Since $\xi.f = fT.\theta$ and i is mono, $\xi.p \in reg(pT)$ which induces unique θ_o with $pT.\theta_o = \xi.p$. $\theta_o.i = iT.\theta$ as pT is epi. We have $(X,\xi) \overset{p}{\to} (I,\theta_o) \overset{i}{\to} (Y,\theta)$ and that θ_o is unique with this property. To complete the proof we have only to show that $(X,\xi) \overset{p}{\to} (I,\theta_o)$ is regular epi in $\mathcal{K}^{\mathbb{T}}$. Let $(X,\xi) \overset{a}{\to} (A,\kappa) \in reg_{\mathbb{T}}(p)$. Suppose $(\zeta,\chi) : B \to X$ are $\mathcal{K}$-morphisms with $\zeta.p = \chi.p$. Let $\tilde{\zeta},\tilde{\chi}$ be the induced homomorphic extensions. Since $\tilde{\zeta}.p$, $\tilde{\chi}.p$ are homomorphisms agreeing on generators, $\tilde{\zeta}.p = \tilde{\chi}.p$. By the hypothesis on a, $\tilde{\zeta}.a = \tilde{\chi}.a$, so restricting to generators we have $\zeta.a = \chi.a$. This proves that $a \in reg_{\mathcal{K}}(p)$. As $X \overset{p}{\to} I$ is a regular epi in $\mathcal{K}$ there exists a unique $\mathcal{K}$-morphism $\tilde{a}$ with $p.\tilde{a} = a$. Since a is a $\mathbb{T}$-homomorphism and pT is epi, $\tilde{a}$ is forced to be a $\mathbb{T}$-homomorphism. []

__2.7 Definition.__ $\mathbb{T}$ is a __regular__ triple in $\mathcal{K}$ if $\mathcal{K}$ is a regular category and if T preserves regular coimage factorizations.

__2.8 Proposition.__ If $\mathbb{T}$ is a regular triple then $\mathcal{K}^{\mathbb{T}}$ is a regular category.

__Proof.__ REG 1, REG 2, REG 3 follow respectively from 2.6, 2.4 and the fact that $U^{\mathbb{T}}$ is faithful. Now let $(X,\xi) \overset{p}{\to} (Y,\theta)$ be a regular epi in $\mathcal{K}^{\mathbb{T}}$. Combining the way the regular commage factorization of p was created at the level $\mathcal{K}$ in 2.6 with 1.2.7 we see that $X \overset{p}{\to} Y$ is regular epi in $\mathcal{K}$. But then REG 4 is clear.

__2.9 Proposition; the precise tripleability theorem (Jon Beck).__ Let $\mathcal{A} \overset{U}{\to} \mathcal{K}$ be a functor. Then U is tripleable iff U has a left adjoint and U creates coequalizers of U-contractible pairs.

__Proof.__ [TM, 1.2.9]. []

__2.10 Definition.__ Let (X,ξ) be a $\mathbb{T}$-algebra and let $A \overset{i}{\rightarrowtail} X$. The __subalgebra of__ (X,ξ) __generated by__ $A =_{dn} \langle A \rangle \rightarrowtail (X,\xi)$, $=_{df}$ the intersection

$$\cap[(D,\alpha) < (X,\xi) : A < D] \rightarrowtail (X,\xi) .$$

When $\langle A \rangle$ exists it is in fact the smallest subalgebra of (X,ξ) containing A.

If $X \xrightarrow{f} Y$ is a $\mathcal{K}$-morphism and if $f = X \overset{p}{\twoheadrightarrow} I \overset{j}{\rightarrowtail} Y$ is a regular coimage factorization of f, $\operatorname{im} f =_{dn} I \overset{j}{\rightarrowtail} Y$. If $A \overset{i}{\rightarrowtail} X \xrightarrow{f} Y$ we also denote $\operatorname{im} i.f$ by "$Af \rightarrowtail Y$". If $X \xrightarrow{f} Y \overset{k}{\leftarrowtail} B$ we denote the pullback of k along f by "$Bf^{-1} \rightarrowtail X$". (That $Bf^{-1} \to X$ is a monomorphism is easily verified).

2.11 Proposition. Let $(X,\xi) \xrightarrow{f} (Y,\theta)$ be a $\mathbb{T}$-homomorphism and let $A \overset{i}{\rightarrowtail} X$, $B \overset{j}{\rightarrowtail} Y$. The following statements are valid.

a. $\langle A \rangle = \operatorname{im} iT.\xi$, providing both exist, and T preserves coimage factorizations.

b. $\langle A \rangle f = \langle Af \rangle$, providing both exist and T preserves coimage factorizations.

c. Bf^{-1}, if it exists, is a subalgebra of (X,ξ).

Proof. a. The diagram

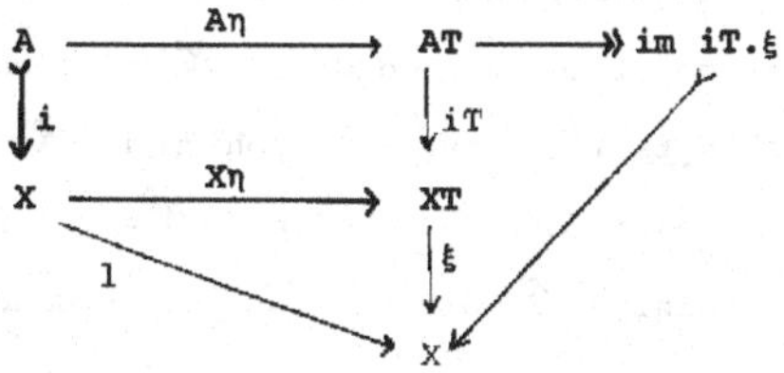

proves that $A \subset \operatorname{im} iT.\xi$. But by 2.6, $\operatorname{im} iT.\xi < (X,\xi)$. Therefore $\langle A \rangle \subset \operatorname{im} iT.\xi$. Conversely, consider the diagram

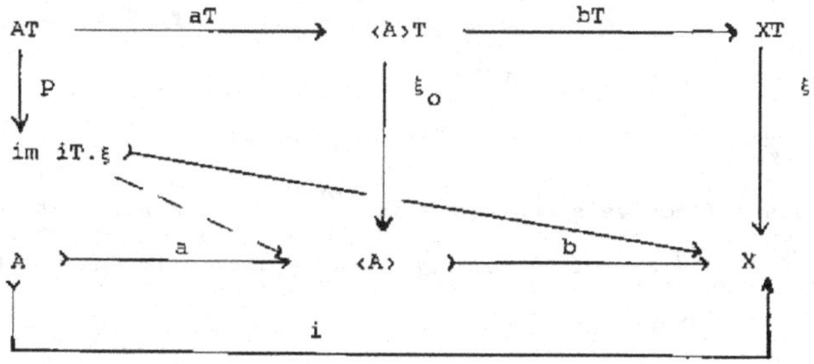

$aT.\xi_o \in \operatorname{reg}(p)$ because b is mono. Therefore, $\operatorname{im} iT.\xi \subset \langle A \rangle$.

b. Suppose $A \overset{p}{\twoheadrightarrow} Af \overset{c}{\rightarrowtail} Y$ is a regular coimage factorization. By hypothesis, pT is a regular epimorphism. We have $\langle Af\rangle = \text{im } \epsilon T.\theta = \text{im } pT.\epsilon T.\theta$ (by 1.2.7) $= \text{im } iT.\xi.f =$ $= \langle A\rangle f$.

c. $Bf^{-1} \to (X,\xi)$ is a monomorphism in $\mathcal{K}^{\Pi}$. But by 2.4, it is clear that the underlying $\mathcal{K}$-morphism of $Bf^{-1} \to (X,\xi)$ is a monomorphism, and hence $Bf^{-1} < (X,\xi)$. Alternate proof: since U preserves Kernel pairs it preserves monos.) [ ]

§ 3. Birkhoff Subcategories.

3.1 Definition. Let $\mathcal{K}$ be a category and let $\mathcal{B}$ be a full subcategory of $\mathcal{K}$. $\mathcal{B}$ is closed under products if every model for a product in $\mathcal{K}$ of a set of $\mathcal{B}$-objects lies in $\mathcal{B}$. $\mathcal{B}$ is closed under subobjects if every monomorphism in $\mathcal{K}$ with range in $\mathcal{B}$ lies in $\mathcal{B}$. Let $\mathcal{C}$ be any subcategory of $\mathcal{K}$. Define $\widehat{\mathcal{C}} =_{df}$ the intersection of all full subcategories of $\mathcal{K}$ containing $\mathcal{C}$ and closed under products and subobjects.

Evidently "$\frown$" is a closure operator on the (large) lattice of subcategories of $\mathcal{K}$, and $\mathcal{C} = \widehat{\mathcal{C}}$ iff $\mathcal{C}$ is closed under products and subobjects.

3.2 Proposition. Let $\mathcal{K}$ be a regular category and let $\mathcal{B} \xrightarrow{i} \mathcal{K}$ be a full subcategory. The following statements are equivalent.

a. $\mathcal{B} = \widehat{\mathcal{B}}$

b. $\mathcal{B}$ is a reflective subcategory of $\mathcal{K}$ in such a way that for every $\mathcal{K}$-object X the reflection $X \xrightarrow{X\eta} X_{\mathcal{B}}$ of X in $\mathcal{B}$ is a regular epimorphism; also $|\mathcal{B}|$ is a union of $\mathcal{K}$-isomorphism classes.

Proof. a implies b. Viewing an isomorphism as a unary product, $|\mathcal{B}|$ is a union of $\mathcal{K}$-isomorphism classes. $\mathcal{B}$ has $\underleftarrow{\lim}$'s and i preserves them. i satisfies the solution set condition by REG 1 and REG 4 . By the adjoint functor theorem it follows that

$\mathcal{B}$ is a reflective subcategory. Let X be a $\mathcal{K}$-object with reflection $X \xrightarrow{X\eta} X_{\mathcal{B}}$. Factor $X\eta = p.k$ through its regular coimage. As k is mono, x is induced with $X\eta.x = p$. By the uniqueness of reflection induced maps, $x.k = 1$. Therefore x is epi and split mono, hence iso, and $X\eta$ is regular epi because p is.

<u>b implies a.</u> Let X be a product in $\mathcal{K}$ of a set of $\mathcal{B}$-objects. Each projection factors through $X\eta$ inducing a map $X_{\mathcal{B}} \xrightarrow{a} X$ such that $X\eta.a = 1_X$. Hence $X\eta$ is split mono; since we assume $X\eta$ is epi, $X\eta$ is an isomorphism. Now suppose X is a $\mathcal{K}$-object admitting a monomorphism j to some object in $\mathcal{B}$. Then j factors through $X\eta$, and hence $X\eta$ is mono. But then $X\eta$ is mono and regular epi, and hence iso. []

For the balance of this section let $\mathbb{T} = (T,\eta,\mu)$ be a regular triple in $\mathcal{K}$.

<u>3.3 Proposition.</u> Let $T \xrightarrow{\lambda} \tilde{T}$ be a pointwise regular epi natural transformation, and suppose further that for every $\mathcal{K}$-object X there exists a $\mathcal{K}$-morphism $X\tilde{\mu}$ such that $X\lambda\lambda.X\tilde{\mu} = X\mu.X\lambda$. Then $\tilde{\mathbb{T}} =_{df} (T,\tilde{\eta},\tilde{\mu})$ (where $\tilde{\eta} =_{df} \eta\lambda$) is a triple in $\mathcal{K}$ and $\lambda\lambda.\tilde{\mu} = \mu.\lambda$.

<u>Proof.</u> The fact that $X\lambda$ is epi yields the unitary axioms. It is also true that $X\lambda\lambda$ and $X\lambda\lambda\lambda$ are epi, e.g. $X\lambda\lambda\lambda = X\lambda TT.X\tilde{T}\lambda T.X\tilde{T}\tilde{T}\lambda$ so use 1.2.8 and the fact that T preserves regular epi's. $X\lambda\lambda$ epi implies $\tilde{\mu}$ is natural, and $X\lambda\lambda\lambda$ epi implies the associativity axiom. The reader may provide the requisite diagrams. []

<u>3.4 The regular triple induced by a $\frown$-closed subcategory.</u> Let $\mathcal{B} \subset \mathcal{K}^{\mathbb{T}}$ be a subcategory such that $\mathcal{B} = \hat{\mathcal{B}}$. By 2.8 $\mathcal{K}^{\mathbb{T}}$ is a regular category, so that by 3.2 $\mathcal{B}$ is a full reflective subcategory with regular epi reflections. In particular, for each $\mathcal{K}$-object X let $(XT,X\mu) \xrightarrow{X\lambda} (X\tilde{T},\xi_X)$ be a regular epi reflection of $(XT,X\mu)$ in $\mathcal{B}$. By the reflection property, each $\mathcal{K}$-morphism $X \xrightarrow{f} Y$ induces unique $f\tilde{T}$ such that $X\lambda.f\tilde{T} = fT.Y\lambda$ which establishes a functor $\mathcal{K} \xrightarrow{\tilde{T}} \mathcal{K}$ and a pointwise regular epi natural transformation $T \xrightarrow{\lambda} \tilde{T}$. For every $\mathcal{K}$-object X, the fact that ξ_X is a $\mathbb{T}$-homomorphism and the reflection property induce unique $X\tilde{\mu}$:

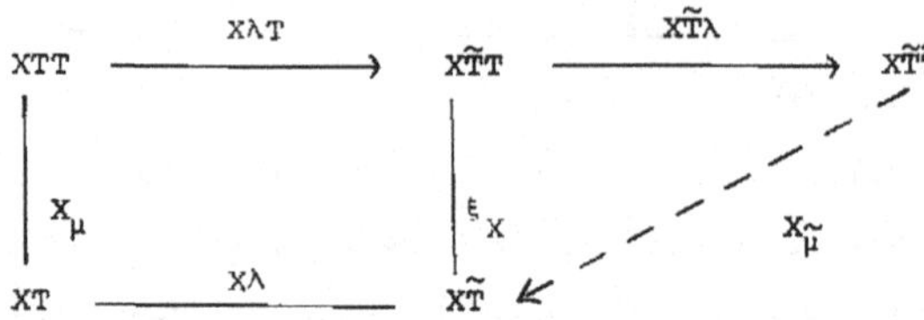

By 3.3, $\widetilde{\mathbb{T}} = (\widetilde{T}, \eta\lambda, \widetilde{\mu})$ is a triple in $\mathcal{K}$ and $\lambda\lambda.\widetilde{\mu} = \mu.\lambda$. $\widetilde{\mathbb{T}}$ is called the <u>regular triple induced by</u> $\mathcal{B}$.

<u>3.5 Definition.</u> A full subcategory $\mathcal{B}$ of $\mathcal{K}^{\mathbb{T}}$ is closed under $U^{\mathbb{T}}$-split epis $=_{df}$ whenever $(X,\xi) \xrightarrow{q} (Q,\alpha) \in \mathcal{K}^{\mathbb{T}}$ with $X \xrightarrow{q} Q$ split epi in $\mathcal{K}$ and $(X,\xi) \in |\mathcal{B}|$ then $(Q,\alpha) \in |\mathcal{B}|$. For each subcategory $\mathcal{C}$ of $\mathcal{K}^{\mathbb{T}}$ define $\widehat{\mathcal{C}} =_{df}$ the intersection of all full subcategories of $\mathcal{K}^{\mathbb{T}}$ closed under products, subalgebras ($=_{df}$ subobjects in $\mathcal{K}^{\mathbb{T}}$) and $U^{\mathbb{T}}$-split epis. A $\widehat{\ }$-closed subcategory of $\mathcal{K}^{\mathbb{T}}$ is called a <u>Birkhoff subcategory of</u> $\mathcal{K}^{\mathbb{T}}$ (because the following theorem is a triple-theoretic version of [B]).

<u>3.6 Proposition.</u> Let $\mathcal{B}$ be a Birkhoff subcategory of $\mathcal{K}^{\mathbb{T}}$ and let $U =_{df}$ the restriction of $U^{\mathbb{T}}$ to $\mathcal{B}$. Then U is tripleable.

<u>Proof.</u> Let $\widetilde{\mathbb{T}} = (\widetilde{T}, \widetilde{\eta}, \widetilde{\mu})$, λ be as in 3.4. We will construct an isomorphism $\mathcal{K}^{\widetilde{\mathbb{T}}} \xrightarrow{\Phi} \mathcal{B}$ such that $\Phi U = U^{\widetilde{\mathbb{T}}}$. If (X,ξ) is a $\widetilde{\mathbb{T}}$-algebra, $(X,\widetilde{\xi})\Phi =_{df} (X, XT \xrightarrow{X\lambda} X\widetilde{T} \xrightarrow{\widetilde{\xi}} X)$. $\widetilde{\eta} = \eta\lambda$ implies $X\eta.X\lambda.\widetilde{\xi} = 1_X$ and $\lambda\lambda.\widetilde{\mu} = \mu.\lambda$ implies $X\mu.X\lambda.\widetilde{\xi} = X\lambda T.\widetilde{\xi}T.X\lambda.\widetilde{\xi}$. If $(X,\widetilde{\xi}) \xrightarrow{f} (Y,\widetilde{\theta})$ is a $\widetilde{\mathbb{T}}$-homomorphism, the naturality of λ guarantees that $(X, X\lambda.\widetilde{\xi}) \xrightarrow{f} (Y, Y\lambda.\widetilde{\theta})$ is a $\widetilde{\mathbb{T}}$-homomorphism. Hence $\mathcal{K}^{\widetilde{\mathbb{T}}} \xrightarrow{\Phi} \mathcal{K}^{\mathbb{T}}$ is a well defined functor with $\Phi U^{\mathbb{T}} = U^{\widetilde{\mathbb{T}}}$. Because λ is pointwise epi, Φ is 1-to-1 on objects. If $(X,\widetilde{\xi})$, $(Y,\widetilde{\theta})$ are $\widetilde{\mathbb{T}}$-algebras and if $(X, X\lambda.\widetilde{\xi}) \xrightarrow{f} (Y, Y\lambda.\widetilde{\theta})$ is a $\mathbb{T}$-homomorphism then $(X,\widetilde{\xi}) \xrightarrow{f} (Y,\widetilde{\theta})$ is a $\widetilde{\mathbb{T}}$-homomorphism because $X\lambda$ is epi. Hence Φ is an isomorphism onto the full subcategory im Φ. We must show im $\Phi = \mathcal{B}$ on objects.

If $(X, X\lambda.\widetilde{\xi}) \in$ im Φ then $(XT, X\widetilde{T}\lambda.X\widetilde{\mu}) \in |\mathcal{B}|$ by the definition of $\widetilde{\mu}$ and $X \xrightarrow{X\widetilde{\eta}} X\widetilde{T}$ is a $\mathcal{K}$-splitting of the $\mathbb{T}$-homomorphism $(X\widetilde{T}, X\widetilde{T}\lambda.X\widetilde{\mu}) \xrightarrow{\widetilde{\xi}} (X, X\lambda.\widetilde{\xi})$, so that $(X, X\lambda.\widetilde{\xi}) \in |\mathcal{B}|$. Conversely, let $(X,\xi) \in |\mathcal{B}|$. By the reflection property there exists a unique $\mathbb{T}$-homomorphism $X\widetilde{T} \xrightarrow{\widetilde{\xi}} X$ with $X\lambda.\widetilde{\xi} = \xi$. Using the fact that $X\lambda\lambda$ is epi it is easy to check that $(X,\tilde{\xi})$ is a $\tilde{\mathbb{T}}$-algebra. The proof is complete. []

§ 4. The category $\mathcal{K}^{(\mathbb{T},\tilde{\mathbb{T}})}$.

For this section, let $\mathbb{T},\tilde{\mathbb{T}}$ be regular triples in $\mathcal{K}$.

4.1 Definition. Define a new category $\mathcal{K}^{(\mathbb{T},\tilde{\mathbb{T}})}$ with objects $(X,\xi,\tilde{\xi})$: $(X,\xi) \in |\mathcal{K}^{\mathbb{T}}|$ and $(X,\tilde{\xi}) \in |\mathcal{K}^{\tilde{\mathbb{T}}}|$ and morphisms $(X,\xi,\tilde{\xi}) \xrightarrow{f} (Y,\theta,\tilde{\theta})$ such that $(X,\xi) \xrightarrow{f} (Y,\theta) \in \mathcal{K}^{\mathbb{T}}$ and $(X,\tilde{\xi}) \xrightarrow{f} (Y,\tilde{\theta}) \in \mathcal{K}^{\tilde{\mathbb{T}}}$ with identities and compositions defined at the level $\mathcal{K}$. $U^{(\mathbb{T},\tilde{\mathbb{T}})} =_{dn}$ the obvious underlying $\mathcal{K}$-object functor.

4.2 Proposition $\mathcal{K}^{(\mathbb{T},\tilde{\mathbb{T}})}$ is a regular category and $U^{(\mathbb{T},\tilde{\mathbb{T}})}$ creates $\varprojlim$'s and regular coimage factorizations.

Proof. That $U^{(\mathbb{T},\tilde{\mathbb{T}})}$ creates $\varprojlim$'s follows easily from 2.4; in particular we have REG 2. REG 3 is clear as $U^{(\mathbb{T},\tilde{\mathbb{T}})}$ is faithful.

Now suppose that $(X,\xi,\tilde{\xi}) \xrightarrow{f} (Y,\theta,\tilde{\theta}) \in \mathcal{K}^{(\mathbb{T},\tilde{\mathbb{T}})}$ is such that f is regular epi in $\mathcal{K}^{\mathbb{T}}$ and in $\mathcal{K}^{\tilde{\mathbb{T}}}$. Then f is regular epi in $\mathcal{K}^{(\mathbb{T},\tilde{\mathbb{T}})}$. To prove it, it is enough to let g be in $\text{reg}_{(\mathbb{T},\tilde{\mathbb{T}})}(f)$ and show that g is in $\text{reg}_{\mathbb{T}}(f)$. Let $(a,b) : (A,\alpha) \to (X,\xi) \in \mathcal{K}^{\mathbb{T}}$ such that $af = bf$. Let $(t,u) : P \to X$ be a pullback of f with itself in $\mathcal{K}$. Since $U^{(\mathbb{T},\tilde{\mathbb{T}})}$ creates $\varprojlim$'s , t,u lift to $\mathcal{K}^{(\mathbb{T},\tilde{\mathbb{T}})}$-morphisms $(P,\gamma,\tilde{\gamma}) \to (X,\xi,\tilde{\xi})$. Since $U^{\mathbb{T}}$ creates $\varprojlim$'s , $(t,u) : (P,\gamma) \to (X,\xi)$ is the pullback of $(X,\xi) \xrightarrow{f} (Y,\theta)$ with itself, which induces $(A,\alpha) \xrightarrow{h} (P,\gamma)$ such that $ht = a$, $hu = b$. Since g is in $\text{reg}_{(\mathbb{T},\tilde{\mathbb{T}})}(f)$ and $tf = uf$ we have $tg = ug$. Therefore $ag = htg = hug = bg$.

That $U^{(\mathbb{T},\tilde{\mathbb{T}})}$ creates regular coimage factorizations now follows easily from 2.6; in particular, REG 1 is established. Let $(X,\xi,\tilde{\xi}) \xrightarrow{f} (Y,\theta,\tilde{\theta})$ be regular epi in $\mathcal{K}^{(\mathbb{T},\tilde{\mathbb{T}})}$. REG 4 will be clear from 2.8 if we show f is regular epi in $\mathcal{K}^{\mathbb{T}}$. This is immediate from 1.2.7 and the way the regular coimage factorization of f in $\mathcal{K}^{(\mathbb{T},\tilde{\mathbb{T}})}$ was constructured. []

<u>4.3. Definition.</u> Let $\mathcal{B}$ be a full subcategory of $\mathcal{K}^{(\mathbb{T},\tilde{\mathbb{T}})}$. $\mathcal{B}$ is a <u>Birkhoff subcategory</u> if $\mathcal{B}$ is closed under products, subobjects and $U^{(\mathbb{T},\tilde{\mathbb{T}})}$-split epis; (the meaning of "closed under $U^{(\mathbb{T},\tilde{\mathbb{T}})}$-split epis" is clear).

<u>4.4 Proposition.</u> Let $\mathcal{B}$ be a Birkhoff subcategory of $\mathcal{K}^{(\mathbb{T},\tilde{\mathbb{T}})}$ and let $U =_{df}$ the restriction of $U^{(\mathbb{T},\tilde{\mathbb{T}})}$ to $\mathcal{B}$. Then U is tripleable iff U satisfies the solution set condition.

<u>Proof.</u> We use 2.9. It is trivial to check that $U^{(\mathbb{T},\tilde{\mathbb{T}})}$ creates coequalizers of $U^{(\mathbb{T},\tilde{\mathbb{T}})}$-contractible pairs. Now suppose $(X,\xi,\tilde{\xi}) \underset{g}{\overset{f}{\rightrightarrows}} (Y,\theta,\tilde{\theta}) \in \mathcal{B}$ with $X \underset{g}{\overset{f}{\rightrightarrows}} Y$ contractible and $Y \xrightarrow{q} Q = \text{coeq}(f,g)$ in $\mathcal{K}$. Since q is a split epimorphism in $\mathcal{K}$, by 1.8.2, the created coequalizer $(Y,\theta,\tilde{\theta}) \xrightarrow{q} (Q,\alpha,\tilde{\alpha})$ is, in fact, in $\mathcal{B}$. Therefore U creates coequalizers of U-contractible pairs. $\mathcal{B}$ is closed under the $\mathcal{K}^{(\mathbb{T},\tilde{\mathbb{T}})}$-$\varprojlim$'s so that $\mathcal{B}$ has and U preserves $\varprojlim$'s. By the adjoint functor theorem, the proof is complete. []

<u>4.5 Definition.</u> Let $(X,\xi,\tilde{\xi}) \in |\mathcal{K}^{(\mathbb{T},\tilde{\mathbb{T}})}|$. $(X,\xi,\tilde{\xi})$ is a $\mathbb{T}$-$\tilde{\mathbb{T}}$ quasicomposite algebra $=_{df}$ for every $\mathcal{K}$-monomorphism $A \overset{i}{\rightarrowtail} X$, the $\mathcal{K}^{(\mathbb{T},\tilde{\mathbb{T}})}$-subalgebra generated by A is $\langle\langle A\rangle_{\mathbb{T}}\rangle_{\tilde{\mathbb{T}}}$. (What we mean by "subalgebra generated by" is clear). Equivalently, if A is a $\tilde{\mathbb{T}}$-subalgebra, so is $\langle A\rangle_{\mathbb{T}}$.

<u>4.6 Proposition.</u> Let $\mathcal{B}$ be a Birkhoff subcategory of $\mathcal{K}^{(\mathbb{T},\tilde{\mathbb{T}})}$ and let $U =_{df} U^{(\mathbb{T},\tilde{\mathbb{T}})}$ restricted to $\mathcal{B}$. If every $\mathcal{B}$-object is a $\mathbb{T}$-$\tilde{\mathbb{T}}$ quasicomposite algebra then U is tripleable.

<u>Proof.</u> By 4.4 we have only to show that U satisfies the solution set condition. Let K be a $\mathcal{K}$-object. Let $\mathcal{E}_1$ be a representative set of regular epis with domain K . Let $\mathcal{E}_2$ be a representative set of split epis with domain of form LT for some L which is the range of some element of $\mathcal{E}_1$. Let $\mathcal{E}_3$ be a representative set of split epis with domain of form $L\tilde{T}$ for some L which is the range of some element of $\mathcal{E}_2$. Now suppose $(X,\xi,\tilde{\xi})$ is an object in $\mathcal{B}$ and that $K \xrightarrow{f} X$ is a $\mathcal{K}$-morphism. There exists

p in $\mathcal{S}_1$ with $f = K \overset{p}{\twoheadrightarrow} L \overset{i}{\rightarrowtail} X$. There exists a model for $\langle L\rangle_{\mathbb{T}}$ such that the canonical split epi $LT \overset{\theta}{\twoheadrightarrow} \langle L\rangle_{\mathbb{T}}$ is in $\mathcal{S}_2$ (as we can always transport a structure map through a $\mathcal{K}$-isomorphism). Similarly there exists a split epi $\langle L\rangle_{\mathbb{T}}\tilde{T} \twoheadrightarrow \langle\langle L\rangle_{\mathbb{T}}\rangle_{\tilde{\mathbb{T}}}$ in $\mathcal{S}_3$. Hence (see the diagram below) we have

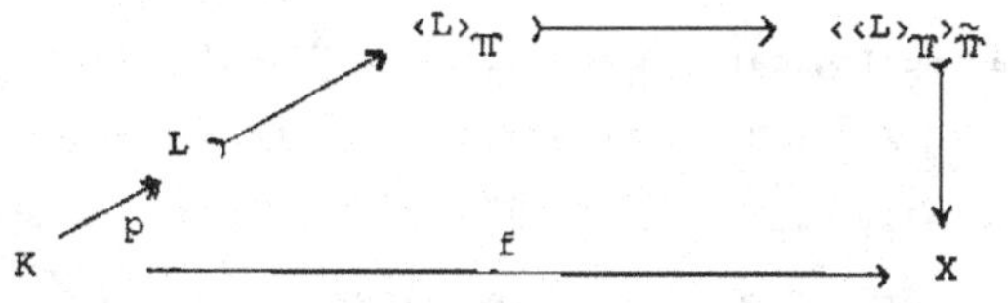

proved that f factors through a set of objects $\{\langle\langle L\rangle_{\mathbb{T}}\rangle_{\tilde{\mathbb{T}}}\}$. The crucial point is our hypothesis which guarantees that each $\langle\langle L\rangle_{\mathbb{T}}\rangle_{\tilde{\mathbb{T}}}$ is in $|\mathcal{B}|$. []

§5. Compact spaces.

5.1 Definition. Let X be a set, $\mathcal{F} \subset 2^X$. $\mathcal{F}^c =_{df} [A \subset X :$ there exists $F \in \mathcal{F}$ with $F \subset A]$. $\mathcal{F}$ is a filter on X if $\mathcal{F} \neq \Phi$, $\Phi \notin \mathcal{F}$, $A,B \in \mathcal{F}$ implies $A \cap B \in \mathcal{F}$ and $\mathcal{F} = \mathcal{F}^c$. An ultrafilter on X is an inclusion maximal filter on X . $X\beta =_{df}$ $[\mathcal{U} : \mathcal{U}$ is an ultrafilter on $X]$. If $A \subset X$, $\mathcal{F} \wedge A =_{df} [F \cap A : F \in \mathcal{F}]$. If $\mathcal{F}$ is a filter on X , it is easy to verify that $A \notin \mathcal{F}$ iff $\mathcal{F} \wedge A'$ is a filter on A' iff $(\mathcal{F} \wedge A')^c$ is a filter on X (where $A' =_{dn}$ the complement of A in X .)

5.2 Lemma. The following statements are valid.

a. For every filter, $\mathcal{F}$, on X , $\mathcal{F} \in X\beta$ iff for every subset A of X either $A \in \mathcal{F}$ or $A' \in \mathcal{F}$.

b. For every filter, $\mathcal{F}$, on X , $\mathcal{F} = \cap[\mathcal{U} \in X\beta : \mathcal{F} \subset \mathcal{U}]$.

Proof. a. If $A \notin \mathcal{F}$, $(\mathcal{F} \wedge A')^c$ is a filter finer than, hence equal to, $\mathcal{F}$. Therefore $A' \in \mathcal{F}$. Conversely, let $\mathcal{G}$ be a filter containing $\mathcal{F}$. If $G \in \mathcal{G}$, $G' \notin \mathcal{F}$ so that $G \in \mathcal{F}$.

b. Let $A \subset X$, $A \notin \mathcal{F}$. $(\mathcal{F} \wedge A')^c$ is a filter on X. By Zorn's Lemma (a nested union of filters is a filter) every filter is contained in an ultrafilter. Hence there exists $\mathcal{U} \in X\beta$ with $(\mathcal{F} \wedge A')^c \subset \mathcal{U}$. We have $\mathcal{F} \subset \mathcal{U}$, $A \notin \mathcal{U}$ proving $A \notin \cap [\mathcal{V} \in X\beta : \mathcal{F} \subset \mathcal{V}]$. []

5.3 Definition Let $(X,\mathcal{S})$ be a topological space, let $\mathcal{F} \subset 2^X$ and let $x \in X$. Recall that $\mathcal{F}$ converges to $x =_{df} \mathcal{F}^c \supset \mathcal{N}_x$ (where $\mathcal{N}_x =_{dn}$ the neighborhood filter of x), $=_{dn} \mathcal{F} \to x$. More generally, if $A \subset X$, $\mathcal{F} \to A =_{dn}$ there exists $x \in A$ with $\mathcal{F} \to x$. If $X \xrightarrow{f} Y$ is a function, $\mathcal{F}f =_{df}$ the filter $[Ff : F \in \mathcal{F}]^c \subset 2^Y$.

5.4 Lemma. The following statements are valid.

a. (Due to Ellis and Gottschalk, see [E+G, lemma 7]. Let $(X,\mathcal{S})$, $(X',\mathcal{S}')$ be topological spaces, let $X \xrightarrow{f} Y$ be a function and let $x \in X$. Then f is continuous at x iff for every $\mathcal{U} \in X\beta$, $\mathcal{U} \to x$ implies $\mathcal{U}f \to xf$.

b. Let $(X,\mathcal{S})$ be a topological space, and let $A \subset X$. Then A is open iff for every $\mathcal{U} \in X\beta$, $\mathcal{U} \to A$ implies $A \in \mathcal{U}$.

c. Let $(X,\mathcal{S})$ be a topological space, and let $X \xrightarrow{f} X'$ be an onto function. Let $\mathcal{S}'$ be the quotient topology induced by f. Then if $(X,\mathcal{S})$ is compact T2 and if $(X,\mathcal{S}) \xrightarrow{f} (X',\mathcal{S}')$ is closed (i.e. maps closed sets to closed sets) then $(X',\mathcal{S}')$ is compact T2.

Proof. a. Let $\mathcal{U} \in X\beta$, $\mathcal{U} \to x$. Let $V \in \mathcal{N}_{xf}$. There exists $W \in \mathcal{N}_x$ with $Wf \subset V$. As $W \in \mathcal{U}$, $V \in \mathcal{U}f$. Now the converse. For every $\mathcal{U} \to x$ we have $\mathcal{U}f \supset \mathcal{N}_{xf}$. $\mathcal{U} \supset \mathcal{U}_{ff^{-1}} \supset \mathcal{N}_{xf}f^{-1}$. By 5.2b, $\mathcal{N}_x = \cap[\mathcal{U} : \mathcal{U} \to x] \supset \mathcal{N}_{xf}f^{-1}$.

b. A is open iff $A \in \bigcap_{x \in A} \mathcal{N}_x = \bigcap_{x \in A} \bigcap_{\mathcal{U} \to x} \mathcal{U} = \bigcap_{\mathcal{U} \to A} \mathcal{U}$.

c. This is standard. See [K, chapter 5, theorem 20, p. 148]. []

5.5 Propositon. Let $\mathcal{C}$ be the category of compact T2 spaces with underlying set functor $\mathcal{C} \xrightarrow{U} \mathcal{S}$. Then U is tripleable.

Proof. A fairly short proof could be given using 2.9. Instead, we offer an independent definition of "compact T2 space" by making the triple explicit. If X is a set with

$x \in X$, $A \subset X$, define $\dot{x} =_{df} [B \subset X : x \in B]$ and $\dot{A} =_{df} [\mathcal{U} \in X\beta : A \in \mathcal{U}]$. The following five statements are trivial to verify: $\dot{x} \in X\beta$, $(\dot{x}) = (\dot{x})$, $\dot{A} \cap \dot{B} = \dot{\overline{A \cap B}}$, $\dot{A'} = \dot{A}'$, $\dot{\Phi} = \Phi$.

Define $\mathbb{B} = (\beta, \eta, \mu)$ by

$$\mathcal{S} \text{---}\ \beta \longrightarrow \mathcal{S}$$

$$X \xrightarrow{f} Y \longmapsto X\beta \xrightarrow{f\beta} Y\beta , \qquad X \xrightarrow{X\eta} X\beta$$

$$\mathcal{U} \longmapsto \mathcal{U}f \qquad\qquad x \longmapsto \dot{x}$$

$$X\beta\beta \xrightarrow{\quad X\mu \quad} X\beta$$

$$\mathcal{A} \longmapsto [A \subset X : \dot{A} \in \mathcal{A}]$$

The proof that $\mathbb{B}$ is a triple in $\mathcal{S}$ will be left to the reader; the details are routine providing one remembers that two ultrafilters are equal if one is contained in the other. (The details are written out in [TM, 2.3.3]). We will construct an isomorphism of categories $\mathcal{C} \xrightarrow{\Phi} \mathcal{S}^{\mathbb{B}}$ such that $\Phi U^{\beta} = U$. Let $(X, \mathcal{J})$ be a compact T2 space. $X\beta \xrightarrow{\xi_{\mathcal{J}}} X =_{df}$ the function sending an ultrafilter to the unique point to which it converges. $X\eta . \xi = 1$ because $\dot{x} \to x$ in all topologies. Now let $\mathcal{A} \in X\beta\beta$. $x =_{df}$ $\langle \mathcal{A}, \xi_{\mathcal{J}}\beta . \xi_{\mathcal{J}} \rangle = [A \subset X : \dot{A} \in \mathcal{A}]^c \xi_{\mathcal{J}}$. To verify the $\xi_{\mathcal{J}}$-associativity axiom we must show that $\langle \mathcal{A}, X\mu \rangle = [A \subset X : \dot{A} \in \mathcal{A}] \to x$. So let $B^{open} \in \mathcal{N}_x$. There exists $\mathcal{L} \in \mathcal{A}$ such that $[\mathcal{U}\xi_{\mathcal{J}} : \mathcal{U} \in \mathcal{L}] \subset B$. Therefore $\mathcal{U} \in \mathcal{L}$ implies $\mathcal{U}\xi_{\mathcal{J}} \in B$ implies there exists $b \in B$ such $\mathcal{U} \to b$. As $B \in \mathcal{N}_b$, $B \in \mathcal{U}$, so $\mathcal{U} \in \dot{B}$. Therefore $\dot{B} \supset \mathcal{L} \in \mathcal{A}$ and $\dot{B} \in \mathcal{A}$, as we wished to show. This defines Φ on objects. Now let $(X, \mathcal{J})$, $(X', \mathcal{J}')$ be compact T2 spaces and let $X \xrightarrow{f} X'$ be a function. f is a $\mathbb{B}$-homomorphism iff $f\beta . \xi_{\mathcal{J}'} = \xi_{\mathcal{J}} . f$ iff for every $\mathcal{U} \in X\beta$ and for every $x \in X$, $\mathcal{U} \to x$ implies $\mathcal{U}f \to xf$ iff f is continuous. Summing up, Φ is a well-defined full and faithful functor such that $\Phi U^{\beta} = U$ and such that Φ is 1-to-1 on objects (using 5.4b for the last statement). To complete the proof we show that Φ is onto on objects.

Let X be a set, and define a topology, $\mathcal{J}_X$, on $X\beta$ by taking $[\dot{A} : A \subset X]$ as a base, which we may do since the $\dot{A}$'s are closed under finite intersections; explicitly, every open set is a union of $\dot{A}$'s and conversely. Let $\mathcal{A} \in X\beta\beta$. $\mathcal{A} \to \mathcal{A}X\mu$, because if

$\mathcal{S}X\mu = [A \subset X : \dot{A} \in \mathcal{S}] \in \dot{B}$ then $B \in [A \subset X : \dot{A} \in \mathcal{S}]$, that is $\dot{B} \in \mathcal{S}$. Moreover if $\mathcal{U} \in X\beta$ is such that $\mathcal{S} \to \mathcal{U}$ it follows that $\mathcal{U} = \mathcal{S}X\mu$, for if $A \in \mathcal{U}$, then $\mathcal{U} \in \dot{A} \in \mathcal{S}$ and hence $A \in [B \subset X : \dot{B} \in \mathcal{S}] = \mathcal{S}X\mu$. This proves that $(X\beta\, \mathcal{S}_X) \in$ obj $\mathcal{C}$ and $(X\beta, \mathcal{S}_X)\Phi = (X\beta, X\mu)$.

Let $\mathcal{L} \overset{i}{\rightarrowtail} X\beta$, and consider the diagram

$$\begin{array}{ccc} \mathcal{L}\beta & \xrightarrow{\quad i \quad} & X_{\beta\beta} \\ \downarrow & & \downarrow X\mu \\ \mathcal{L} & \overset{i}{\rightarrowtail} & X \end{array}$$

One sees immediately that $\mathcal{L}$ is a subalgebra of $(X\beta, X\mu)$ iff every ultrafilter on $\mathcal{L}$ converges in $\mathcal{L}$ iff (applying a well-known theorem of topology) $\mathcal{L}$ is closed.

Now let (X,ξ) be any β-algebra. $(X\beta, X\mu) \overset{\xi}{\to} (X,\xi)$ is a β-homomorphism onto. Let $\mathcal{S}$ be the quotient topology induced by ξ on X. Let $\mathcal{L} \subset X\beta$. $\mathcal{L}$ is closed iff $\mathcal{L} < (X\beta, X\mu)$ implies (by 2.11b) $\mathcal{L}\xi < (X,\xi)$ implies (by 2.11c) $(\mathcal{L}\xi)\xi^{-1} < (X\beta, X\mu)$ iff $(\mathcal{L}\xi)\xi^{-1}$ is closed in $(X\beta, \mathcal{S}_X)$ iff $\mathcal{L}$ is closed in $(X, \mathcal{S})$. Therefore ξ is a closed mapping. By 5.4c, $(X, \mathcal{S}) \in$ obj $\mathcal{C}$. Finally, for $\mathcal{U} \in X\beta$ we show $\mathcal{U} \underset{\mathcal{S}}{\to} \mathcal{U}\xi$. Let $\mathcal{U}\xi \in A \in \mathcal{S}$. There exists $B \subset X$ with $\mathcal{U} \in \dot{B} \subset A\xi^{-1}$. For all $b \in B$, $b = \dot{b}\xi \in A\xi^{-1}\xi = A$. Therefore $A \supset B \in \mathcal{U}$ and $A \in \mathcal{U}$. []

<u>5.6 Remarks.</u> a. Let (X,ξ) be a β-algebra and let $A \subset X$. Then A is a subalgebra iff A is closed.

b. Free β-algebras are totally disconnected.

c. We can easily prove the Tychonoff theorem in the weak form "the cartesian product of compact T2 spaces is compact".

<u>Proof.</u> To prove (a), use the argument given for free algebras in the proof of 5.5. (b) is easy using the properties of "•"; notice that the class of clopen subsets of $(X\beta, X\mu)$ is precisely $[\dot{A} : A \subset X]$. For the third statement, construct the product in $\mathcal{S}^{\beta}$:

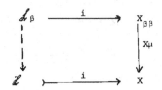

$$\begin{array}{ccc} (\Pi X_i)\mu & \xrightarrow{\quad pr_i \quad} & X_i\beta \\ \Big\downarrow{\scriptstyle \xi} & & \Big\downarrow{\scriptstyle \xi_i} \\ \Pi X_i & \xrightarrow{\quad pr_i \quad} & X_i \end{array}$$

Now observe that the diagram says that an ultrafilter on the product converges iff it converges pointwise, a characterization of the cartesian product topology of any family of topological spaces. []

§ 6. Operations.

For this section fix a triple $\mathbb{T} = (T,\eta,\mu)$ in $\mathcal{S}$.

6.1 Proposition. $\mathbb{T}$ is a regular triple.

Proof. That $\mathcal{S}$ is a regular category is well known; ordinary image factorizations provide the regular coimage factorizations. T preserves all epimorphisms and all monomorphisms with non-empty domain since these are split. To complete the proof we must show that for each set X, $(\Phi \overset{i}{\rightarrowtail} X)T$ is mono. This is clear if $\Phi T = \Phi$. Otherwise there exists a function $X \overset{f}{\to} \Phi T$. By freeness, $\mathbb{T}$-homomorphisms from $(\Phi T,\Phi\mu)$ to any $\mathbb{T}$-algebra (A,α) are in bijective correspondence with functions from Φ to A. Hence $\Phi T \overset{iT}{\to} XT \overset{fT}{\to} \Phi TT \overset{\Phi\mu}{\to} \Phi T$ is the identity map and iT is (split) mono. []

6.2 Definition. Let n be a set. "Raising to the nth power" is a functor:

$$\mathcal{S} \xrightarrow{\quad 1^n \quad} \mathcal{S}$$

$$X \xrightarrow{\ f\ } Y \longmapsto X^n \xrightarrow{\ f^n\ } X^n$$

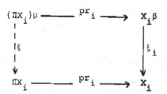

$\mathbb{T}(n) =_{df} [g : g$ is a natural transformation from 1^n to $T]$. For (X,ξ) a $\mathbb{T}$-algebra and $g \in \mathbb{T}(n)$, $\xi^g =_{df}$ the function $X^n \xrightarrow{Xg} XT \xrightarrow{\xi} X$. ξ^g is called an <u>n-ary operation</u> of (X,ξ) and the set of all such $=_{dn} \mathcal{O}_n(X,\xi)$.

6.3 and 6.4 below are indications that $\mathbb{T}$-algebras are characterized by their operations. See [TM, §2.2] for further details.

<u>6.3 Proposition.</u> Let (X,ξ), (Y,θ) be $\mathbb{T}$-algebras, and let $X \xrightarrow{f} Y$ be a function. The following statements are pairwise equivalent.

a. f is a $\mathbb{T}$-homomorphism.

b. For every set n and for every $g \in \mathbb{T}(n)$ the diagram

$$\begin{array}{ccc} X^n & \xrightarrow{f^n} & Y^n \\ {\scriptstyle \xi^g}\downarrow & (*)_g & \downarrow{\scriptstyle \theta^g} \\ X & \xrightarrow{f} & Y \end{array}$$

commutes.

c. $(*)_g$ commutes for every $g \in \mathbb{T}(X)$.

<u>Proof. a implies b.</u>

$$\begin{array}{ccccc} X^n & \xrightarrow{Xg} & XT & \xrightarrow{\xi} & X \\ \downarrow{\scriptstyle f^n} & & \downarrow{\scriptstyle fT} & & \downarrow{\scriptstyle f} \\ Y^n & \xrightarrow{Yg} & YT & \xrightarrow{\theta} & Y \end{array}$$

<u>b implies c.</u> This is obvious.

<u>c implies a.</u> Consider the diagram of "a implies b" with $n = X$. Let $x \in XT$. By the Yoneda Lemma there exists $g \in \mathbb{T}(X)$ with $\langle 1_X, Xg\rangle = x$. We have $\langle x, \xi.f\rangle = \langle 1_X, \xi^g.f\rangle = \langle 1_X, Xg.fT.\theta\rangle = \langle x, fT.\theta\rangle$. []

<u>6.4 Proposition.</u> Let (X,ξ) be a $\mathbb{T}$-algebra and let $A \overset{i}{\rightarrowtail} X$ be a subset. Then $A = \langle A\rangle$ iff for every $g \in \mathbb{T}(A)$, $A^A \xrightarrow{i^A} X^A \xrightarrow{\xi^g} X$ factors through A.

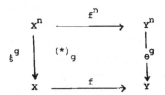

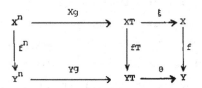

<u>Proof.</u> If $(A,\xi_o) < (X,\xi)$, $\xi_o{}^g$ is the desired factorization. Conversely, consider the diagram:

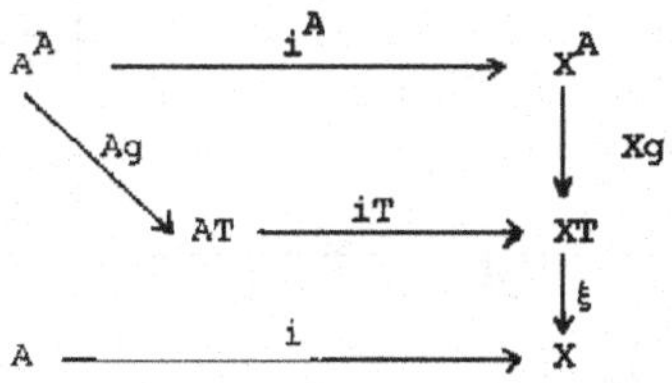

Let $x \in AT$. By the Yoneda Lemma there exists $g \in \mathbb{T}(A)$ with $\langle 1_A, Ag\rangle = x$. Hence as im $i^A.\xi^g \subset A$ by hypothesis, $\langle x, iT.\xi\rangle = \langle 1_A, i^A.\xi^g\rangle \in A$. Therefore $iT.\xi$ factors through i and $A = \langle A\rangle$. []

<u>6.5 Lemma.</u> Let n,m be sets, $g \in \mathbb{T}(n)$, $(X,\xi) \in |\mathcal{S}^{\mathbb{T}}|$, $(X^m,\dot{\xi}) =_{df} (X,\xi)^m$ and let $(X^n)^m \overset{\chi}{\to} (X^m)^n$ be the canonical bijection. Then

$$(X^n)^m \xrightarrow{(\xi^g)^m} X^m = (X^n)^m \xrightarrow{\chi} (X^m)^n \xrightarrow{\dot{\xi}^g} X^m .$$

<u>Proof.</u>

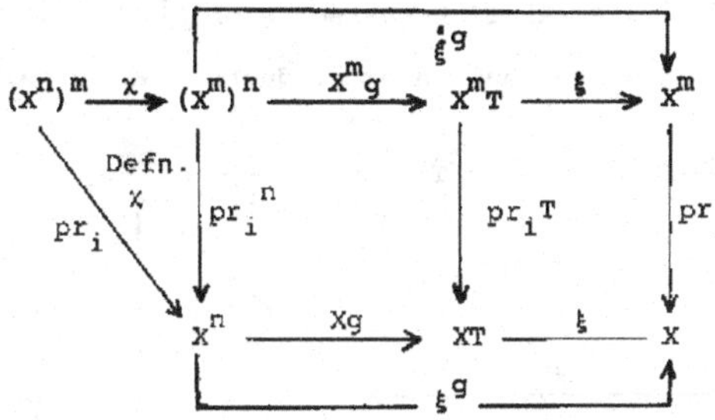

. []

For the balance of this section fix another triple $\tilde{\mathbb{T}} = (T,\tilde{\eta},\tilde{\mu})$ in $\mathcal{S}$.

<u>6.6 Proposition.</u> Let $(X,\xi,\tilde{\xi}) \in |\mathcal{S}^{(\mathbb{T},\tilde{\mathbb{T}})}|$. The following statements are equivalent.

a. For all sets n , for all $g \in \mathbb{T}(n)$, ξ^g is a $\tilde{\mathbb{T}}$-homomorphsim.

b. For all sets m , for all $h \in \tilde{\mathbb{T}}(n)$, $\tilde{\xi}^h$ is a $\mathbb{T}$-homomorphism.

<u>Proof.</u> Use 6.3, 6.5 and the symmetry of the diagram

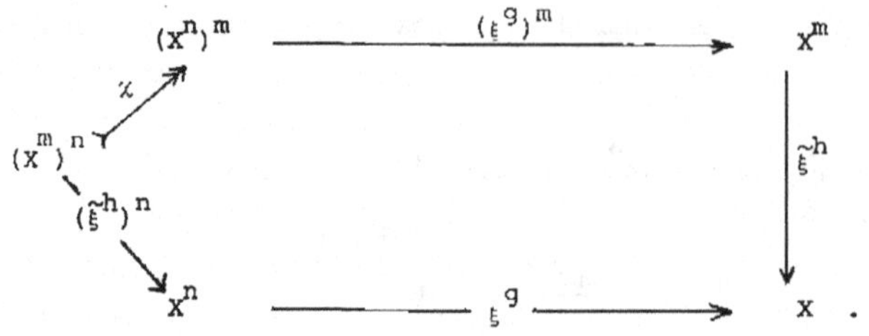

[]

<u>6.7 Definition.</u> $(X,\xi,\tilde{\xi}) \in |\mathcal{S}^{(\mathbb{T},\tilde{\mathbb{T}})}|$ is a $\mathbb{T}$-$\tilde{\mathbb{T}}$ <u>bialgebra</u> if it satisfies either of the equivalent conditions of 6.6. The full subcategory of $\mathbb{T}$-$\tilde{\mathbb{T}}$ bialgebras $=_{dn} \mathcal{S}^{[\mathbb{T},\tilde{\mathbb{T}}]}$ and the restriction of $U^{(\mathbb{T},\tilde{\mathbb{T}})}$ to $\mathcal{S}^{[\mathbb{T},\tilde{\mathbb{T}}]} =_{dn} U^{[\mathbb{T},\tilde{\mathbb{T}}]}$. If $U^{[\mathbb{T},\tilde{\mathbb{T}}]}$ is tripleable, the resulting triple, $=_{dn} \mathbb{T}\otimes\tilde{\mathbb{T}}$, $=_{df}$ the tensor product of $\mathbb{T}$ and $\tilde{\mathbb{T}}$. It is an open question whether or not $\mathbb{T}\otimes\tilde{\mathbb{T}}$ always exists. A constructive proof can be given if both $\mathbb{T}$ and $\tilde{\mathbb{T}}$ have a rank (in the sense of [TM, 2.2.6])by generalizing Freyd's proof in [F]. By 4.4 and 6.8 below the problem reduces to showing that $U^{[\mathbb{T},\tilde{\mathbb{T}}]}$ satisfies the solution set condition.

<u>6.8 Proposition.</u> $\mathcal{S}^{[\mathbb{T},\tilde{\mathbb{T}}]}$ is a Birkhoff subcategory of $\mathcal{S}^{(\mathbb{T},\tilde{\mathbb{T}})}$.

<u>Proof.</u> The diagram of 6.5 shows "closed under products". Now consider the diagram:

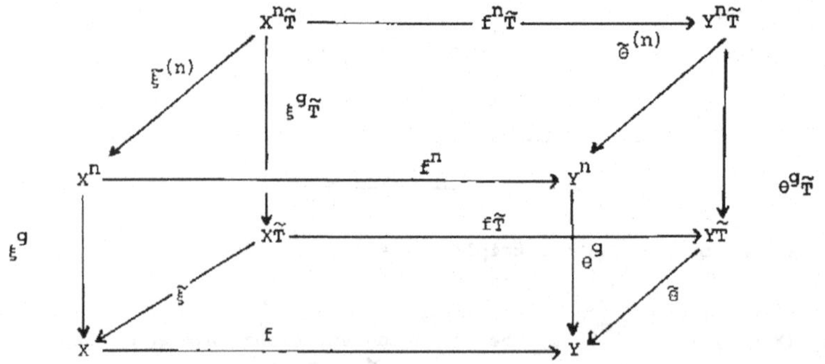

If $(X,\xi,\tilde{\xi}) \xrightarrow{f} (Y,\theta,\tilde{\theta}) \in \mathcal{S}^{(\mathbb{T},\tilde{\mathbb{T}})}$, all commutes except possibly the left and right faces. Hence if f is mono then right implies left; if f is epi then so is $f^n\tilde{T}$ so left implies right. []

<u>6.9 Definition.</u> Let (X,ξ) be a $\mathbb{T}$-algebra, let A be a subset of X and let n be a set. Consider the factorization $A \overset{i}{\rightarrowtail} \langle A\rangle \overset{j}{\rightarrowtail} X$. Because $\langle A\rangle^n \overset{j^n}{\rightarrowtail} X^n < X^n$ we have a factorization

$$\begin{array}{ccccccc} A^n & \xrightarrow{\;k\;} & \langle A^n\rangle & \xrightarrow{\;m\;} & \langle A\rangle^n & \overset{j^n}{\rightarrowtail} & X^n . \\ & \multicolumn{3}{c}{\underrightarrow{\qquad\quad i^n \qquad\quad}} & \uparrow & & \end{array}$$

Say that <u>subalgebras commute with powers in</u> $\mathcal{S}^{\mathbb{T}}$ if m is always an isomorphism for all (X,ξ), A, n .

<u>6.10 Proposition.</u> Suppose that subalgebras commute with powers in $\mathcal{S}^{\tilde{\mathbb{T}}}$. Then every $\mathbb{T}$-$\tilde{\mathbb{T}}$ bialgebra is a $\mathbb{T}$-$\tilde{\mathbb{T}}$ quasicomposite algebra, and hence $\mathbb{T}\otimes\tilde{\mathbb{T}}$ exists.

<u>Proof.</u> Let $(X,\xi,\tilde{\xi})$ be a $\mathbb{T}$-$\tilde{\mathbb{T}}$ bialgebra and let $(A,\xi_o) < (X,\xi)$. For each $g \in \mathbb{T}(\langle A\rangle_{\tilde{\mathbb{T}}})$ consider the diagram

$$\begin{array}{ccccccc} A^{\langle A\rangle_{\tilde{\mathbb{T}}}} & \rightarrowtail & \langle A^{\langle A\rangle_{\tilde{\mathbb{T}}}}\rangle & = & \langle A\rangle_{\tilde{\mathbb{T}}}^{\langle A\rangle_{\tilde{\mathbb{T}}}} & \rightarrowtail & X^{\langle A\rangle_{\tilde{\mathbb{T}}}} \\ \Big\downarrow \xi^g & & & & & & \Big\downarrow \xi^g \\ A & \rightarrowtail & & \langle A\rangle_{\tilde{\mathbb{T}}} & & \rightarrowtail & X \end{array}$$

Since ξ^g is a $\tilde{\mathbb{T}}$-homomorphism it follows from 2.11b that ξ^g maps $\langle A^{\langle A\rangle}\tilde{\mathbb{T}}\rangle$ into $\langle A\rangle_{\tilde{\mathbb{T}}}$. Since $\langle A^{\langle A\rangle}\tilde{\mathbb{T}}\rangle = \langle A\rangle_{\tilde{\mathbb{T}}}^{\langle A\rangle}\tilde{\mathbb{T}}$, ξ^g maps $\langle A\rangle_{\tilde{\mathbb{T}}}^{\langle A\rangle}\tilde{\mathbb{T}}$ into $\langle A\rangle_{\tilde{\mathbb{T}}}$. It follows from 6.4 that $\langle A\rangle_{\tilde{\mathbb{T}}}$ is a $\mathbb{T}$-subalgebra. The last statement is immediate from 6.8 and 4.6. []

<u>§ 7. Compact algebras.</u>

<u>7.1 Proposition.</u> For every triple, $\mathbb{T}$, in $\mathcal{S}$, every $\mathbb{T}$-$\boldsymbol{\beta}$ bialgebra is a $\mathbb{T}$-$\boldsymbol{\beta}$ quasicomposite algebra. In particular, $\mathbb{T}\otimes\boldsymbol{\beta}$ always exists.

<u>Proof.</u> A well-known theorem of topology is: "the product of the closures is the closure of the product." Using the Tychonoff theorem and 5.6a we have that subalgebras commute

with powers in $\mathcal{S}^{\beta}$. Now use 6.10. []

A $\mathbf{T}\otimes\beta$ algebra is, by definition, a $\mathbf{T}$-algebra whose underlying set is provided with a compact T2 topology in such a way that $\mathbf{T}$-operators are continuous. Hence $\mathbf{T}\otimes\beta$-algebras deserve to be —and are— called <u>compact $\mathbf{T}$-algebras.</u>

<u>7.2 Example; discrete actions with compact phase space.</u> Let G be a discrete monoid with associated triple $\mathbf{G}$. If $g \in \mathbf{G}(n)$ it is easy to check, using the Yoneda Lemma, that for each G-set (X,α), α^g factors as a projection map followed by the "transition" map induced from X to X by the action of some element of G. Hence $\mathcal{S}^{\mathbf{G}\otimes\beta}$ is the category of compact T2 transformation semigroups with phase semigroup G, that is since G is discrete $X \times G \xrightarrow{\alpha} X$ is continuous iff each transition $X \xrightarrow{\alpha^g} X$ is continuous.

<u>7.3 Proposition.</u> Compact topological dynamics is tripleable. More precisely, let G be a monoid with associated triple $\mathbf{G}$. Let $\mathscr{S}$ be any topology whatever on the underlying set of G. $\mathcal{B} =_{df}$ the full subcategory of $\mathcal{S}^{\mathbf{G}\otimes\beta}$ generated by objects (X,ξ,α) such that $(X,\xi) \times (G,\mathscr{S}) \xrightarrow{\alpha} (X,\xi)$ is continuous. Then $\mathcal{B}$ is a Birkhoff subcategory of $\mathcal{S}^{\mathbf{G}\otimes\beta}$, and in particular $\mathcal{B}$ is tripleable. (Compact topological dynamics is recovered by insisting that $\mathscr{S}$ be compatible with G.)

<u>Proof.</u> Consider a product of $\mathcal{B}$-objects, $(X,\alpha,\xi) = \Pi(X_i,\alpha_i,\xi_i)$. At the level of sets we have

$$\begin{array}{ccc} X \times G & \xrightarrow{\quad\alpha\quad} & X \\ \Big\downarrow{\scriptstyle pr_i \times 1} & & \Big\downarrow{\scriptstyle pr_i} \\ X_i \times G & \xrightarrow{\quad\alpha_i\quad} & X_i \end{array}$$

$(X,\xi) = \Pi(X_i,\xi_i)$ in $\mathcal{S}^{\beta}$ by 4.2, and hence in the category of all topological spaces by the Tychonoff theorem. Therefore α is continuous because each $\alpha . pr_i$ is.

Next, let $(A,\alpha_o,\xi_o) \overset{i}{\rightarrowtail} (X,\alpha,\xi)$ be a $\mathbf{G}\otimes\beta$-subalgebra with $(X,\alpha,\xi) \in |\mathcal{B}|$. We have $\alpha_o . i = (i \times 1).\alpha$. Now all monomorphisms in $\mathcal{S}^{\beta}$ become relative subspaces when viewed in the category of all spaces because every algebraic monomorphism is an isomorphism into. Therefore α_o is continuous because $\alpha_o . i$ is.

To show that $\mathcal{B}$ is closed under quotients it suffices to prove the following topological lemma: Consider the situation

$$\begin{array}{ccc} X \times H & \longrightarrow & X \\ \Big\downarrow {\scriptstyle f \times 1} & & \Big\downarrow {\scriptstyle f} \\ Y \times H & \xrightarrow{\ b\ } & Y \end{array}$$

where X, H, Y are topological spaces with X compact and Y T2 and where a is continuous and f is continuous onto. Then b is continuous. To prove it we use 5.4a. Let $\mathcal{U}$ be an ultrafilter on $Y \times H$ such that $\mathcal{U} \rightarrow (y,h) \in Y \times H$. Because f is onto, $U(f \times 1)^{-1} \neq \emptyset$ for all $U \in \mathcal{U}$. Hence there exists an ultrafilter $\mathcal{V}$ on $X \times H$ with $\mathcal{V} \supset \mathcal{U}(f \times 1)^{-1}$. $U \supset U(f \times 1)^{-1}(f \times 1)$ proves $\mathcal{U} = \mathcal{V}(f \times 1)$. Since X is compact there exists $x \in X$ such that $\mathcal{V} pr_X \rightarrow x$. $\mathcal{V} pr_H = \mathcal{V}(f \times 1) pr_H = \mathcal{U} pr_H \rightarrow h$. If $N \in \mathcal{N}_x, M \in \mathcal{N}_h$ there exist $V, W \in \mathcal{V}$ with $N \supset V\, pr_X, M \supset W\, pr_H$ and then $N \times M \supset (V \cap W) pr_X \times (V \cap W) pr_H \supset V \cap W \in \mathcal{V}$ proves that $\mathcal{V} \rightarrow (x,h)$. Since $\mathcal{U} pr_Y = \mathcal{V}(f \times 1) pr_Y = \mathcal{V} pr_X\, f \longrightarrow (x,h) pr_X\, f = xf$ and Y is $T2, xf = y$. Therefore $\mathcal{U} b = \mathcal{V}(f \times 1)\, b = \mathcal{V} a\, f \rightarrow (x,h)\, a\, f = (x,h)(f \times 1)\, b = (y,h)\, b$ as desired. []

Proposition 7.3 says that a compact T2 topological transformation group may equally well be viewed as a set with algebraic structure. Certain results of Ellis [Ea] [Eb] can be conveniently proved by this approach, and certain questions originating in topological dynamics may be asked in $\mathcal{S}^{\mathbf{T}}$. See [TM, §2.4, 2.5] for further details.

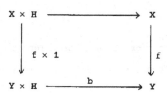

REFERENCES

[TM]	Ernest Manes	A triple miscellany: Some aspects of the theory of algebra over a triple, dissertation, Wesleyan Un. 1967
[B]	G.D. Birkhoff,	On the structure of abstract algebras, proc. Cambridge Phil. Soc. vol. 31 (1935), 433-454.
[E + G]	Robert Ellis and W.H. Gottschalk	Homomorphisms of transformation groups, Trans. Amer. Math. Soc. vol. 94 (1960), 258-271.
[Ea]	Robert Ellis	A semigroup associated with a transformation group, Trans. Amer. Math. Soc. vol. 94 (1960), 272-281.
[Eb]	Robert Ellis,	Universal minimal sets, Proc. Amer. Math. Soc. vol 11 (1960), 540-543.
[F]	Peter Freyd	Algebra valued functors in general and tensor products in particular, Colloq. Math. vol. 14 (1966), 89-106.
[GA]	R. Godement	Théorie des faisceaux Paris, Hermann, 1958, appendix.
[K]	J.L. Kelley	General topology New York, Van Nostrand, 1955.
[M]	Barry Mitchell	Theory of categories, New York, Acad. press, 1965.

DISTRIBUTIVE LAWS

Jon Beck

The usual distributive law of multiplication over addition, $(x_0 + x_1)(y_0 + y_1) \rightarrow x_0y_0 + x_0y_1 + x_1y_0 + x_1y_1$, combines the mathematical structures of abelian groups and monoids to produce the more interesting and complex structure of rings. From the point of view of "triples," a distributive law provides a way of interchanging two types of operations and making the functorial composition of two triples into a more complex triple.

The main formal properties and different ways of looking at distributive laws are given in §1. §2 is about algebras over composite triples. These are found to be objects with two structures, and the distributive law or interchange of operations appears in its usual form as an equation which the two types of operations must obey. §3 is about some frequently-occurring diagrams of adjoint functors which are connected with distributive laws. §4 is devoted to Examples. There is an Appendix on compositions of adjoint functors.

I should mention that many properties of distributive laws, some of them beyond the scope of this paper, have also been developed by Barr, Linton and Manes. In particular, one can refer to Barr's paper "Composite cotriples" in this volume. Since Barr's paper is available, I omitted almost all references to cotriples.

I would like to acknowledge the support of an NAS - NRC (AFOSR) Postdoctoral Fellowship at the E.T.H., Zürich, while this paper was being prepared, as well as the hospitality of the Mathematics Institute of the University of Rome, where some of the commutative diagrams were found.

One general fact about triples will be used. If $\varphi : \mathbf{S} \rightarrow \mathbf{T}$ is a map of triples in $\underline{A}$, the functor $\underline{A}^{\varphi}: \underline{A}^{\mathbf{S}} \leftarrow \underline{A}^{\mathbf{T}}$ usually has a left adjoint, for which there is a coequalizer formula:

$$ASF^{T} \underset{A\bar{\varphi}}{\overset{\sigma F^{T}}{\rightrightarrows}} AF^{T} \longrightarrow (A,\sigma)\otimes_{\mathbf{S}} F^{T} .$$

Here (A,σ) is an $\mathbb{S}$ - algebra and the coequalizer is calculated in $\underline{A}^{\mathbb{T}}$. The natural operation $\bar{\varphi}$ of $\mathbb{S}$ on F^{T} is the composition

$$(AST, AS\mu^{T}) \xrightarrow{\varphi T} (ATT, AT\mu^{T}) \xrightarrow{\mu^{T}} (AT, A\mu^{T}) \ .$$

The notation $(\ )\otimes_{\mathbb{S}}F^{T}$ for the left adjoint is justifiable. Later on the symbol $\underline{A}^{\varphi}$ is replaced by a Hom notation. The adjoint pair $(\ )\otimes_{\mathbb{S}}F^{T}$,$\underline{A}^{\varphi}$ is always tripleable.

1. <u>Distributive laws</u>, <u>composite and lifted triples</u>. A <u>distributive law of</u> $\mathbb{S}$ <u>over</u> $\mathbb{T}$ is a natural transformation ℓ: TS → ST such that

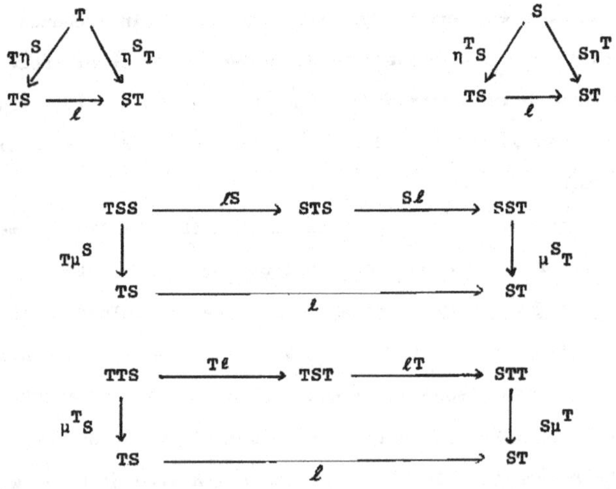

commute.

The <u>composite triple</u> defined by ℓ is $\mathbb{S}\mathbb{T} = (ST, \eta^{S}\eta^{T}, S\ell T.\mu^{S}\mu^{T})$. That is, the composite functor ST : $\underline{A} \to \underline{A}$, with unit and multiplication

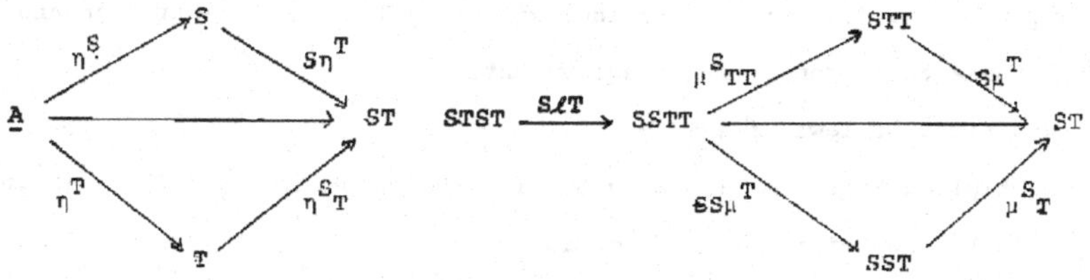

is a triple in $\underline{A}$. The units of $\mathbb{S}$ and $\mathbb{T}$ give triple maps

$$S\eta^T : \mathbb{S} \longrightarrow \mathbb{S}\mathbb{T} ,$$

$$\eta^S T : \mathbb{T} \longrightarrow \mathbb{S}\mathbb{T} .$$

The proofs of these facts are just long naturality calculations. Note that the composite triple should be written $(\mathbb{S}\mathbb{T})_\ell$ to show its dependence on ℓ , but that is not usually observed.

In addition to the composite triple, ℓ defines a <u>lifting</u> of the triple $\mathbb{T}$ into the category of $\mathbb{S}$-algebras. This is the triple $\tilde{\mathbb{T}}$ in $\underline{A}^{\mathbb{S}}$ defined by

$$\tilde{\mathbb{T}} = \begin{cases} \tilde{T} : \quad (A,\sigma)\tilde{T} = (AT, A\ell.\sigma T) , \\ \tilde{\eta} : (A,\sigma)\tilde{\eta} = A\eta : (A,\sigma) \to (A,\sigma)\tilde{T} , \\ \tilde{\mu} : (A,\sigma)\tilde{\mu} = A\mu : (A,\sigma)\tilde{T}\tilde{T} \to (A,\sigma)\tilde{T} . \end{cases}$$

It follows from compatibility of ℓ with $\mathbb{S}$ that $A\ell.\sigma T$ is an $\mathbb{S}$-algebra structure, and from compatibility of ℓ with $\mathbb{T}$ that $\tilde{\eta},\tilde{\mu}$ are maps of $\mathbb{S}$-algebras.

That $\tilde{\mathbb{T}}$ is a <u>lifting</u> of $\mathbb{T}$ is expressed by the commutativity relations

$\tilde{T}U^S = U^S T$, $\tilde{\eta}U^S = U^S\eta$, $\tilde{\mu}U^S = U^S\mu$.

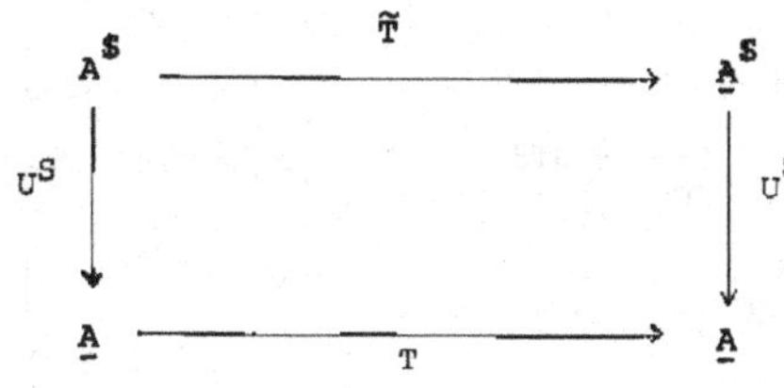

<u>Proposition.</u> Not only do distributive laws give rise to composite triples and liftings, but in fact these three concepts are equivalent:

(1) distributive laws $\ell : TS \to ST$,

(2) multiplications $m : STST \to ST$ with the properties : $(\mathbb{S}\mathbb{T})_m = (ST, \eta^S\eta^T, m)$ is a triple in $\underline{A}$, the natural transformations

$$\mathbb{S} \xrightarrow{S\eta^T} \mathbb{S}\mathbb{T} \xleftarrow{\eta^S T} \mathbb{T}$$

are triple maps, and the <u>middle unitary law</u>

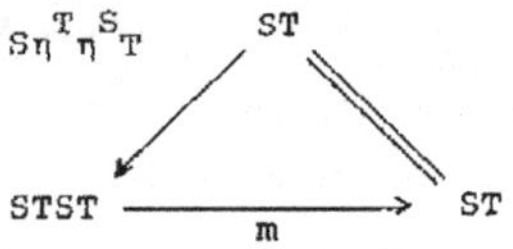

holds,

(3) liftings $\tilde{\mathbb{T}}$ of the triple $\mathbb{T}$ into $\underline{A}^{\mathbb{S}}$.

<u>Proof</u>. Maps (1) → (2), (1) → (3) have been constructed above. It remains to construct their inverses and prove that they are equivalences.

(2) → (1). Given m, define ℓ as the composition

$$TS \xrightarrow{\eta^S TS\eta^T} STST \xrightarrow{m} ST$$

Compatibility of ℓ with the units of $\mathbb{S}$ and $\mathbb{T}$ is trivial. As to compatibility with the multiplication in $\mathbb{T}$,

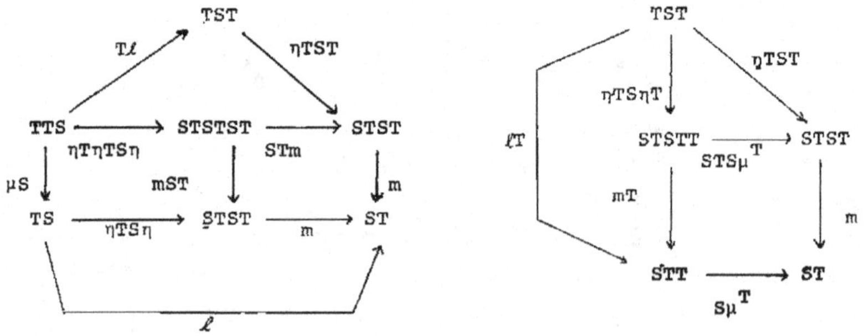

commute, the second because $\mathbb{T} \to (\mathbb{S}\mathbb{T})_m$ is a triple map. This reduces the problem to showing that an associative law holds between μ^T and m:

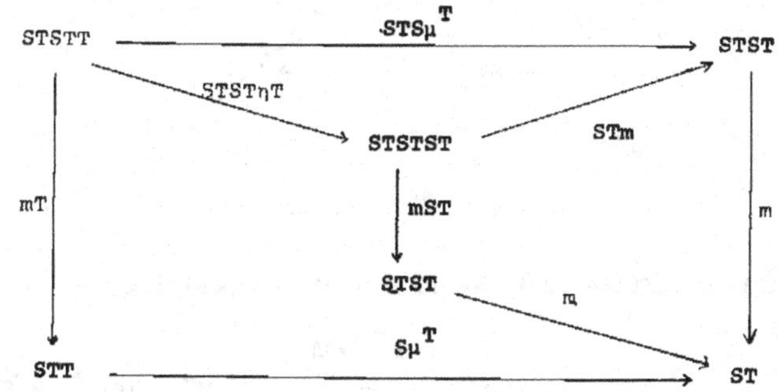

This commutes since $S\mu^T = ST\eta^S T.m$, as follows from the fact that $\mathbb{T} \to (\mathbb{S}\mathbb{T})_m$ is a triple map, and also the middle unitary law:

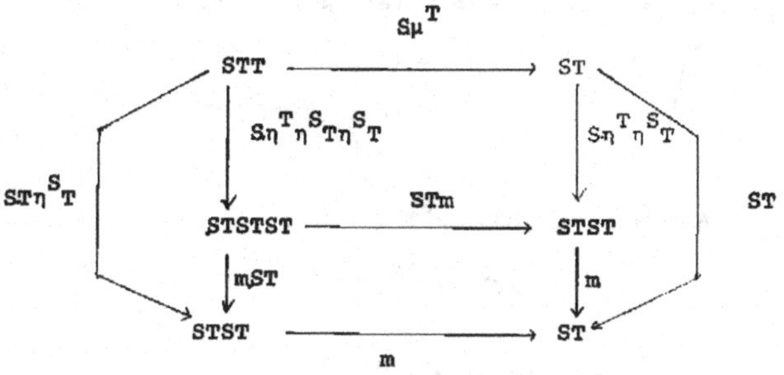

The proof that ℓ is compatible with multiplication in $\mathbb{S}$ is similar; it uses the associative law

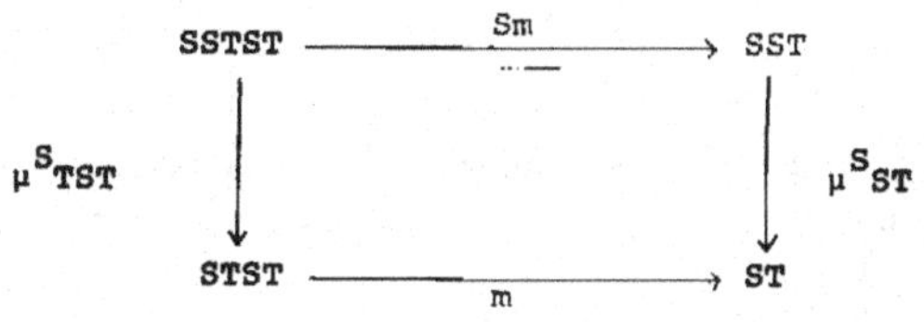

The composition (1) → (2) → (1) is clearly the identity. (2) → (1) → (2) is the

identity because of

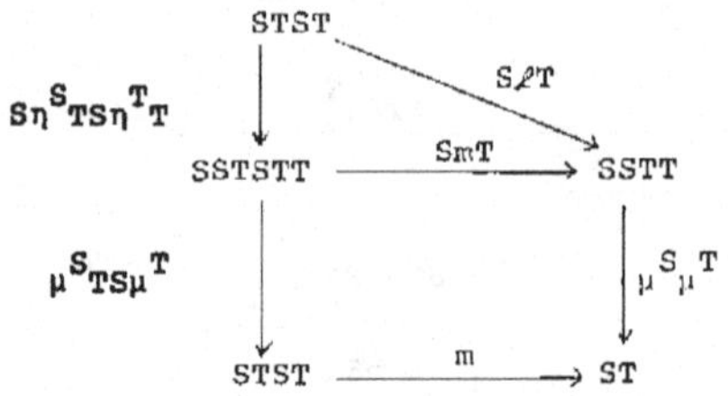

(3) → (1) . If $\tilde{\mathbb{T}}$ is a lifting of $\mathbb{T}$ define ℓ as the composition

$$TS \xrightarrow{\eta^S TS} STS = F^S U^S TS = F^S \tilde{T} U^S (FU)^S \xrightarrow{F^S \tilde{T}(\epsilon U)^S} F^S \tilde{T} U^S = (FU)^S T = ST,$$

where the abbreviation $(FU)^S$ stands for $F^S U^S$ . .... (1) → (3) → (1) is the identity. If we write $\ell \to \tilde{\mathbb{T}} \to \ell'$, then $\ell' = \ell$:

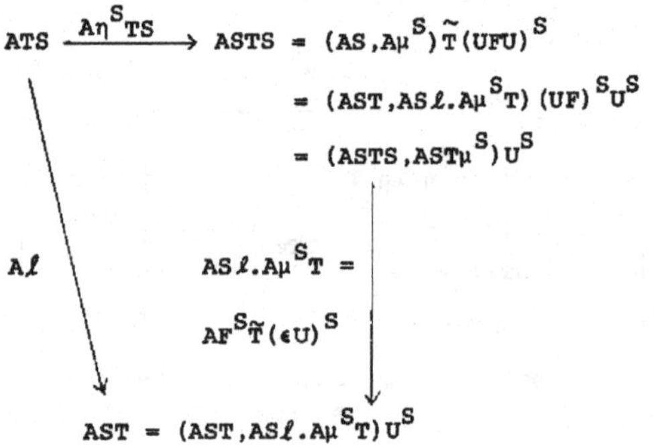

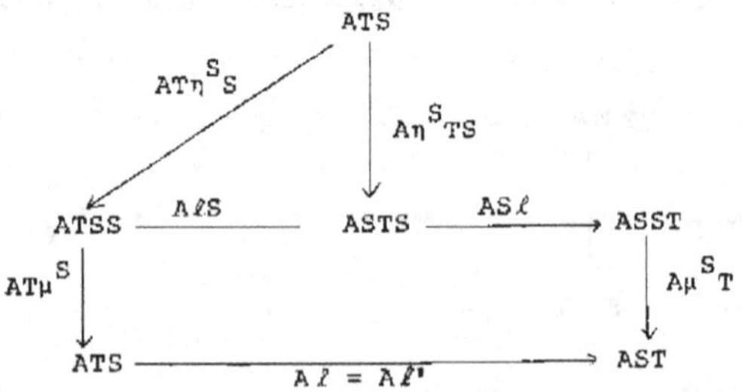

(3) → (1) → (3) is the identity. Let us write $\tilde{\mathbb{T}} \to \mathcal{L} \to \tilde{\tilde{\mathbb{T}}}$ and prove $\tilde{\mathbb{T}} = \tilde{\tilde{\mathbb{T}}}$.

Any lifting $\tilde{\mathbb{T}}$ of $\mathbb{T}$ can be written $(A,\sigma)\tilde{T} = (AT,(A,\sigma)\tilde{\sigma})$, where $\tilde{\sigma} : U^{S}TS \to U^{S}T$ is a natural $\mathbb{S}$-structure on $U^{S}T$. Restricting the lifting to free $\mathbb{S}$-algebras, $AF^{S}\tilde{T} = (AST, A\sigma_0)$, where $\sigma_0 = F^{S}\tilde{\sigma} : STS \to ST$ is a natural $\mathbb{S}$-structure on ST, which in addition satisfies an internal associativity relation involving μ^{S} :

$$\begin{array}{ccc} SSTS & \xrightarrow{\ S\sigma_0\ } & SST \\ {\scriptstyle \mu^{S}TS}\downarrow & & \downarrow{\scriptstyle \mu^{S}T} \\ STS & \xrightarrow[\ \sigma_0\ ]{} & ST \end{array}$$

This follows from the fact that $A\mu^{S} : ASF^{S} \to AF^{S}$ is an $\mathbb{S}$-algebra map.

Similarly, write $(A,\sigma)\tilde{\tilde{T}} = (AT,(A,\sigma)\tilde{\tilde{\sigma}}), AF^{S}\tilde{\tilde{T}} = (AST, A\sigma_1)$.

We must show that $\tilde{\sigma} = \tilde{\tilde{\sigma}}$. This is done first for free $\mathbb{S}$-algebras, i.e., $\sigma_0 = \sigma_1$, and then the result is deduced for all $\mathbb{S}$-algebras by means of the canonical epimorphism $\sigma : AF^{S} \to (A,\sigma)$. If $\tilde{\mathbb{T}} \to \mathcal{L}$, then $\mathcal{L}$ is the composition

$$\begin{array}{lcl} ATS \xrightarrow{\ \eta^{S}TS\ } & ASTS & = AF^{S}\tilde{T}U^{S}F^{S}U^{S} \\ & & = (AST, A\sigma_0)U^{S}F^{S}U^{S} \\ \quad\searrow{\scriptstyle \ell} & & = (ASTS, AST\mu^{S})U^{S} \\ & & \quad\downarrow \sigma_0 \\ & AST & = (AST, A\sigma_0)U^{S} \end{array}$$

Now, $(AS, A\mu^{S})\tilde{\tilde{T}} = (AST, A\sigma_1)$. But since $\mathcal{L} \to \tilde{\tilde{\mathbb{T}}}$,

$$\begin{aligned} (AS, A\mu^{S})\tilde{\tilde{T}} &= (AST, AS\ell . A\mu^{S}T) \\ &= (AST, AS\eta^{S}TS . AS\sigma_0 . A\mu^{S}T) \\ &= (AST, A\sigma_0). \end{aligned}$$

Thus $\sigma_1 = \sigma_0$. Applying $\tilde{\tilde{T}}$ and $\tilde{T}$ to the canonical epimorphism of the free algebra,

$$\begin{array}{ccccc} AF^{S}\tilde{\tilde{T}} = (AST, A\sigma_1) & \xrightarrow{\ \sigma T\ } & (AT,(A,\sigma)\tilde{\tilde{\sigma}}) & = & (A,\sigma)\tilde{\tilde{T}} \\ \| \qquad\qquad \| & & & & \\ AF^{S}\tilde{T} = (AST, A\sigma_0) & \xrightarrow{\ \sigma T\ } & (AT,(A,\sigma)\tilde{\sigma}) & = & (A,\sigma)\tilde{T} \end{array}$$

But $A\eta^{S}T.\sigma T = AT$. A general fact in any tripleable category is that if $f : (A,\sigma) \to (A',\sigma')$, $f : (A,\sigma) \to (A',\sigma'')$, and f is a split epimorphism in $\underline{A}$, then $\sigma' = \sigma''$. Thus $\tilde{\sigma} = \tilde{\tilde{\sigma}}$.

Of course, $\tilde{\tilde{\eta}} = \tilde{\eta}$, $\tilde{\tilde{\mu}} = \tilde{\mu}$, since these are just the unique liftings of η, μ into $\underline{A}^{\mathbb{S}}$ and do not depend on ℓ. Thus $\tilde{\mathbb{T}} = \tilde{\tilde{\mathbb{T}}}$,

q.e.d.

2. <u>Algebras over the composite triple.</u> Let $\ell : TS \to ST$ be a distributive law, and $\mathbb{ST}, \tilde{\mathbb{T}}$ the corresponding composite and lifted triples.

<u>Proposition</u>. Let (A,ξ) be an $\mathbb{ST}$-algebra. Since $\mathbb{S}, \mathbb{T} \to \mathbb{ST}$ are triple maps, the compositions

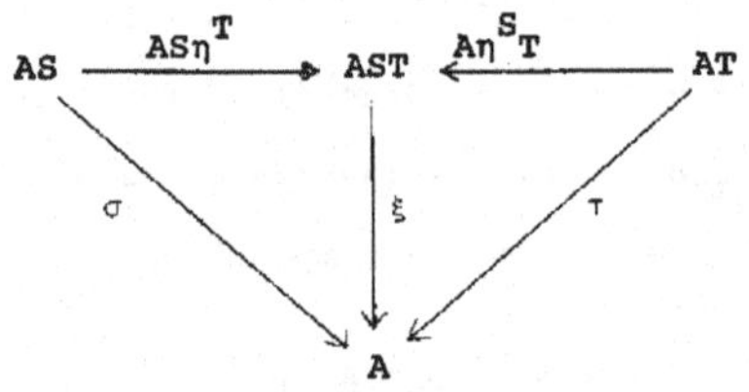

are $\mathbb{S}$- and $\mathbb{T}$-structures on A, and it turns out that σ is "ℓ-distributive" over τ :

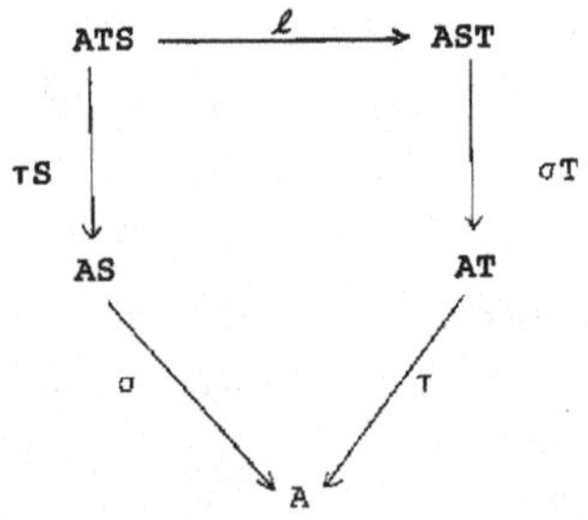

A $\tilde{\mathbb{T}}$-algebra in $\underline{A}^{\mathbb{S}}$ consists of an $\mathbb{S}$-algebra (A,σ) with a $\tilde{\mathbb{T}}$-structure $\tau : (A,\sigma)\tilde{T} \to (A,\sigma)$. Thus τ must be both a $\mathbb{T}$-structure $AT \to A$ and an $\mathbb{S}$-algebra map. The latter condition is equivalent to ℓ-distributivity of σ over τ. The above therefore defines a functor

$$(\underline{A}^{\mathbb{S}})^{\tilde{\mathbb{T}}} \xleftarrow{\ \Phi^{-1}\ } \underline{A}^{\mathbb{S}\mathbb{T}}$$

Finally, the triple induced in $\underline{A}$ by the composite adjoint pair below is exactly the ℓ-composite $\mathbb{S}\mathbb{T}$. The "semantical comparison functor" Φ is an isomorphism of categories, with the above Φ^{-1} as inverse.

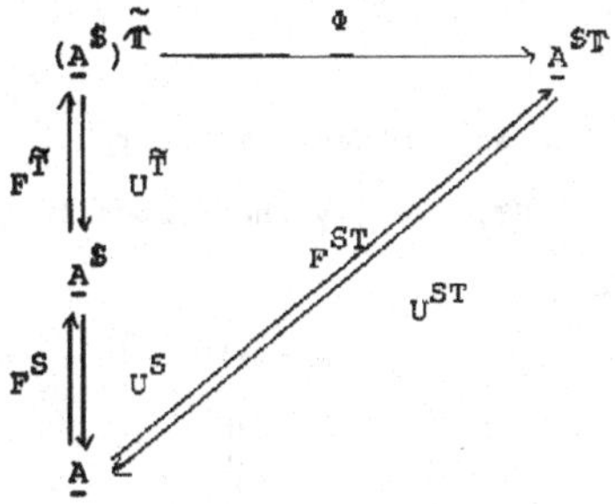

The formula for Φ is $(A,\sigma,\tau)\Phi = (A,\sigma T.\tau)$ in this context.

<u>Proof.</u> First, distributivity holds between σ,τ .

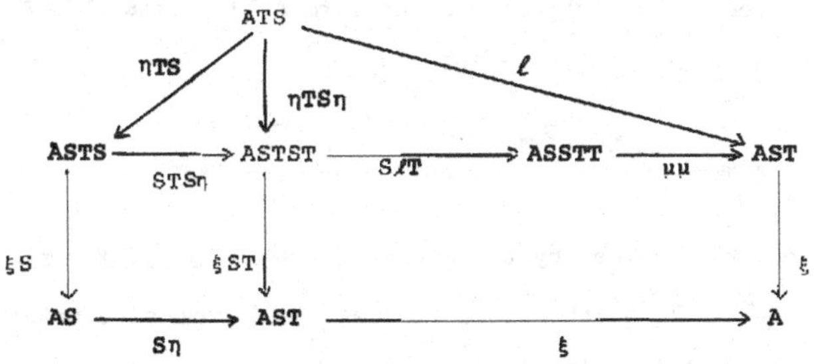

commutes, so we only need to show that $\xi = \sigma T.\tau$. (This is also the essential part in proving that $\Phi\Phi^{-1} = \Phi^{-1}\Phi = \text{id}$.)

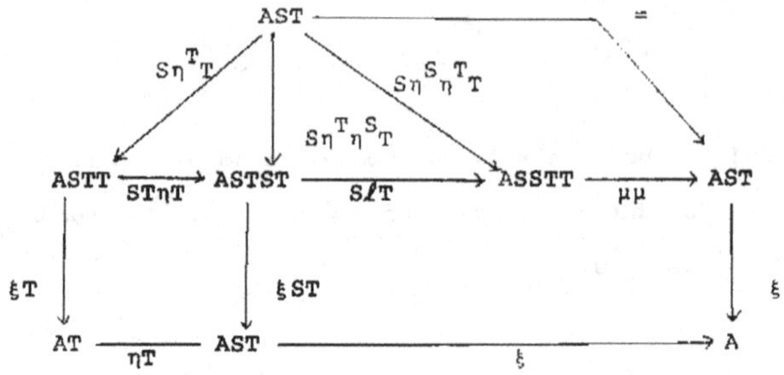

Now compute the composite adjointness. The formula for $F^{\tilde{T}}$ is $(A,\sigma) \to (AT, A\ell.\sigma T, A\mu^T)$. Thus $F^S F^{\tilde{T}} U^{\tilde{T}} U^S = ST$. Clearly the composite unit is $\eta^S \eta^T$. As for the counit, that is the contraction

$$\begin{array}{ccccc}
(A,\sigma,\tau)U^{\tilde{T}}U^S F^S F^{\tilde{T}} & \longrightarrow & (A,\sigma,\tau)U^{\tilde{T}}F^{\tilde{T}} & \longrightarrow & (A,\sigma,\tau) \\
\| & & \| & & \| \\
(AST, AS\ell.A\mu^S T, A\mu^T) & \xrightarrow[\sigma T]{} & (AT, A\ell.\sigma T, A\mu^T) & \xrightarrow[\tau]{} & (A,\sigma,\tau)
\end{array}$$

The multiplication in a triple induced by an adjoint pair is always the value of the counit on free objects, here objects of the form $AF^S F^{\tilde{T}} = (AST, AS\ell.A\mu^S T, AS\mu^T)$. Thus the multiplication in the induced triple is

$$ASTST \xrightarrow{AS\ell T.A\mu^S TT} ASTT \xrightarrow{AS\mu^T} AST$$

which is exactly that defined by the given distributive law ℓ. The composite adjointness $\underline{A} \to (\underline{A}^{\mathbf{S}})^{\tilde{\mathbf{T}}} \to \underline{A}$ therefore induces the ℓ-composite triple $\mathbf{ST}$.

By the universal formula for Φ and the above counit formula,

$$\begin{aligned}
(A,\sigma,\tau)\Phi &= (A, (A,\sigma,\tau)((U^{\tilde{T}}\epsilon^S F^{\tilde{T}})\epsilon^{\tilde{T}})U^{\tilde{T}}U^S) \\
&= (A, \sigma T.\tau).
\end{aligned}$$

q.e.d.

3. <u>Distributive laws and adjoint functors.</u> A distributive law enables four pairs of adjoint functors to exist, all of which are tripleable.

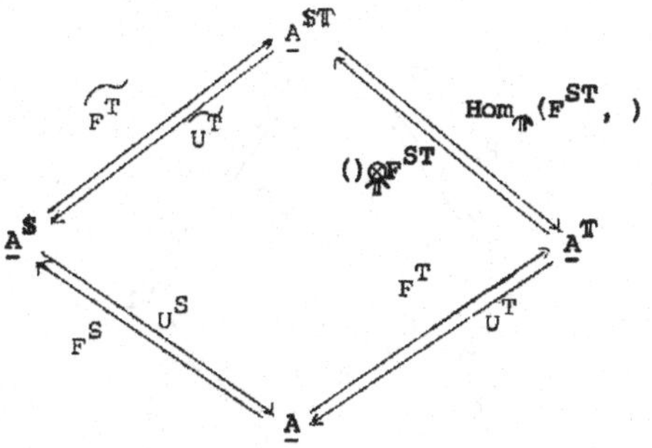

Here $\widetilde{F^T} = \widetilde{F^T}\Phi$, $\widetilde{U^T} = \Phi^{-1}\widetilde{U^T}$ are the liftings of F^T, U^T into $\underline{A}^{\mathbb{S}}$ given by the Proposition, §2. $(\ )\otimes_{\mathbb{T}} F^{ST}$ and its adjoint are induced by the triple map $\eta^S T : \mathbb{T} \to \mathbb{S}\mathbb{T}$ as described in the Introduction. $\widetilde{F^T}$ could be written $(\ )\otimes_{\mathbb{S}} F^{ST}$, of course.

Since the composite underlying $\underline{A}$-object functors $\underline{A}^{\mathbb{S}\mathbb{T}} \to \underline{A}$ are equal, the natural map e described in the Appendix is induced. It is a functorial equality

$$U^S F^T \xrightarrow{\ =\ } \widetilde{F^T}.\mathrm{Hom}_{\mathbb{T}}(F^{ST},\ ).$$

The above functorial equality, or isomorphism in general, will be referred to as "distributivity". I now want to demonstrate a converse, to the effect that if an adjoint square is commutative and distributive, then distributive laws hold between the triples and cotriples that are present.

<u>Proposition.</u> Let

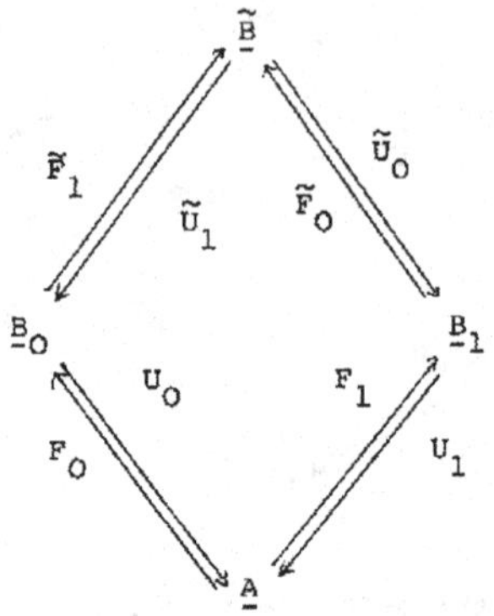

be an adjoint square which commutes by virtue of adjoint natural isomorphisms $u : \tilde{U}_1 U_0 \xrightarrow{\sim} \tilde{U}_0 U_1$, $f : F_1\tilde{F}_0 \xrightarrow{\sim} F_0\tilde{F}_1$. Let $T_0, T_1, \tilde{G}_1, \tilde{G}_0$ be the triples in $\underline{A}$ and cotriples in $\underline{B}$ which are induced. Let $e : U_0F_1 \to \tilde{F}_1\tilde{U}_0$, $e' : U_1F_0 \to \tilde{F}_0\tilde{U}_1$ be defined as in the Appendix. The adjoint square is <u>distributive</u>(in an asymmetrical sense,0 overl) if e is an isomorphism. Assume this. Then with φ,ψ the isomorphisms defined in the proof (and induced by e), we have that

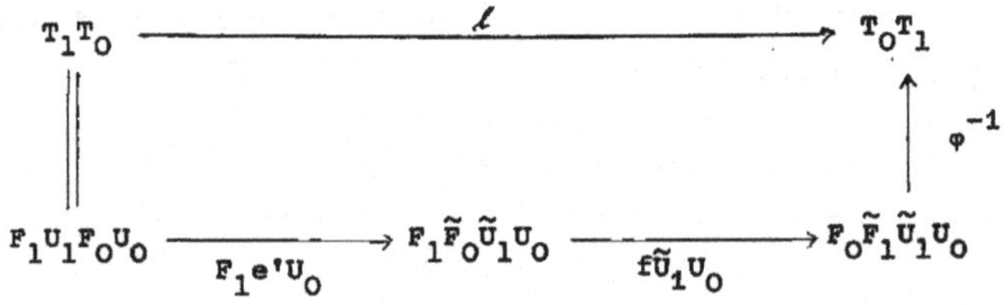

$$\begin{array}{ccccc}
\tilde{G}_1\tilde{G}_0 & & \xrightarrow{\quad\lambda\quad} & & \tilde{G}_0\tilde{G}_1 \\
\downarrow{\scriptstyle \psi^{-1}} & & & & \| \\
\tilde{U}_1U_0F_0\tilde{F}_1 & \xrightarrow[uF_0\tilde{F}_1]{} & \tilde{U}_0U_1F_0\tilde{F}_1 & \xrightarrow[\tilde{U}_0e'F_1]{} & \tilde{U}_0\tilde{F}_0\tilde{U}_1\tilde{F}_1
\end{array}$$

are distributive laws of T_0 over T_1 , $\widetilde{\mathbb{G}}_1$ over $\widetilde{\mathbb{G}}_0$.

If the adjoint square is produced by a distributive law TS $\to$ ST as described at the start of §3 , so that $\mathbb{S}$ corresponds to T_0 and $\mathbb{T}$ to T_1 , then the distributive law given by the above formula is the original one.

<u>Proof.</u> Let

$$\mathbb{T} = \begin{cases} T = F_0\tilde{F}_1\tilde{U}_1U_0 \ : \ \underline{A} \to \underline{A} \\ \eta = \eta_0(F_0\ \tilde{\eta}_1U_0) \ : \underline{A} \to T \\ \mu = F_0\tilde{F}_1((\tilde{U}_1\epsilon_0\tilde{F}_1)\tilde{\epsilon}_1)\tilde{U}_1U_0 \ : TT \to T \end{cases}$$

be the total triple induced by the left hand composite adjointness. e,u induce a natural isomorphism

$$\begin{array}{ccccc}
T_0T_1 & & \xrightarrow{\quad\varphi\quad} & & T \\
\| & & & & \| \\
F_0U_0F_1U_1 & \xrightarrow[F_0eU_1]{} & F_0\tilde{F}_1\tilde{U}_0U_1 & \xrightarrow[F_0\tilde{F}_1u^{-1}]{} & F_0\tilde{F}_1\tilde{U}_1U_0
\end{array}$$

By transfer of structure, any functor isomorphic to a triple also has a triple structure. Thus we have an isomorphism of triples $\varphi : (T_0T_1)_m = (T_0T_1, \eta_0\eta_1, m) \to \mathbb{T}$.

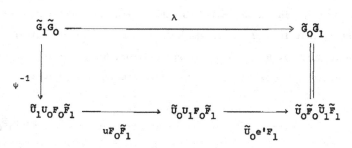

Actually, the diagrams in the Appendix show that φ transfers units as indicated, and m is the quantity that is defined via the isomorphism. A short calculation also shows that m is middle-unitary. By (1) $\longleftrightarrow$ (2), Proposition, §1 , m is induced by the distributive law $(\eta_0 T_1 T_0 \eta_1)m : T_1 T_0 \to T_0 T_1$. Now, consider the diagram

$$
\begin{array}{c}
F_1 U_1 F_0 U_0 \xrightarrow{\qquad F_1 e^{\prime} U_0 \qquad} F_1 \tilde{F}_0 \tilde{U}_1 U_0 \\
\end{array}
$$

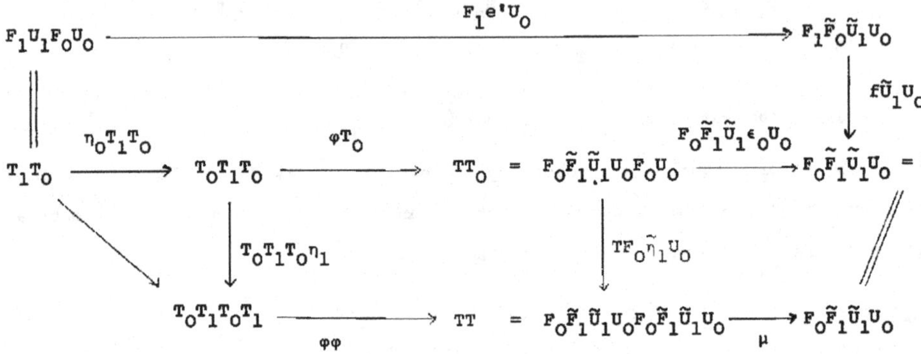

The upper figure commutes, by expanding φ , and naturality. Its top line is $\ell\varphi$. Thus
$$\ell = (\eta_0 T_1 T_0 \eta_1)(\varphi\varphi)\mu\varphi^{-1} = (\eta_0 T_1 T_0 \mu_1)m \text{ is a distributive law.}$$

The proof that λ is a distributive law is dual. One defines the total cotriple

$$
\mathbb{G} = \left\{
\begin{array}{l}
\tilde{G} = \tilde{U}_1 U_0 F_0 \tilde{F}_1 \quad : \tilde{\underline{B}} \to \tilde{\underline{\underline{B}}} \\[2mm]
\tilde{\epsilon} = (\tilde{U}_1 \epsilon_0 \tilde{F}_1)\tilde{\epsilon}_1 \quad : \tilde{G} \to \tilde{\underline{B}} \\[2mm]
\delta = \tilde{U}_1 U_0 (\eta_0 (F_0 \tilde{\eta}_1 U_0)) F_0 \tilde{F}_1 : \tilde{G} \to \tilde{G}\tilde{G}
\end{array}
\right.
$$

and uses the isomorphism

$$
\begin{array}{c}
\tilde{G}_1 \tilde{G}_0 \xleftarrow{\qquad \psi \qquad} \tilde{G} \\
\end{array}
$$

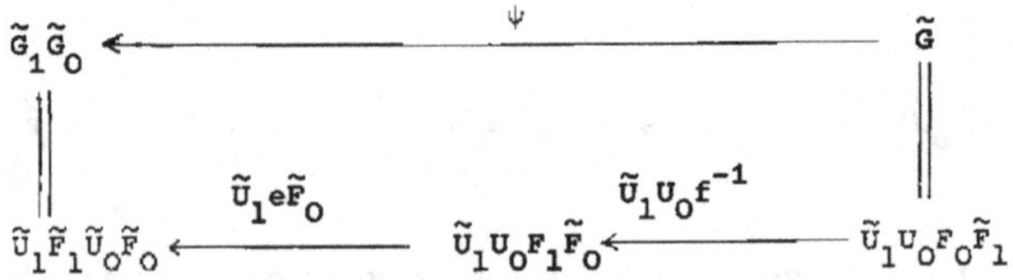

to induce a similar isomorphism of cotriples $\psi : \mathbb{G} \to (\mathbb{G}_1 \mathbb{G}_0)_d$.

Finally, if the original adjoint square is produced by a distributive law ℓ : TS → ST, the Proposition, §2, shows that the total triple is the ℓ-composite $\mathbb{ST}$, and φ is an identity map.

q.e.d.

One can easily obtain distributive laws of mixed type, for example, $\tilde{\mathbb{T}}_1\mathbb{G}_0 \to \mathbb{G}_0\tilde{\mathbb{T}}_1$.

<u>Remark on structure-semantics of distributive laws.</u> Triples in $\underline{A}$ give rise to adjoint pairs over $\underline{A}$, $\underline{A} \to \underline{A}^{\mathbb{T}} \to \underline{A}$, and adjoint pairs $\underline{A} \to \underline{B} \to \underline{A}$ give rise to triples in $\underline{A}$. This yields the structure-semantics adjoint pair for triples:

$$\text{Ad }\underline{A} \underset{\sigma}{\overset{\check{\sigma}}{\rightleftarrows}} (\text{Trip }\underline{A})^* .$$

This adjoint pair is a reflection ($\sigma\check{\sigma}$ = id.) and the comparison functor Φ : id. → $\check{\sigma}\sigma$ is the unit.

Something similar can be done for distributive adjoint situations over $\underline{A}$ and distributive laws.

Define a <u>distributive law in</u> $\underline{A}$ to be a triple ($\mathbb{S}$,$\mathbb{T}$,ℓ) where $\mathbb{S}$,$\mathbb{T}$ are triples and ℓ : TS → ST is a distributive law. A <u>map</u> (φ,ψ) : ($\mathbb{S}$,$\mathbb{T}$,ℓ) → ($\mathbb{S}'$,$\mathbb{T}'$,ℓ') is a pair of triple maps $\mathbb{S}$ → $\mathbb{S}'$, $\mathbb{T}$ → $\mathbb{T}'$ which is compatible with ℓ, ℓ' . Let <u>Dist</u>($\underline{A}$) be this category.

A <u>distributive adjoint situation over</u> $\underline{A}$ means a diagram

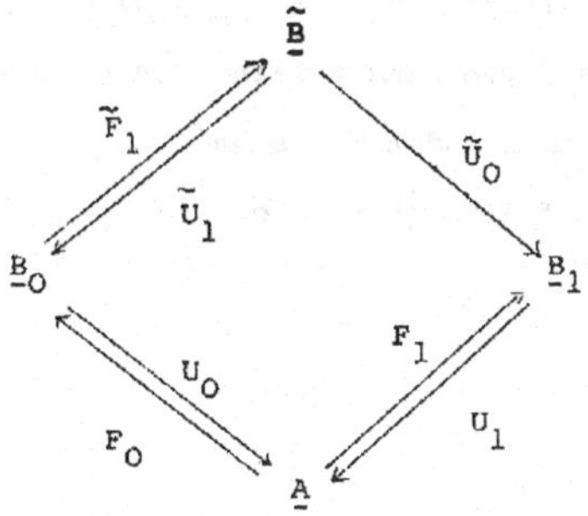

where $(F_0,U_0),(\tilde{F}_1,\tilde{U}_1)(F_1,U_1)$ are adjoint pairs, $\tilde{U}_1\tilde{U}_0 = U_0U_1$, and the natural map $U_0F_1 \to \tilde{F}_1\tilde{U}_0$ is an isomorphism. A map of such adjoint situations consists of functors $\underline{B}_0 \to \underline{B}'_0$, $\underline{B}_1 \to \underline{B}'_1$, $\underline{\tilde{B}} \to \underline{\tilde{B}}'$ commuting with the underlying object functors $\tilde{U}_1$, $U_0,\tilde{U}_0,U_1$, . . .

Distributive laws give rise to distributive adjoint situations over $\underline{A}$, and vice versa (note that $\ell = (\eta_0T_1T_0\eta_1)m$ and m does not involve $\tilde{F}_0$). Thus we have an adjoint pair

$$\underline{\text{Distributive Adj}}\ \underline{A} \underset{\sigma}{\overset{\check{\sigma}}{\rightleftarrows}} (\underline{\text{Dist}}\ \underline{A})^* .$$

The structure functor $\check{\sigma}$ is left adjoint to the semantics functor σ, $\sigma\check{\sigma} = \text{id.}$, and the unit is a combination of Φ's. This is the correct formulation of the above Proposition.

4. Examples. (1) Multiplication and addition. Let $\underline{A}$ be the category of sets, let $\$$ be the free monoid triple in $\underline{A}$, and T the free abelian group triple. Then $\underline{A}^{\$}$ is the category of monoids and $\underline{A}^{T}$ is the category of abelian groups. For every set X the usual interchange of addition and multiplication

$$\prod_{i=0}^{m} \sum_{j_i=0}^{n_i} x_{ij_i} \longrightarrow \sum_{j_0=0}^{n_0} \cdots \sum_{j_m=0}^{n_m} \prod_{i=0}^{m} x_{ij_i}$$

can be interpreted as a natural transformation $XTS \xrightarrow{\ell} XST$ and is a distributive law of multiplication over addition, that is, of $\$$ over T, in the formal sense.

The composite $\$T$ is the free ring triple. XST is the polynomial ring Z[X] with the elements of X as noncommuting indeterminates.

The canonical diagram of adjoint functors is:

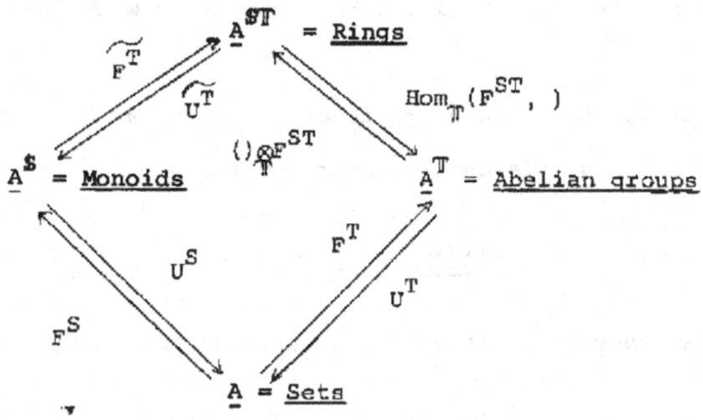

$\widetilde{F^T}$ is the free abelian group functor lifted into the category of monoids, and is known as the "monoid ring" functor. $Hom_{\mathbb{T}}(F^{ST},)$ is the forgetful functor. If A is an abelian group, the value of the left adjoint, $A\otimes_{\mathbb{T}}F^{ST}$, is the Z-tensor ring generated by A, namely Z + A + A⊗A + ... The natural map $U^S F^T \to \widetilde{F^T}.Hom_{\mathbb{T}}(F^{ST},)$ is the identity, that is, distributivity holds. Both compositions give the free abelian groups generated by the elements of monoids.

The scheme is : the distributive law ℓ: TS → ST produces the adjoint square, which, being distributive (§3), induces a distributive law $\lambda : \mathbb{G}_{Ab}\mathbb{G}_{Mon} \to \mathbb{G}_{Mon}\mathbb{G}_{Ab}$, where $G_{Mon} = \widetilde{U^T}\widetilde{F^T}$, $G_{Ab} = Hom_{\mathbb{T}}(F^{ST},)\otimes_{\mathbb{T}}F^{ST}$. This λ is that employed by Barr in his "Composite cotriples", this volume (Theorem(4.6)).

A distributive law ST → TS would have the air of a universal solution to the problem of factoring polynomials into linear factors. This suggests that the composite TS has little chance of being a triple.

(2) <u>Constants</u>. Any set C can be interpreted as a triple in the category of sets, $\underline{A}$, via the coproduct injection and folding map X → C + X, C + C + X → C + X. $\underline{A}^{C+()}$ is the category of sets with C as constants. For example, if C=1, $\underline{A}^{1+()}$ is the category of pointed sets.

Let $\mathbb{T}$ be any triple in $\underline{A}$. A natural map ℓ : C + XT → (C + X)T is defined in an obvious way, using Cη. ℓ is a distributive law of C + () over $\mathbb{T}$. The composite triple $C + \mathbb{T}$ has as algebras $\mathbb{T}$-algebras furnished with the set C as constants.

(3) Group actions. Let π be a monoid or group. π can be interpreted as a triple in $\underline{A}$, the category of sets, via cartesian product:

$$X \xrightarrow{(x,1)} X \times \pi, \qquad X \times \pi \times \pi \xrightarrow{(x,\sigma_1\sigma_2)} X \times \pi.$$

$\underline{A}^{\pi}$ is the category of π-sets. If T is any functor $\underline{A} \to \underline{A}$, there is a natural map

$$XT \times \pi \xrightarrow{\ \ell\ } (X \times \pi)T.$$

Viewing $XT \times \pi$ as a π-fold coproduct of XT with itself, ℓ has the value

$$XT \xrightarrow{(X,\sigma)T} (X \times \pi)T$$

on the σ-th cofactor, if (X,σ) is the map $x \to (x,\sigma)$. If $\mathbb{T} = (T,\eta,\mu)$ is a triple in $\underline{A}$, ℓ is a distributive law of π over $\mathbb{T}$. The algebras over the composite triple $\pi\mathbb{T}$ are $\mathbb{T}$-algebras equipped with π-operations. The elements of π act as $\mathbb{T}$-homomorphisms.

Example (3) can be combined with (2) to show that any triple $\mathbb{S}$ generated by constants and unary operations has a canonical distributive law over any triple $\mathbb{T}$ in $\underline{A}$. The $\mathbb{S}\mathbb{T}$-algebras are $\mathbb{T}$-linear automata.

(4) No new equations in the composite triple. It is known that if $\mathbb{T}$ is a consistent triple in sets, then the unit $X\eta^T : X \to XT$ is a monomorphism for every X. And every triple in sets, as a functor, preserves monomorphisms. Thus if $\mathbb{S}$, $\mathbb{T}$ are consistent triples, and $\ell : TS \to ST$ is a distributive law, then the triple maps $\mathbb{S}, \mathbb{T} \to \mathbb{S}\mathbb{T}$ are monomorphisms of functors.

This means that the operations of $\mathbb{S}$ and of $\mathbb{T}$ are mapped injectively into operations of $\mathbb{S}\mathbb{T}$, and no new equations hold among them in the composite.

The triples excluded as "inconsistent" are the terminal triple and one other:

(a) $XT = 1$ for all X,

(b) $oT = o$, $XT = 1$ for all $X \neq o$.

(5) Distributive laws on rings as triples in the category of abelian groups.

Let S and T be rings. S and T can be interpreted as triples $\mathbb{S}$ and $\mathbb{T}$ in the category of abelian groups, $\underline{A}$, via tensor product:

$$A \xrightarrow{a\otimes 1} A\otimes S, \quad A\otimes S\otimes S \xrightarrow{a\otimes s_0 s_1} A\otimes S.$$

$\underline{A}^{\mathbb{S}}$ and $\underline{A}^{\mathbb{T}}$ are the categories of S- and T- modules. The usual interchange map of the tensor product, $\ell : T\otimes S \to S\otimes T$, gives a distributive law of $\mathbb{S}$ over $\mathbb{T}$. This is just what is needed to make the composite $S\otimes T$ into a ring:

$$S\otimes T\otimes S\otimes T \xrightarrow{S\otimes \ell\otimes T} S\otimes S\otimes T\otimes T \xrightarrow{\text{mult.} \otimes \text{mult.}} S\otimes T.$$

This ring multiplication is the multiplication in the composite triple $\mathbb{S}\mathbb{T}$.

The interchange map is adjoint to a distributive law between the adjoint cotriples Hom(S,), Hom(T,). This is a general fact about adjoint triples.

The identities for a distributive law are especially easy to check in this example, as are certain conjectures about distributive laws.

Let $\underline{A}$ be the category of graded abelian groups, and let S and T be graded rings. Then two obvious transpositions of the graded tensor product, $T\otimes S \to S\otimes T$, exist :

$$t\otimes s \to s\otimes t,$$

$$t\otimes s \to (-1)^{\dim s.\, \dim t}\, s\otimes t$$

Both are distributive laws of the triple $\mathbb{S} = (\ )\otimes S$ over $\mathbb{T} = (\ )\otimes T$. They give different graded ring structures on $S\otimes T$ and different composite triples $\mathbb{S}\mathbb{T}$.

Finally, note the following ring multiplication in $S\otimes T$:

$$(s_0 \otimes t_0)(s_1 \otimes t_1) = (-1)^{\dim s_0 \,.\, \dim t_1}\, s_0 s_1 \otimes t_0 t_1.$$

The maps

$$S \xrightarrow{s\otimes 1} S\otimes T \xleftarrow{1\otimes t} T$$

are still ring homomorphisms, but the "middle unitary" law described in §1 does not hold.

A number of problems are open, in such areas as homology, the relation of composites to tensor products of triples, and possible extension of the distributive law formalism to non-tripleable situations, for example, the following, suggested by Knus-Stammbach:

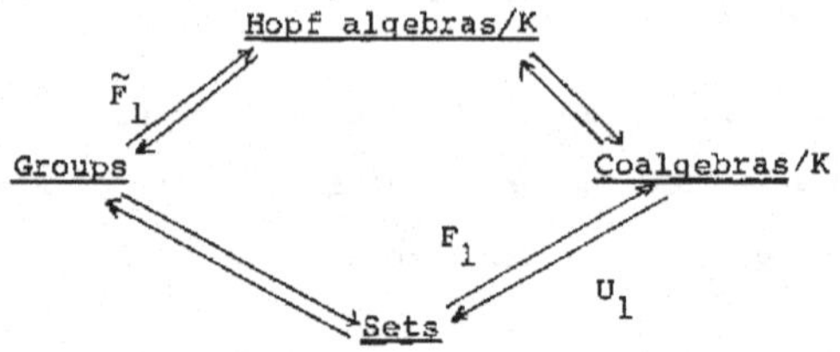

($XF_1 = K(X)$, CU_1 = all $c \in C$ such that $\Delta(c) = c \otimes c$, and $\pi\tilde{F}_1$ = the group algebra $K(\pi)$.)

5. __Appendix__. If there are adjoint pairs of functors

$$\underline{A} \underset{U}{\overset{F}{\rightleftarrows}} \underline{B} \underset{\tilde{U}}{\overset{\tilde{F}}{\rightleftarrows}} \underline{\tilde{B}}\,, \qquad \begin{array}{l} \eta : \underline{A} \to FU,\ \tilde{\eta} : \underline{B} \to \tilde{F}\tilde{U}, \\ \epsilon : UF \to \underline{B},\ \tilde{\epsilon} : \tilde{U}\tilde{F} \to \underline{\tilde{B}}, \end{array}$$

then there is a __composite__ adjoint pair

$$\underline{A} \underset{\tilde{U}U}{\overset{F\tilde{F}}{\rightleftarrows}} \underline{\tilde{B}}$$

whose unit and counit are

$$\underline{A} \xrightarrow{\eta(F\tilde{\eta}U)} F\tilde{F}\tilde{U}U\,, \quad \tilde{U}UF\tilde{F} \xrightarrow{(\tilde{U}\epsilon\tilde{F})\tilde{\epsilon}} \underline{\tilde{B}}\,.$$

Given adjoint functors

$$\underline{A} \underset{U}{\overset{F}{\rightleftarrows}} \underline{B}\,, \quad \underline{A} \underset{U'}{\overset{F'}{\rightleftarrows}} \underline{B}$$

then the diagrams

$$\begin{array}{ccc} F' & \xrightarrow{f} & F \\ {\scriptstyle \eta F'}\downarrow & & \uparrow{\scriptstyle F\epsilon'} \\ FUF' & \xrightarrow[FuF']{} & FU'F' \end{array} \qquad\qquad \begin{array}{ccc} U & \xrightarrow{u} & U' \\ {\scriptstyle U\eta'}\downarrow & & \uparrow{\scriptstyle \epsilon U'} \\ UF'U' & \xrightarrow[UfU']{} & UFU' \end{array}$$

establish a 1-1 correspondence between morphisms $u : U \to U'$, $f : F' \to F$. Corresponding morphisms are called adjoint.

Let

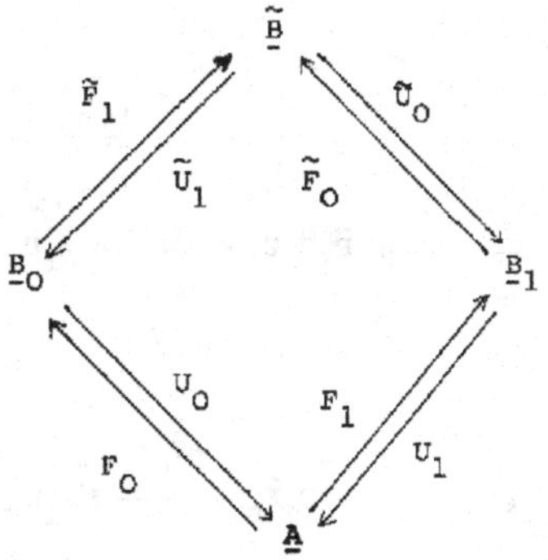

be adjoint pairs of functors, and let $u : \tilde{U}_1 U_0 \to \tilde{U}_0 U_1$, $f : F_1 \tilde{F}_0 \to F_0 \tilde{F}_1$ be adjoint natural transformations. Then u,f induce a natural transformation e which plays a large role in §3 :

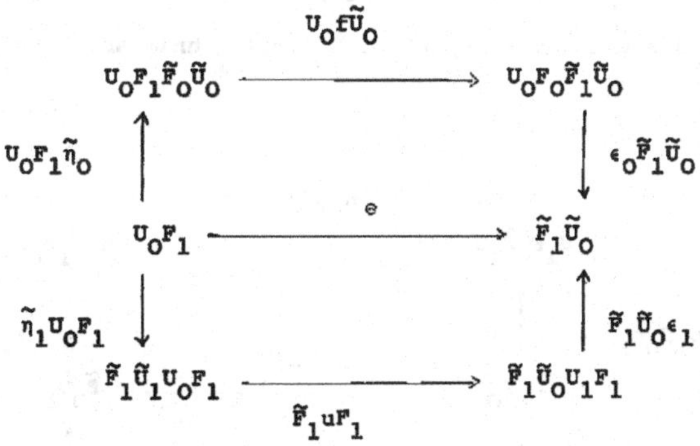

The following diagrams commute:

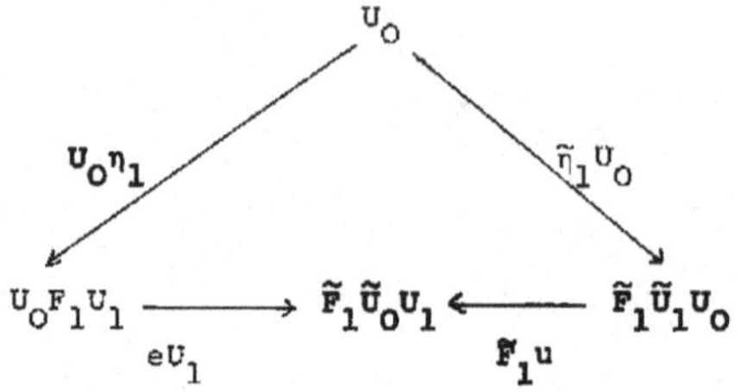

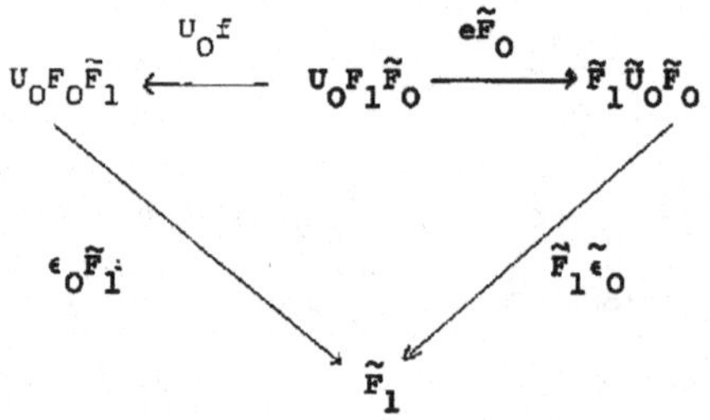

If in addition adjoint maps $u^{-1} : \tilde{U}_0 U_1 \to \tilde{U}_1 U_0$, $f^{-1} : F_0\tilde{F}_1 \to F_1\tilde{F}_0$ are available, they induce a natural map e' with similar unit and counit properties:

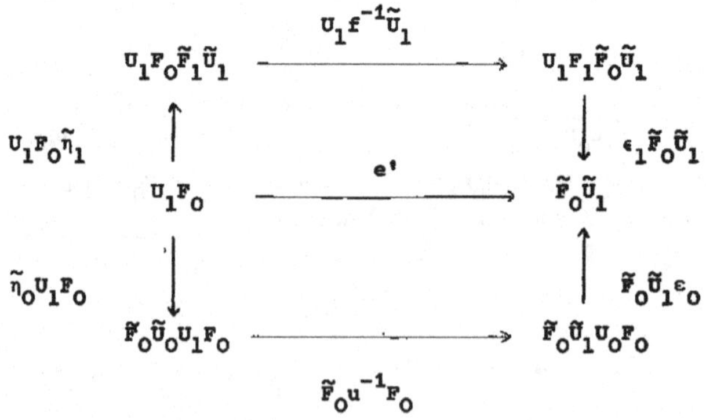

ORDINAL SUMS AND EQUATIONAL DOCTRINES

by F. William Lawvere

Our purpose is to describe some examples and to suggest some directions for the study of categories with equational structure. To equip a category $\underline{A}$ with such a structure means roughly to give certain "$\underline{C}$-tuples of $\underline{D}$-ary operations"

$$\underline{A}^{\underline{D}} \xrightarrow{\theta} \underline{A}^{\underline{C}}$$

for various categories $\underline{D}$ and $\underline{C}$, in other words, "operations" in general operate (functorially or naturally) on diagrams in $\underline{A}$, not only on n-tuples, and may be subjected to equations involving both composition of natural transformations and Godement multiplication of natural transformations and functors. By an equational doctrine we mean an invariant form of a system of indices and conditions which specifies a particular species of structure of the general type just described. Thus equational doctrines bear roughly the same relation to the category of categories which algebraic theories bear to the category of sets. Further development will no doubt require contravariant operations (to account for closed categories) and "weak algebras" (to allow for even the basic triple axioms holding "up to isomorphism"), but in this article we limit ourselves to strong standard constructions in the category of categories.

Thus, for us an <u>equational doctrine</u> will consist of the following data: 1) a Rule $\mathcal{D}$ which assigns to every category $\underline{B}$ another category $\underline{B}\mathcal{D}$ and to every pair of categories $\underline{B}$ and $\underline{A}$, a functor

$$\underline{A}^{\underline{B}} \xrightarrow{\mathcal{D}} (\underline{A}\mathcal{D})^{(\underline{B}\mathcal{D})}$$

2) a rule η which assigns to every category $\underline{B}$ a functor

$$\underline{B} \xrightarrow{\underline{B}\eta} \underline{B}\mathcal{D}$$

3) a rule μ which assigns to every category $\underline{B}$ a functor

$$(\underline{B}\mathcal{D})\mathcal{D} \xrightarrow{\underline{B}\mu} \underline{B}\mathcal{D}$$

These data are subject to seven axioms, expressing that $\mathcal{D}$ is strongly functorial, η , μ strongly natural, and that together they form a standard construction (= monad = triple). For example, part of the functoriality of $\mathcal{D}$ is expressed by the commutativity of

$$\begin{array}{ccc}
\underline{B}^{\underline{C}} \times \underline{A}^{\underline{B}} & \xrightarrow{\mathcal{D}\times\mathcal{D}} & (\underline{B}\mathcal{D})^{(\underline{C}\mathcal{D})} \times (\underline{A}\mathcal{D})^{(\underline{B}\mathcal{D})} \\
\text{compn.} \downarrow & & \swarrow \text{compn.} \\
\underline{A}^{\underline{C}} & \xrightarrow{\mathcal{D}} & (\underline{A}\mathcal{D})^{(\underline{C}\mathcal{D})}
\end{array}$$

while the naturality of μ is expressed by the commutativity of

$$\begin{array}{ccc}
\underline{A}^{\underline{B}} & \xrightarrow{\mathcal{D}^2} & (\underline{A}\mathcal{D}\mathcal{D})^{\underline{B}\mathcal{D}\mathcal{D}} \\
\mathcal{D} \downarrow & & \downarrow (\underline{A}\mu)^{(\underline{B}\mathcal{D}\mathcal{D})} \\
(\underline{A}\mathcal{D})^{(\underline{B}\mathcal{D})} & \xrightarrow[(\underline{A}\mathcal{D})^{(\underline{B}\mu)}]{} & (\underline{A}\mathcal{D})^{(\underline{B}\mathcal{D}\mathcal{D})}
\end{array}$$

and the associativity of μ by the commutativity of

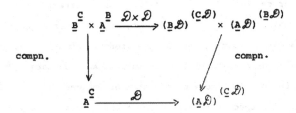

$$\begin{array}{ccc} \underline{B}\mathcal{D}\mathcal{D}\mathcal{D} & \xrightarrow{(\underline{B}\mathcal{D})\mu} & (\underline{B}\mathcal{D})\mathcal{D} \\ {\scriptstyle \ulcorner \underline{B}\mu \urcorner \mathcal{D}}\downarrow & & \downarrow{\scriptstyle \underline{B}\mu} \\ (\underline{B}\mathcal{D})\mathcal{D} & \xrightarrow[\underline{B}\mu]{} & \underline{B}\mathcal{D} \end{array}$$

In the last diagram the left column denotes the value of the functor

$$(\underline{B}\mathcal{D})^{(\underline{B}\mathcal{D}\mathcal{D})} \xrightarrow{\quad \mathcal{D} \quad} (\underline{B}\mathcal{D})\mathcal{D}^{(\underline{B}\mathcal{D}\mathcal{D})\mathcal{D}}$$

at the object $\ulcorner \underline{B}\mu \urcorner$ of its domain which corresponds to the functor $\underline{B}\mu$.

An algebra (sometimes called a "theory") over the given doctrine means a category $\underline{A}$ with a functor $\underline{A}\mathcal{D} \xrightarrow{\alpha} \underline{A}$ subject to the usual two conditions. Homomorphisms between algebras are also defined as usual, although probably "weak homomorphisms will have to be considered later too.

For examples of doctrines, consider any category $\underline{D}$ and let $\mathcal{D} : \underline{B} \rightsquigarrow \underline{B}^{\underline{D}}$ with η,μ defined diagonally. Or let $\mathcal{D} : \underline{B} \rightsquigarrow \underline{D}^{(\underline{D}^{\underline{B}})}$ with obvious (though complicated) η,μ. Clearly a strongly adjoint equational doctrine is determined by a category $\underline{M} = \underline{1}\mathcal{D}$ equipped with a strictly associative functorial multiplication $\underline{M} \times \underline{M} \longrightarrow \underline{M}$ with unit.

One of several important operations on doctrines is the formation of the opposite doctrine

$$\mathcal{D}^* : \underline{B} \rightsquigarrow ((\underline{B}^{op})\mathcal{D})^{op}$$

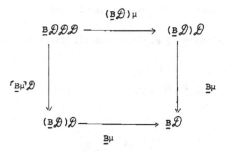

(Note that $(\ )^{op}$, while covariant, is not a strong endofunctor of cat; however it operates on the strong endofunctors in the manner indicated)

Denoting by $\mathrm{Cat}^{\mathcal{D}}$ the category of algebras (or theories) over the doctrine $\mathcal{D}$, we define

$$\mathrm{Hom}_{\mathcal{D}} : (\mathrm{Cat}^{\mathcal{D}})^{\mathrm{op}} \times \mathrm{Cat}^{\mathcal{D}} \longrightarrow \mathrm{Cat}$$

by the <u>equalizer</u> diagram

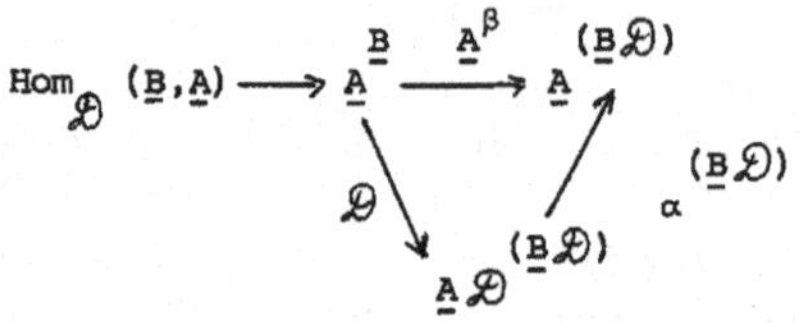

where β,α denote the algebra structures on $\underline{B},\underline{A}$ respectively. That is if $\underline{B} \overset{f}{\underset{g}{\rightrightarrows}} \underline{A}$ are two algebra homomorphisms and if $\varphi : f \longrightarrow g$ is a natural transformation, then φ is considered to belong to the <u>category</u> Hom iff it also satisfies under Godement multiplication the same equation which defines the notion of homomorphism:

$$(\ulcorner\varphi\urcorner\mathcal{D})\alpha = \beta\varphi$$

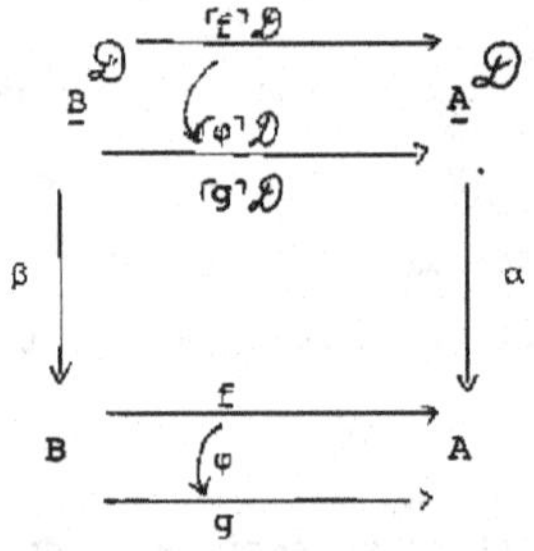

$\mathrm{Hom}_{\mathcal{D}}(\underline{B},\underline{A})$ may or may not be a full subcategory of $\underline{A}^{\underline{B}}$, depending on $\mathcal{D}$.

In particular

$$\mathrm{Hom}_{\mathcal{D}}(\underline{1}\mathcal{D},-) : \mathrm{Cat}^{\mathcal{D}} \longrightarrow \mathrm{Cat}$$

is the underlying functor, which has a strong left adjoint together with which it resolves $\mathscr{D}$.

For a given $\mathscr{D}$-algebra $\underline{A},\alpha$ the functor

$$\mathrm{Hom}\ (-,\underline{A})\ :\ (\mathrm{Cat}^{\mathscr{D}})^{op} \longrightarrow \mathrm{Cat}$$

might be called "$\mathscr{D}$-semantics with values in $\underline{A}$". It has a strong left adjoint, given by $\underline{C} \rightsquigarrow \underline{A}^{\underline{C}}$. (That $\underline{A}^{\underline{C}}$ is a $\mathscr{D}$-algebra for an abstract category $\underline{C}$ and $\mathscr{D}$-algebra $\underline{A}$ is seen by noting that

$$\underline{C} \xrightarrow{ev} \underline{A}^{(\underline{A}^{\underline{C}})} \longrightarrow (\underline{A}\mathscr{D})^{(\underline{A}^{\underline{C}})\mathscr{D}}$$

corresponds by symmetry to a functor

$$(\underline{A}^{\underline{C}})\mathscr{D} \longrightarrow (\underline{A}\mathscr{D})^{\underline{C}}$$

which when followed by $\alpha^{\underline{C}}$ gives the required $\mathscr{D}$-structure on $\underline{A}^{\underline{C}}$). We thus obtain by composition a new doctrine $\mathscr{D}_{\underline{A}}$, the "dual doctrine of $\mathscr{D}$ in the $\mathscr{D}$-algebra $\underline{A}$". Explicitly,

$$\mathscr{D}_{\underline{A}}\ :\ \underline{C} \rightsquigarrow \mathrm{Hom}_{\mathscr{D}}(\underline{A}^{\underline{C}},\underline{A}).$$

the comparison functor Φ in

$$\begin{array}{ccc} (\mathrm{Cat}^{\mathscr{D}})^{op} & \xrightarrow{\quad\Phi\quad} & \mathrm{Cat}^{\mathscr{D}_{\underline{A}}} \\ & {\scriptstyle \mathrm{Hom}_{\mathscr{D}}(-,\underline{A})} \searrow \quad \swarrow & \\ & \mathrm{Cat}^{\circlearrowleft \mathscr{D}_{\underline{A}}} & \end{array}$$

then has a left adjoint given by

$$\Phi^{\vee}\ :\ \underline{C} \rightsquigarrow \mathrm{Hom}_{\mathscr{D}_{\underline{A}}}(\underline{C},\underline{1}\mathscr{D}\Phi)$$

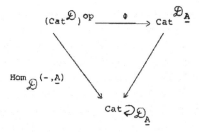

Actually $\underline{1}\mathcal{D}\Phi = \underline{A}$ as a category, but with the induced $\mathcal{D}_{\underline{A}}$-structure, rather than the given $\mathcal{D}$-structure.

For a trivial example, note that if 1 denotes the identity doctrine, then $\mathrm{Hom}_1(\underline{B},\underline{A}) = A^{\underline{B}}$ and $1_{\underline{A}} : C \rightsquigarrow \underline{A}^{(\underline{A}^{\underline{C}})}$. The dual $1_{\underline{A}}$ of the identity doctrine in $\underline{A}$ thus might be called the full 2-clone of $\underline{A}$; it takes on a somewhat less trivial aspect if we note that giving $\underline{A}$ a structure α over any doctrine $\mathcal{D}$ induces a morphism

$$\mathcal{D} \xrightarrow{\tilde{\alpha}} 1_{\underline{A}}$$

of doctrines, since

$$\underline{A}^{\underline{C}} \xrightarrow{\mathcal{D}} \underline{A}\mathcal{D}^{\underline{C}\mathcal{D}}$$

yields by symmetry a functor which can be composed with $\alpha^{(\underline{A}^{\underline{C}})}$. The image $\mathcal{D}/_{\langle\underline{A},\alpha\rangle}$ of $\tilde{\alpha}$, if it could be defined in general, would then be the doctrine of "$\mathcal{D}$-algebras in which hold all equations valid in $\langle\underline{A},\alpha\rangle$". In a particular case Kock has succeeded in defining such an image doctrine, and put it to good use in the construction of the doctrine of colimits (see below).

For a more problematic example of the dual of a doctrine, let $\underline{S}_o$ denote the skeletal category of <u>finite</u> sets, and let $[\underline{S}_o, \ulcorner\underline{B}\urcorner]$ denote the category whose objects are arbitrary

$$n \xrightarrow{\underline{B}} \underline{B}, \quad n \in \underline{S}_o$$

and whose morphisms are given by pairs,

$$n \xrightarrow{\sigma} n' \quad , \quad B \xrightarrow{b} \sigma B'$$

Then $\underline{B} \rightsquigarrow [\underline{S}_o, \ulcorner\underline{B}\urcorner]$ becomes a doctrine by choosing a strictly associative sum operation in $\underline{S}_o$ with help of which to define μ. The algebras over the resulting doctrine are arbitrary categories equipped with strictly associative finite coproducts. Algebras over the opposite doctrine $\mathcal{D}$ are then categories equipped with strictly associative finite <u>products</u>. By choosing a suitable version (<u>not</u> skeletal) of the category $\underline{S}$ of small sets, it can be made into a particular algebra $A,\alpha\rangle = \langle\underline{S},\times\rangle$ over $\mathcal{D}$. Then $\mathrm{Hom}_{\mathcal{D}}(-,\underline{S})$ is seen to include by restriction the usual functorial semantics of algebraic theories. Thus in particular <u>every algebraic category</u> $\underline{C}$ has canonically the structure of a $\mathcal{D}_{\underline{S}}$– algebra, $\mathcal{D}_{\underline{S}}$ denoting the dual doctrine $\underline{C} \rightsquigarrow \mathrm{Hom}_{\mathrm{prod}}(\underline{S}^{\underline{C}},\underline{S})$. The latter doctrine is very rich, having as operations arbitrary $\varprojlim$, directed $\varinjlim$ and probably more (?). Thus if $\underline{C}$ is the category of algebras over a small theory, $\underline{C}\Phi^{\vee} = \mathrm{Hom}_{\mathcal{D}_{\underline{S}}}(\underline{C},\underline{S})$ must consist of functors which are representable by finitely generated algebras. Thus if one could further see that a sufficient number of coequalizers were among the $\mathcal{D}_{\underline{S}}$ operations (meaning that the representing algebras would have to be projective) we would have a highly natural method of obtaining all the information about an algebraic theory which could possibly be recovered from its category of algebras alone, namely the method of the dual doctrine (which goes back at least to M. H. Stone in the case of sets)

Another construction possible for any doctrine $\mathcal{D}$ is that of $\underline{B}_{\mathcal{D}}$: <u>the category of all possible $\mathcal{D}$-structures on the category</u> $\underline{B}$. It is defined as the $\varprojlim$ of the following finite diagram in Cat:

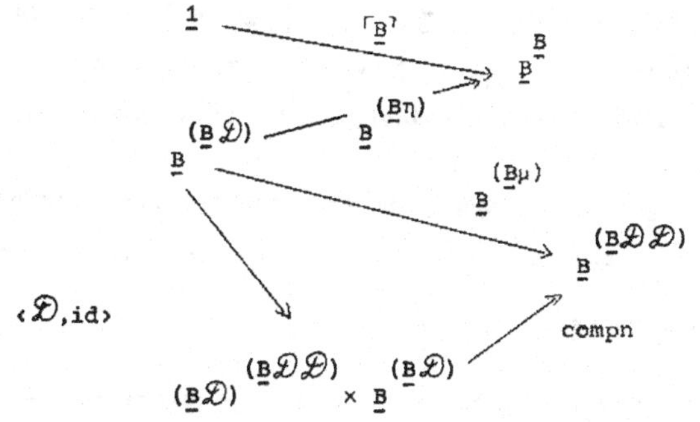

Thus the notion of <u>morphism</u> between different $\mathcal{D}$-structures on the same category $\underline{B}$ is defined by imposing the same equations on <u>natural transformations</u> which are imposed on functors in defining the individual structures. For example, with the appropriate $\mathcal{D} = (\ )\times \underline{\Delta}$ defined below, $\underline{B}_{\mathcal{D}} = \mathrm{Trip}(\underline{B})$ = the usual category of all standard constructions in $\underline{B}$. Incidentally, we might call a doctrine $\mathcal{D}$ <u>categorical</u> if for any $\underline{B}$, any two objects in $\underline{B}_{\mathcal{D}}$ are uniquely isomorphic; this would <u>not</u> hold for the doctrine of standard constructions, but would for various doctrines of limits or colimits such as those whose development has been begun by Kock ($\underline{B}_{\mathcal{D}}$ will of course be $\underline{0}$ for many $\underline{B}$).

By the <u>ordinal sum</u> of two categories we mean the pushout

$$\begin{array}{ccc} \underline{A} \times |\underline{2}| \times \underline{B} & \longrightarrow & \underline{A} \times \underline{2} \times \underline{B} \\ \downarrow & & \downarrow \\ \underline{A} + \underline{B} & \dashrightarrow & \underline{A} +_{\sigma} \underline{B} \end{array}$$

in which the left vertical arrow takes $\langle a,i,b\rangle \rightsquigarrow a$ if $i = 0$, $\rightsquigarrow b$ if $i = 1$. Thus $\underline{A} +_{\sigma} \underline{B}$ may be visualized as $\underline{A} + \underline{B}$ with exactly one morphism $A \longrightarrow B$ adjoined for every $A \in \underline{A}$, $B \in \underline{B}$. Actually what we have just defined is the ordinal sum over $\underline{2}$; we could also consider the ordinal sum over any category $\underline{C}$ of any family $\{\underline{A}_C\}$ of categories indexed by the objects of $\underline{C}$. For example, with help of the ordinal sum over $\underline{3}$ we see that $+_{\sigma}$ is an associative bifunctor on Cat; it has the empty category $\underline{0}$ as neutral object. Also $\underline{1} +_{\sigma} \underline{1} = \underline{2}$, $\underline{1} +_{\sigma} \underline{2} = \underline{3}$, etc. One has $\underline{1} +_{\sigma} \omega \cong \omega$ but $\omega +_{\sigma} 1 \not\cong \omega$, showing that $+_{\sigma}$ is <u>not</u> commutative; it is not even commutative when applied to <u>finite</u> ordinals, if we consider what it does to morphisms.

Now $\underline{B} \rightsquigarrow \underline{1} +_{\sigma} \underline{B}$ may be seen to be the doctrine whose algebras are categories equipped with an initial object, while its opposite doctrine $\underline{B} \rightsquigarrow \underline{B} +_{\sigma} \underline{1}$ is the doctrine of terminal objects.

Consider the category - with - a - strictly - associative - multiplication (denoted by juxtaposition) generated as such by an object T and two morphisms

$$T^2 \xrightarrow{\mu} T \xleftarrow{\eta} 1$$

subject to the three laws familiar from the definition of standard construction. Denote this (finitely - presented!) category with multiplication by $\underline{\Delta}$. Clearly then $(\ )\times \underline{\Delta}$ is a doctrine whose algebras are precisely standard constructions. To obtain a concrete representation of $\underline{\Delta}$, define a functor

$$\underline{\Delta} \longrightarrow \text{Cat}$$

by sending $1 \rightsquigarrow \underline{0}$, $T \rightsquigarrow \underline{1}$, $T^2 \rightsquigarrow \underline{2}$, and noting that since all diagrams ending in $\underline{1}$ commute there is a unique extension to a <u>functor</u> which takes juxtaposition in $\underline{\Delta}$ into <u>ordinal sum</u> in Cat. For example, $T\eta, \eta T \rightsquigarrow \partial_0, \partial_1$. Clearly the categories which are values of our functor are just all the finite ordinal numbers (including $\underline{0}$): we claim that the functor is actually <u>full</u> and <u>faithful</u>. For suppose

$$\underline{n} \xrightarrow{\sigma} \underline{m}$$

is any functor (order-preserving map) between finite ordinals. Then

$$\underline{m} = \sum_{i\in \underline{m}}^{\sigma} \underline{1}$$

and denoting by $\underline{n}_i$ the inverse image of i by σ, we actually have that σ itself is an ordinal sum

$$\sigma = \sum_{i\in \underline{m}}^{\sigma} \sigma_i$$

where $\sigma_i : \underline{n}_i \longrightarrow \underline{1}$. Since such σ_i is unique we need only show that $\underline{n} \longrightarrow \underline{1}$ can be somehow expressed using composition and juxtaposition in terms of T, η, μ. For this define $\mu_n : T^n \longrightarrow T$ by

$\mu_0 = \eta$ (corresponding to an empty fiber $\underline{n}_i$)

$\mu_1 = T$ (corresponding to a singleton fiber $\underline{n}_i$)

$$\mu_2 = \mu \qquad \text{(corresponding to a two-point fiber } n_i\text{)}$$

$$\mu_{n+2} = \mu_{n+1}T.\mu$$

Thus $\mu_{n+2} = \mu T^n.\mu T^{n-1} \ . \ . \ . \ . \ \mu T.\mu$ and every map is a juxtaposition (ordinal sum) of the μ's. Furthermore any calculation involving the order-preserving maps can be carried out using only the triple laws. Thus $\underline{\Delta}$ could also be given the usual (infinite) presentation as a pure category with generators

$$d_i = T^i \eta T^{n-i} : T^n \longrightarrow T^{n+1} \qquad i = o, \ . \ . \ . \ , n$$

$$s_i = T^i \mu T^{n-i} : T^{n+2} \longrightarrow T^{n+1} \qquad i = o, \ . \ . \ . \ , n$$

if desired, although the finite presentation using ordinal sums and the triple laws seems much simpler.

It results in particular that the category $\underline{\Delta}$ of finite ordinals (including $\underline{0}$) and order preserving maps carries itself a canonical standard construction $\underline{n} \rightsquigarrow \underline{1} +_{\sigma} \underline{n}$ (just the restriction of the doctrine of initial objects from Cat to $\underline{\Delta}$). Denote by $A\underline{\Delta}$ the category of algebras for this standard construction, which is easily seen to have as objects all non-zero ordinals and as morphisms the order preserving maps <u>which preserve first element</u>. By construction $\underline{A\Delta}$ carries a standard <u>co</u>-construction. But it also has a $\underline{\Delta}$-structure because it is a <u>self-dual</u> category. Namely, since a finite ordinal is a complete category, and since on such a functor preserving initial objects preserves all colimits, we have the isomorphism "taking right adjoints":

$$(A\underline{\Delta})^{op} \xrightarrow[\approx]{adj.} \underline{\Delta A}$$

where $\underline{\Delta A}$ denotes the category of maps preserving last element. But now the covariant operation $(\ )^{op}$ on Cat restricts to $\underline{\Delta}$ and takes $\underline{\Delta A}$ into $A\underline{\Delta}$. Thus composing these two processes we obtain the claimed isomorphism

$$(A\underline{\Delta})^{op} \xrightarrow{\ \sim\ } A\underline{\Delta}$$

and hence a standard construction on $A\underline{\Delta}$.

Now let $\underline{A}$ be any category equipped with a standard construction $\underline{T}$, which we interpret as a category with a given action of $\underline{\Delta}$. Then

$$\mathrm{Hom}_{\underline{\Delta}}(A\underline{\Delta},\underline{A}) \cong \underline{A}^{\underline{T}}$$

the Eilenberg-Moore category of $\langle \underline{A},\underline{T} \rangle$. Since the latter carries canonically an action of $\underline{\Delta}^*$, we see that $A\underline{\Delta}$ has in another sense the co-structure of a standard co-construction, and get an adjoint pair

$$\mathrm{Cat}^{\underline{\Delta}} \underset{\otimes_{\underline{\Delta}^*} A\underline{\Delta}}{\overset{\mathrm{Hom}_{\underline{\Delta}}(A\underline{\Delta},-)}{\rightleftarrows}} \mathrm{Cat}^{\underline{\Delta}^*}$$

in which the lower assigns to every standard co-construction the associated Kleisli category of free coalgebras. For ease in dealing with these relationships it may be useful to use the following notation for $A\underline{\Delta}$, in which A is just a symbol:

$$\begin{array}{ccccccc} 1+3 & & 1+2 & & 1+1 & & 1+o \\ \ldots\ldots\; AT^3 & \rightrightarrows & AT^2 & \overset{\xi T}{\underset{A\mu}{\rightrightarrows}} & AT & \overset{\xi}{\underset{A\eta}{\rightleftarrows}} & A \end{array}$$

Clearly one can also obtain the doctrine of adjoint triples, describing a simultaneous action of $\underline{\Delta}$ and $\underline{\Delta}^*$ related by given adjunction maps. The writer does not know of a simple concrete representation of the resulting category $\tilde{\underline{\Delta}}$ with strictly associative multiplication. The same could be asked for the doctrine of Frobenius standard constructions, determined by the monoid in Cat presented as follows

$$1 \underset{\epsilon}{\overset{\eta}{\rightleftarrows}} T \underset{\mu}{\overset{\delta}{\rightleftarrows}} T^2$$

Triple laws for η,μ and cotriple laws for ϵ,δ are required to hold, as are the following four equations:

$$\delta T.T\mu.T\epsilon = \mu ,$$

$$T\delta.\mu T.\epsilon T = \mu ,$$

$$\eta T.\delta T.T\mu = \delta ,$$

$$T\eta.T\delta.\mu T = \delta .$$

An algebra over this doctrine has an underlying triple and an underlying cotriple whose associated free and cofree functors are the same. For example, if G is a finite group, then in any abelian category $\underline{A}$, $AT = \bigoplus_G A$ has such a structure. The characteristic property from group representation theory actually carries over to the general case: There is a "quadratic form" $\beta = \mu.\epsilon : T^2 \longrightarrow 1$ which is "associative" $T\mu.\beta = \mu T.\beta$ and "non-singular" i.e. there is $\alpha = \eta.\delta : 1 \longrightarrow T^2$ quasi-inverse to β (i.e. they are adjunctions for $T \dashv T$.)

In order to construct doctrines whose algebras are categories associatively equipped with colimits, Kock considers categories Cat_o of categories and functors which are "regular" in the sense that the total category of a fibration belongs to Cat_o whenever the base and every fiber belong to Cat_o. In order to make $\underline{B} \rightsquigarrow [Cat_o, \ulcorner\underline{B}\urcorner] = Dir_{Cat_o}(\underline{B})$ into a strict standard construction in Cat, Kock found it necessary to construe Cat_o as having for each of its objects $\underline{C}$ a given well ordering on the set of objects of $\underline{C}$ and on each hom set of $\underline{C}$. Then with considerable effort he is able to choose a version of the Grothendieck process (taking $\underline{C} \xrightarrow{R} Cat_o$ for $\underline{C} \in Cat_o$ to the associated op-fibration over $\underline{C}$ in Cat_o) which gives rise to a strictly-associative

$$Dir_{Cat_o}(Dir_{Cat_o}(\underline{B})) \xrightarrow{\underline{B}\mu} Dir_{Cat_o}(\underline{B})$$

One then defines the colimits - over - indexcategories - in - Cat_o doctrine $\mathfrak{R}$ by

$$\mathfrak{R} = Dir_{Cat_o}(-) \Big/ \langle \underline{S}, \underrightarrow{\lim} \rangle$$

showing first, also with some effort, that there does exist an equivalent version $\underline{S}$ of the category of small (relative to Cat_o) sets which can be equipped with a strictly associative $\varinjlim$ i.e. a colimit assigment which is also an algebra structure

$$Dir_{Cat_o}(\underline{S}) \xrightarrow{\varinjlim} \underline{S}$$

for the "precolimit" doctrine Dir_{Cat_o}.

By choosing the appropriate Cat_o and by making use of the "opposite" doctrine construction, one then sees that the notions of a category equipped with small $\varinjlim$, finite $\varprojlim$, or countable products, etc, etc, are all essentially doctrinal. Hence presumably, given an understanding of free products, quotients, Kronecker products, distributive laws, etc for doctrines, so are the notions of abelian category, $\mathcal{S}$-topos, ab-topos (the latter two without the usual "small generating set" axiom) also doctrinal. (In order to view, for example, the distributive axiom for topos as a distributive law in the Barr-Beck sense, it may be necessary to generalize the notion of equational doctrine to allow the associative laws for μ or α to hold only up to isomorphism (?).)

The value of knowing that a notion of category - with-structure is equationally doctrinal should be at least as great as knowing that a category is triplable over sets. We have at the moment however no intrinsic characterization of those categories enriched over Cat which are of the form $Cat^{\mathcal{D}}$ for some equational doctrine . However the Freyd Hom-Tensor Calculus would seem to extend easily from theories over sets to doctrines over Cat to give the theorem: any strongly left adjoint functor

$$Cat^{\mathcal{D}_1} \longrightarrow Cat^{\mathcal{D}_2}$$

is given by $(-)\otimes_{\mathcal{D}_1}\underline{A}$ where $\underline{A}$ is a fixed category equipped with a $\mathcal{D}_2$-structure and a $\mathcal{D}_1$-costructure. For example, consider the (doctrinal) notion of $\underline{2}$-topos, meaning a partially ordered set with small sups and finite infs which distribute over the sups (morphisms to preserve just the mentioned structure). Then of course the

Sierpinski space represents the "open sets" functor

$$\underline{\mathrm{Top}}^{op} \longrightarrow \underline{2}\text{-}\underline{\mathrm{Topos}}$$

Consider on the other hand the functor

$$\mathcal{S}\text{-}\underline{\mathrm{Topos}} \longrightarrow \underline{2}\text{-}\underline{\mathrm{Topos}}$$

which assigns to every $\mathcal{S}$-topos the set of all subobjects of the terminal object; this is represented by the $\mathcal{S}$-topos $\underline{E}$ with one generator X subject to $X \xrightarrow{\approx} X \times X$, hence has a strong left adjoint $- \otimes \underline{E}$ which, when restricted to $\underline{\mathrm{Top}}^{op}$ is just the assigment of the category of sheaves to each space. Or again consider the functor "taking abelian group objects"

$$\mathcal{S}\text{-}\underline{\mathrm{Topos}} \longrightarrow \text{ab-}\underline{\mathrm{Topos}}$$

Since this is $\underline{F} \rightsquigarrow \mathrm{Hom}_{\text{fin prod}}(\underline{Z},\underline{FU})$ where $\underline{Z}$ is the category of finitely generated free abelian groups and $\underline{FU}$ denotes the category with finite products underlying the topos $\underline{F}$, we see that our functor is represented by $\underline{A}$ = the relatively free topos over the category $\underline{Z}$ with finite products. Hence there is a strong left adjoint $- \otimes \underline{A}$ which should be useful in studying the extent to which an arbitrary Grothendieck category differs from the abelian sheaves on some $\mathcal{S}$-site.

Bibliography

Kock, Anders Limit Monads in Categories
Aarhus Universitet Math Preprint
Series 1967/68 No.6

Freyd, Peter Algebra-Valued-Functors in General and Tensor Products in Particular, Colloq. Math. 14 (1966), 89-106

Bénabou, Jean Thesis

CATEGORIES WITH MODELS

H. Appelgate and M. Tierney (1)

1. Introduction

General remarks. A familiar process in mathematics is the creation of "global" objects from given "local" ones. The "local" objects may be called "models" for the process, and one usually says that the "global" objects are formed by "pasting together" the models. The most immediate example is perhaps given by manifolds. Here the "local" objects are open subsets of Euclidean space, and as one "pastes together" by homeomorphisms, diffeomorphisms, etc., one obtains respectively topological, differentiable, etc., manifolds. Our object in this paper is to present a coherent, categorical treatment of this "pasting" process.

The general plan of the paper is the following. In § 2, we consider the notion of a category $\underline{A}$ with models $I : \underline{M} \longrightarrow \underline{A}$, and give a definition of what we mean by an "$\underline{M}$-object of $\underline{A}$". Our principal tool for the study of $\underline{M}$-objects is the theory of cotriples, and its connection with $\underline{M}$-objects is developed in § 3. There we show that the $\underline{M}$-objects of $\underline{A}$ can be identified with certain coalgebras over a "model-induced" cotriple. § 4 exploits this identification by using it to prove equivalence theorems relating the category of $\underline{M}$-objects to categories of set-valued functors. § 5 consists of a detailed study of several important examples, and § 6 is concerned with the special case where the model-induced cotriple is idempotent.

In subsequent papers, we will consider model-induced adjoint functors, and models in closed (autonomous) categories.

---

(1) The second named author was partially supported by the NSF under Grant GP 6783.

It is a pleasure to record here our indebtedness to Jon Beck for numerous instructive conversations and many useful suggestions. Thanks are also due F.W. Lawvere for suggesting the example of G-spaces, and, in general, for being a patient listener.

Notation. Let $\underline{A}$ be a category. If A_1 and A_2 are objects of $\underline{A}$, the set of morphisms $f : A_1 \longrightarrow A_2$ will be denoted by (A_1, A_2) . If there is a possibility of confusion, we will use $\underline{A}(A_1, A_2)$ to emphasize $\underline{A}$. Similarly, if $\underline{A}$ and $\underline{B}$ are categories and $F_1, F_2 : \underline{A} \longrightarrow \underline{B}$ are functors, (F_1, F_2) denotes the class of natural transformations $\eta : F_1 \longrightarrow F_2$. The value of η at $A \in \underline{A}$ will be written $\eta A : F_1 A \longrightarrow F_2 A$ unless the situation is complicated, in which case we write $\eta . A : F_1 . A \longrightarrow F_2 . A$. Composition of morphisms will be denoted by juxtaposition (usual order). Diagram means commutative diagram unless otherwise specified, and all functors are covariant.

Let

$$\underline{A} \xrightarrow{F} \underline{B} \underset{G_2}{\overset{G_1}{\rightrightarrows}} \underline{C} \xrightarrow{H} \underline{D}$$

be a collection of categories and functors, and let $\eta : G_1 \longrightarrow G_2$ be a natural transformation. Then

$$H\eta F : HG_1 F \longrightarrow HG_2 F$$

is the natural transformation given by

$$H\eta F.A = H(\eta(FA))$$

for $A \in \underline{A}$. When $\underline{A} = \underline{B}$ and $F = 1_{\underline{A}}$ $(\underline{C} = \underline{D}$ and $H = 1_{\underline{C}})$ we write $H\eta$ (ηF) . Godement's 5 rules of functorial calculus for this situation are used without comment.

The category of sets will be denoted by $\mathfrak{S}$, and we say a category $\underline{A}$ is small, if $|\underline{A}|$ (the class of objects of $\underline{A}$) is a set.

$\underline{A}^*$ is the dual of $\underline{A}$.

We shall use primarily the following criterion for adjoint functors. Namely, given functors

$$\underline{A} \underset{R}{\overset{L}{\rightleftarrows}} \underline{C}$$

L is <u>coadjoint</u> to R (or R is <u>adjoint</u> to L) iff there are natural transformations

$$\eta : 1_{\underline{A}} \longrightarrow RL$$

called the <u>unit</u>, and

$$\varepsilon : LR \longrightarrow 1_{\underline{C}}$$

called the <u>counit</u>, such that

$$L \xrightarrow{L\eta} LRL \xrightarrow{\varepsilon L} L$$

and

$$R \xrightarrow{\eta R} RLR \xrightarrow{R\varepsilon} R$$

are the respective identities 1_L and 1_R. We abbreviate this in the notation

$$(\varepsilon,\eta) : L \longrightarrow\!\!\!\!| \; R .$$

Let

$$D : \underline{J} \longrightarrow \underline{A}$$

be a functor. Any $A \in \underline{A}$ defines an obvious constant functor

$$A : \underline{J} \longrightarrow \underline{A} .$$

A pair (C,γ) consisting of an object $\underline{C} \in \underline{A}$ and a natural transformation

$$\gamma : D \longrightarrow C$$

will be called a <u>colimit</u> of D iff for each $A \in \underline{A}$, composition with γ induces a 1-1 correspondence between $\underline{A}$-morphisms $C \longrightarrow A$, and natural transformations $D \longrightarrow A$. In other words, C is an object of $\underline{A}$, and γ is a universal family of morphisms

$$\gamma j : Dj \longrightarrow C \qquad \text{for } j \in \underline{J}$$

such that if $\alpha : j \longrightarrow j'$ is in $\underline{J}$, then

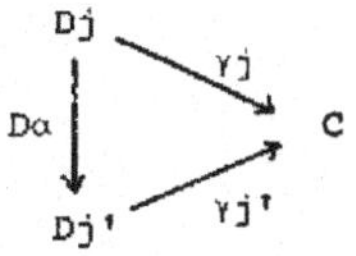

commutes. Universal means that if

$$\gamma' j : Dj \longrightarrow C' \qquad j \in \underline{J}$$

is any such family, then there is a unique morphism $f : C \longrightarrow C'$ in $\underline{A}$ such that

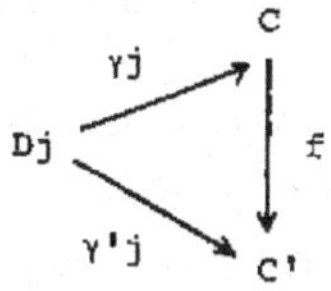

commutes for all $j \in \underline{J}$. Clearly any two colimits of D are isomorphic in the obvious sense. A choice of colimit, when it exists, will be denoted by $\varinjlim D$, and the natural transformation will be understood. The <u>limit</u> of D is defined dually, and denoted by $\varprojlim D$. We say $\underline{A}$ has <u>small colimits</u> if for every functor

$$D : \underline{J} \longrightarrow \underline{A}$$

with $\underline{J}$ is small, there exists a colimit of D in $\underline{A}$.

2. Categories with models

<u>Singular and realization functors</u>. A category $\underline{A}$ together with a functor

$$I : \underline{M} \longrightarrow \underline{A}$$

where $\underline{M}$ is small, is called a <u>category with models</u>. $\underline{M}$ will be called the <u>model</u>

<u>category</u>, and $\underline{A}$ is sometimes called the <u>ambient category</u>.

Given a category $\underline{A}$ with models, I defines a <u>singular functor</u>

$$s : \underline{A} \longrightarrow (\underline{M}^*, \mathfrak{S})$$

as follows: for $A \in \underline{A}$,

$$sA : \underline{M}^* \longrightarrow \mathfrak{S}$$

is the functor given by

$$sA.M = (IM,A)$$
$$sA.\alpha = (I\alpha,A)$$

for M an object and α a morphism in $\underline{M}$. If $f : A \longrightarrow A'$ is a morphism in $\underline{A}$, then

$$sf : sA \longrightarrow sA'$$

is the obvious natural transformation induced by composition with f .

The following example, to be discussed later in greater detail, may help to motivate the terminology and future definitions. Let $\underline{A} = \underline{Top}$, the category of topological spaces and continuous maps, and let $\underline{\Delta}$ be the standard simplicial category. Let

$$I : \underline{\Delta} \longrightarrow \underline{Top}$$

be the functor which assigns to each simplex [n] the standard geometric simplex Δ_n , and to each simplicial morphism $\alpha : [m] \longrightarrow [n]$ the uniquely determined affine map $\Delta_\alpha : \Delta_m \longrightarrow \Delta_n$. Then, $(\underline{\Delta}^*, \mathfrak{S})$ is the category of simplicial sets, and

$$s : \underline{Top} \longrightarrow (\underline{\Delta}^*, \mathfrak{S})$$

is the usual singular functor of homology theory.

Let us assume now that $\underline{A}$, our category with models, has small colimits. Then we can construct a coadjoint

$$r : (\underline{M}^*, \mathcal{S}) \longrightarrow \underline{A}$$

to s as follows. (This construction, when $\underline{M}$ is $\underline{\Delta}$, seems to be due originally to Kan [5].) Let $F : \underline{M}^* \longrightarrow \mathcal{S}$ and consider the category (Y,F) whose objects are pairs (M,x) where $M \in \underline{M}$ and $x \in FM$, and whose morphisms $(M,x) \longrightarrow (M',x')$ are morphisms $\alpha : M \longrightarrow M'$ in $\underline{M}$ such that $F\alpha(x') = x$. (As the notation indicates, this is Lawvere's comma category where $Y : \underline{M} \longrightarrow (\underline{M}^*, \mathcal{S})$ is the Yoneda embedding.) There is an obvious functor

$$\partial_o : (Y,F) \longrightarrow \underline{M}$$

given by $\partial_o(M,x) = M$, $\partial_o\alpha = \alpha$. Consider the composite of ∂_o with I ,

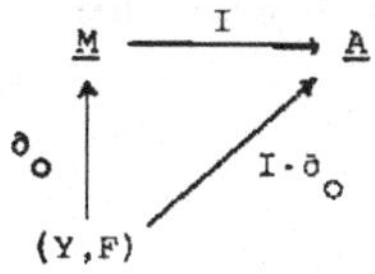

and put

$$rF = \varinjlim I\cdot\partial_o .$$

Let us denote the components of the universal natural transformation $i : I\cdot\partial_o \longrightarrow rF$ by

$$i(M,x) : IM \longrightarrow rF .$$

(In what follows, we will often omit the M and write simply $i_x : IM \longrightarrow rF$ where $x \in FM$.) The functoriality of r is determined, for $\gamma : F \longrightarrow F'$ a natural transformation, by the diagram

$$\begin{array}{ccc} IM & & \\ {\scriptstyle i_x}\downarrow & \searrow^{i_{\gamma M(x)}} & \\ rF & \xrightarrow{r\gamma} & rF' \end{array}$$

The unit $\eta : 1 \longrightarrow sr$ is given by

$$(\eta F.M)(x) = i(M,x)$$

for $F \in (\underline{M}^*, \mathfrak{S})$, $M \in \underline{M}$, and $x \in FM$. The counit $\varepsilon : rs \longrightarrow 1$ is determined by the diagram

$$\begin{array}{ccc} IM & & \\ {\scriptstyle i_\varphi}\downarrow & \searrow^{\varphi} & \\ rsA & \xrightarrow{\varepsilon A} & A \end{array}$$

for $A \in \underline{A}$, $M \in \underline{M}$, and $\varphi \in sA.M$. It is now trivial to verify that η and ε are natural, and that the composites

$$s \xrightarrow{\eta s} srs \xrightarrow{s\varepsilon} s$$

$$r \xrightarrow{r\eta} rsr \xrightarrow{\varepsilon r} r$$

are the respective identites 1_s and 1_r . Thus, we have $(\varepsilon,\eta) : r \longrightarrow\!\!| \; s$. We shall call r a <u>realization functor</u>, since in the previous example r is the geometric realization of Milnor [6].

Often, the categories that we work with come equipped with a colimit preserving underlying set functor

$$U : \underline{A} \longrightarrow \mathfrak{S} .$$

In this case, the underlying set of rF admits an easy description; that is, one can describe easily the colimit of the composite functor

$$(Y,F) \xrightarrow{I \cdot \partial_o} \underline{A} \xrightarrow{U} \mathfrak{S} ,$$

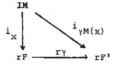

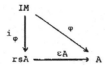

which is, by assumption, the underlying set of rF. Namely, consider the set $\mathcal{T}$ of all triples (M,x,m) where $(M,x) \in (Y,F)$ and $m \in UIM$. Let $\equiv$ be the equivalence relation on $\mathcal{T}$ generated by the relation: $(M,x,m) \sim (M',x',m')$ iff there is $\alpha : (M,x) \longrightarrow (M',x')$ in (Y,F) such that $UI\alpha(m) = m'$ (i.e. $(M,F\alpha(x'),m) \sim (M',x',UI\alpha(m))$). Let $|M,x,m|$ denote the equivalence class containing (M,x,m). (Again, we will often drop the M, writing simply $|x,m|$.) It is easy to see that the set $\tilde{\mathcal{T}}$ of equivalence classes, together with the family of functions

$$i'(M,x) : UIM \longrightarrow \tilde{\mathcal{T}}$$

given by $i'(M,x)(m) = |M,x,m|$, is a colimit of $UI_F \cdot \partial_o$.

<u>Atlases</u>

Let $\underline{A}$ be a category with models

$$I : \underline{M} \longrightarrow \underline{A} ,$$

and let $A \in \underline{A}$. A subfunctor $\mathfrak{G} \hookrightarrow sA$ of the singular functor sA will be called an <u>$\underline{M}$-preatlas</u> for A. $\varphi : IM \longrightarrow A$ is said to be an <u>$\mathfrak{G}$-morphism</u> if $\varphi \in \mathfrak{G}M$.

Let $\mathcal{A}$ be a set of $\underline{A}$-morphisms of the form $\varphi : IM \longrightarrow A$ for $M \in \underline{M}$ and A a fixed object of $\underline{A}$. If $\mathfrak{G} \hookrightarrow sA$ is an $\underline{M}$-preatlas for A, we say $\mathfrak{G}$ <u>contains</u> $\mathcal{A}$ if each $\varphi \in \mathcal{A}$ is an $\mathfrak{G}$-morphism. Since there is at least one preatlas containing $\mathcal{A}$ - namely sA itself - and since the intersection of any family of preatlases containing $\mathcal{A}$ is a preatlas containing $\mathcal{A}$, we can define the $\underline{M}$-preatlas <u>generated by</u> $\mathcal{A}$ to be the smallest preatlas containing $\mathcal{A}$. Clearly, this consists of all morphisms $\psi : IM' \longrightarrow A$ such that ψ can be factored as

$$\begin{array}{ccc} IM' & \xrightarrow{\ \psi\ } & A \\ {\scriptstyle I\alpha}\downarrow & \nearrow_{\varphi} & \\ IM & & \end{array}$$

where α is a morphism in $\underline{M}$ and $\varphi \in \mathcal{A}$.

Assuming $\underline{A}$ has small colimits, we have the realization

$$r : (\underline{M}^*, \mathcal{S}) \longrightarrow \underline{A}$$

with $(\varepsilon, \eta) : r \dashv s$. Let $\mathfrak{G}$ be an $\underline{M}$-preatlas for A, and let

$$j : \mathfrak{G} \longrightarrow sA$$

be the inclusion. Let $e : r\mathfrak{G} \longrightarrow A$ be the composite of

$$r\mathfrak{G} \xrightarrow{rj} rsA \xrightarrow{\varepsilon A} A .$$

e can be characterized as that $\underline{A}$-morphism $r\mathfrak{G} \longrightarrow A$ such that

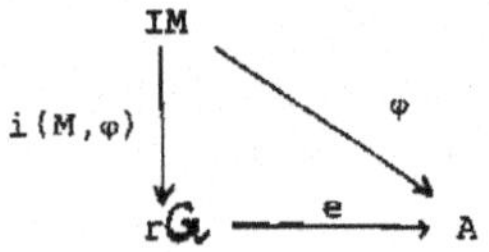

commutes for $M \in \underline{M}$, and $\varphi \in \mathfrak{G}M$.

An object $A \in \underline{A}$ having a preatlas $\mathfrak{G} \longrightarrow sA$ for which e is an isomorphism will be called an $\underline{M}$-object with atlas $\mathfrak{G}$. Intuitively, e epic means the $\mathfrak{G}$-morphisms φ cover A, e monic means they are compatible, and the full isomorphism condition means that, in addition, the $\underline{A}$-structure (e.g. the topology) of A is determined by the φ's.

An $\underline{M}$-object A is thus isomorphic to a small colimit of models, i.e. a colimit of a functor with small domain that factors through I. Namely,

$$\varinjlim I \cdot \partial_0 = r\mathfrak{G} \xrightarrow[\sim]{e} A .$$

This suggests that all such colimits are $\underline{M}$-objects, and this is confirmed in

<u>Proposition 2.1</u>

Let I' be a functor with small domain that factors through I :

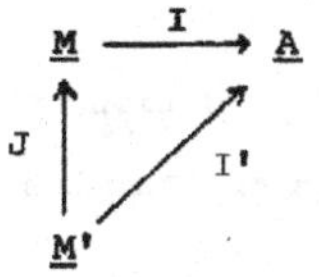

Let $A = \varinjlim I'$ with universal family

$$\gamma M' : I'M' \longrightarrow A .$$

Let $\mathfrak{G} \hookrightarrow sA$ be the $\underline{M}$-preatlas for A generated by the set

$$\mathcal{A} = \{\gamma M' : IJM' = I'M' \longrightarrow A\} .$$

Then A is an $\underline{M}$-object with atlas $\mathfrak{G}$; i.e. $e : r\mathfrak{G} \longrightarrow A$ is an isomorphism.

<u>Proof</u>: Let $e' : A \longrightarrow r\mathfrak{G}$ be the $\underline{A}$-morphism determined by

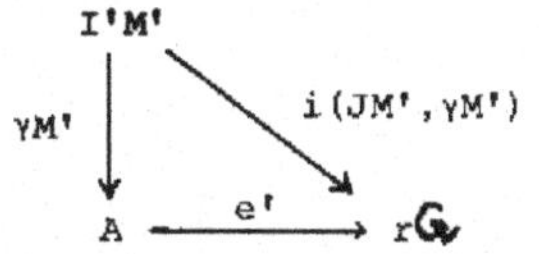

This makes sense, since $\mathcal{A}$ is contained in $\mathfrak{G}$. But then

$$ee'\gamma M' = ei(JM',\gamma M') = \gamma M'$$

so $ee' = 1_A$. On the other hand, for any $\varphi \in \mathfrak{G}M$ we have

$$e'ei(M,\varphi) = e'\varphi .$$

But $\varphi = \gamma M' \cdot I\alpha$ for some $\alpha : M \longrightarrow JM'$ in $\underline{M}$. Therefore,

$$e'\varphi = e'\gamma M' \cdot I\alpha = i(JM',\gamma M')I\alpha = i(M,\gamma M' \cdot I\alpha) = i(M,\varphi) ,$$

and $e'e = 1_{r\mathfrak{G}}$.

By 2.1 then, the $\underline{M}$-objects of $\underline{A}$ are exactly those objects of $\underline{A}$ that are "pasted together" from models; i.e. that are colimits of models. These are the "global" objects referred to in the introduction.

The remainder of this section is devoted to a technical condition on $\underline{M}$-objects and a lemma on pullbacks. These will be needed in the section on examples (§ 5).

<u>Regular</u> <u>$\underline{M}$-objects</u>. Let $\mathcal{G} \hookrightarrow sA$ be an $\underline{M}$-preatlas for A with generating set $\mathcal{A}$. $\mathcal{A}$ is said to be a <u>regular</u> <u>generating</u> <u>set</u> iff for each pair

$$\varphi_1 : IM_1 \longrightarrow A$$

$$\varphi_2 : IM_2 \longrightarrow A$$

of morphisms in $\mathcal{A}$, there are morphisms

$$\alpha_1 : M \longrightarrow M_1$$

$$\alpha_2 : M \longrightarrow M_2$$

in $\underline{M}$ such that

$$\begin{array}{ccc} IM & \xrightarrow{I\alpha_1} & IM_1 \\ {\scriptstyle I\alpha_2}\downarrow & & \downarrow{\scriptstyle \varphi_1} \\ IM_2 & \xrightarrow[\varphi_2]{} & A \end{array}$$

is a pullback diagram in $\underline{A}$. An $\underline{M}$-object A is called <u>regular</u> if it has an atlas $\mathcal{G} \longrightarrow sA$ with a regular generating set.

<u>Lemma 2.2</u>

Let $f : A \longrightarrow A'$ be a morphism in $\underline{A}$, and

$$\begin{array}{ccc} P & \xrightarrow{k_1} & A \\ {\scriptstyle k_2}\downarrow & & \downarrow{\scriptstyle f} \\ A & \xrightarrow[f]{} & A' \end{array}$$

a pullback diagram.

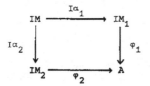

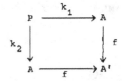

Then f is monic iff at least one of k_1 , k_2 is monic.

Proof: If f is monic, then $k_1 = k_2$, and P being a pullback says their common value is monic. Suppose, say, k_1 is monic. Then $\exists | \varphi : P \to P$ such that $k_1\varphi = k_1$ and $k_2\varphi = k_1$. But $k_1\varphi = k_1 = k_1 1_P$, so $\varphi = 1_P$, $k_2 = k_1$, and f is monic. We are grateful to H.B. Brinkmann for pointing out this short proof to us.

Note that if A is a regular $\underline{M}$-object, and all $I\alpha : IM \longrightarrow IM'$ are monic in $\underline{A}$, 2.2 implies that all $\varphi : IM \longrightarrow A$ in a regular generating set for an atlas for A are monic.

3. The model induced cotriple

Here we shall first briefly recall some facts from the theory of cotriples and coalgebras [1]. We then show how $I : \underline{M} \longrightarrow \underline{A}$ induces a cotriple $\mathbb{G}$ in a category $\underline{A}$ with models, and discuss the relation of $\mathbb{G}$ to the singular and realization functors of § 2. Finally, we identify $\underline{M}$-objects with certain coalgebras over $\mathbb{G}$.

So, recall that in a category $\underline{A}$, a cotriple $\mathbb{G} = (G, \varepsilon, \delta)$ consists of a functor $G : \underline{A} \longrightarrow \underline{A}$ together with natural transformations $\varepsilon : G \longrightarrow 1_{\underline{A}}$ and $\delta : G \longrightarrow G^2$ satisfying the following diagrams:

$$\begin{array}{ccc} G & \xrightarrow{\ \delta\ } & G^2 \\ & {\scriptstyle 1_G}\searrow & \downarrow{\scriptstyle \varepsilon G} \\ & & G \end{array} \qquad\qquad \begin{array}{ccc} G & \xrightarrow{\ \delta\ } & G^2 \\ & {\scriptstyle 1_G}\searrow & \downarrow{\scriptstyle G\varepsilon} \\ & & G \end{array}$$

and

$$\begin{array}{ccc} G & \xrightarrow{\delta} & G^2 \\ {\scriptstyle\delta}\downarrow & & \downarrow{\scriptstyle\delta G} \\ G^2 & \xrightarrow[G\delta]{} & G^3 \end{array} .$$

Given any adjoint pair of functors

$$\underline{A} \underset{F}{\overset{U}{\rightleftarrows}} \underline{B} \quad \text{with} \quad (\varepsilon,\eta) : F \longrightarrow\!\!| \; U ,$$

it is easy to see that $\mathbb{G} = (FU,\varepsilon,F\eta U)$ defines a cotriple in $\underline{A}$. Furthermore, as shown in [1], any cotriple $\mathbb{G} = (G,\varepsilon,\delta)$ in $\underline{A}$ arises in this way by considering the category of $\mathbb{G}$-coalgebras $\underline{A}_{\mathbb{G}}$. An object of $\underline{A}_{\mathbb{G}}$ is a pair (A,θ) where $A \in \underline{A}$, and $\theta : A \longrightarrow GA$ is a morphism in $\underline{A}$ satisfying the following two diagrams:

$$\begin{array}{ccc} A & \xrightarrow{\theta} & GA \\ & {\scriptstyle 1_A}\searrow & \downarrow{\scriptstyle\varepsilon A} \\ & & A \end{array} \quad , \qquad \begin{array}{ccc} A & \xrightarrow{\theta} & GA \\ {\scriptstyle\theta}\downarrow & & \downarrow{\scriptstyle G\theta} \\ GA & \xrightarrow{\delta A} & G^2A \end{array} .$$

A morphism $f : (A',\theta') \longrightarrow (A,\theta)$ in $\underline{A}_{\mathbb{G}}$ is a morphism $f : A' \longrightarrow A$ in $\underline{A}$ such that

$$\begin{array}{ccc} A' & \xrightarrow{f} & A \\ {\scriptstyle\theta'}\downarrow & & \downarrow{\scriptstyle\theta} \\ GA' & \xrightarrow{Gf} & GA \end{array}$$

commutes. Now one defines functors $L : \underline{A}_{\mathbb{G}} \longrightarrow \underline{A}$ and $R : \underline{A} \longrightarrow \underline{A}_{\mathbb{G}}$ by $L(A,\theta) = A$, $RA = (GA,\delta A)$. Then, we have $\varepsilon : LR \longrightarrow 1$, and we obtain $\beta : 1 \longrightarrow RL$ by setting

$$\beta(A,\theta) = \theta : (A,\theta) \longrightarrow (GA,\delta A) .$$

It is immediate that these definitions give $(\varepsilon,\beta) : L \longrightarrow\!\!| \; R$ and $\mathbb{G} = (G,\varepsilon,\delta) = (LR,\varepsilon,L\beta R)$.

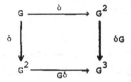

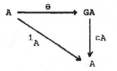

Now let $\underline{A}$ be a category with models

$$I : \underline{M} \longrightarrow \underline{A} .$$

(For the moment, let us not require $\underline{M}$ small.) If $\underline{A}$ has sufficient colimits, I defines a cotriple in $\underline{A}$ as follows. Let $A \in \underline{A}$, and consider the comma category (I,A) . Thus, an object in (I,A) is a pair (M,φ) where $M \in \underline{M}$ and $\varphi : IM \longrightarrow A$ is a morphism in $\underline{A}$. A morphism $\alpha : (M',\varphi') \longrightarrow (M,\varphi)$ is a morphism $\alpha : M' \longrightarrow M$ in $\underline{M}$ such that

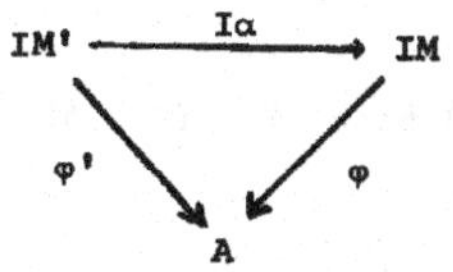

commutes. We have the obvious functor

$$I \cdot \partial_o : (I,A) \longrightarrow \underline{A}$$

as in § 2 . Assuming it exists, we set

$$GA = \varinjlim I \cdot \partial_o ,$$

and denote the (M,φ)-th component of the universal family by

$$i_\varphi : IM \longrightarrow GA .$$

G is a functor if we give, for $f : A' \longrightarrow A$ in $\underline{A}$, $Gf : GA' \longrightarrow GA$ by requiring

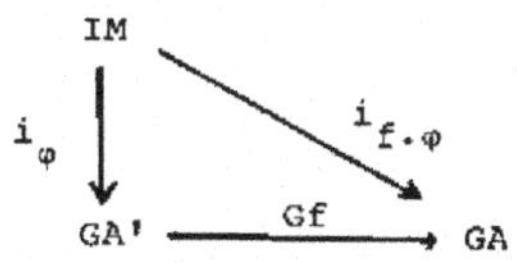

for each $\varphi : IM \longrightarrow A'$. $\varepsilon A : GA \longrightarrow A$ is determined by

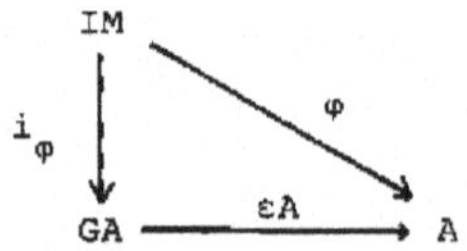

for each $\varphi : IM \longrightarrow A$, and $\delta A : GA \longrightarrow G^2A$ by

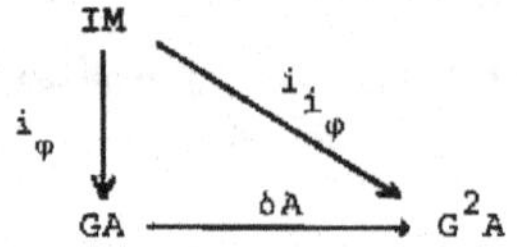

It is easy to check that ε and δ are natural, and that $\mathbb{G} = (G,\varepsilon,\delta)$ is a cotriple on $\underline{A}$. We call $\mathbb{G}$ the <u>model induced cotriple</u>.

Having $\mathbb{G}$, we can form the category $\underline{A}_{\mathbb{G}}$ of $\mathbb{G}$-coalgebras, and we obtain a diagram

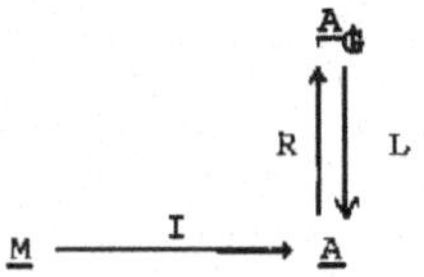

We show first that there is a functor $\bar{I} : \underline{M} \longrightarrow \underline{A}_{\mathbb{G}}$ such that $L\bar{I} = I$. That is, we exhibit a coalgebra structure on each IM , which is functorial with respect to morphisms in $\underline{M}$. So, let

$$\theta_M : IM \longrightarrow GIM$$

be the morphism $i_{1_{IM}}$. By definition of ε , we have

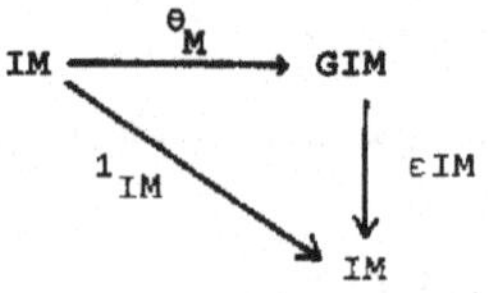

In the diagram

$$\begin{array}{ccc} IM & \xrightarrow{\theta_M} & GIM \\ {\scriptstyle\theta_M}\downarrow & & \downarrow{\scriptstyle G\theta_M} \\ GIM & \xrightarrow{\delta IM} & G^2IM \end{array}$$

the common diagonal is i_{θ_M} . Thus θ_M is a coalgebra structure for IM . Suppose $\alpha : M' \longrightarrow M$ is a morphism in $\underline{M}$. Then, in the diagram

$$\begin{array}{ccc} IM' & \xrightarrow{I\alpha} & IM \\ {\scriptstyle\theta_{M'}}\downarrow & & \downarrow{\scriptstyle\theta_M} \\ GIM' & \xrightarrow{GI\alpha} & GIM \end{array} ,$$

the common diagonal is $i_{I\alpha}$, and θ_M is functorial in M . Thus, if we set

$$\bar{I}M = (IM, \theta_M)$$

$$\bar{I}\alpha = I\alpha$$

for $M \in \underline{M}$ and $\alpha : M' \longrightarrow M$, we obtain the required functor $\bar{I}$.

Let us now suppose that we are in a models situation

$$I : \underline{M} \longrightarrow \underline{A}$$

where $\underline{M}$ is small, and $\underline{A}$ has small colimits. Then, as in § 2, we have the adjoint pair

$$\underline{A} \underset{r}{\overset{s}{\rightleftarrows}} (\underline{M}^*, \mathcal{S}) \quad \text{with} \quad (\varepsilon, \eta) : r \longrightarrow\!\!| \; s \quad .$$

Moreover, examining the definition of r , one sees immediately that

$$\mathbb{G} = (rs, \varepsilon, r\eta s) \quad .$$

(We gave a direct definition of $\mathbb{G}$ since it may occur that $\mathbb{G}$ exists, although rF

does not for arbitrary F . This can happen, for example, when $\underline{M}$ is not small or $\underline{A}$ does not have all small colimits.) Thus, in the terminology of [1], the above adjointness generates the cotriple $\mathbb{G}$. Therefore, again by [1], there is a canonical functor

$$\bar{r} : (\underline{M}^*, \mathcal{S}) \longrightarrow \underline{A}_{\mathbb{G}}$$

given by $\bar{r}F = (rF, r\eta F)$ and $\bar{r}\gamma = r\gamma$ for $F \in (\underline{M}^*, \mathcal{S})$ and $\gamma : F' \longrightarrow F$ a natural transformation. We have, of course, also the lifted singular functor

$$\bar{s} : \underline{A}_{\mathbb{G}} \longrightarrow (\underline{M}^*, \mathcal{S})$$

associated with the lifted models

$$\bar{I} : \underline{M} \longrightarrow A_{\mathbb{G}} \ .$$

That is, $\bar{s}$ is the functor defined by

$$\bar{s}(A,\theta)\cdot M = \underline{A}_{\mathbb{G}}((IM,\theta_M),(A,\theta)) \ .$$

for $(A,\theta) \in \underline{A}_{\mathbb{G}}$ and $M \in \underline{M}$. Since each $\underline{A}_{\mathbb{G}}$-morphism is also an $\underline{A}$-morphism, we have an inclusion of functors

$$j : \bar{s} \longrightarrow sL \ .$$

Proposition 3.1

For each $(A,\theta) \in \underline{A}_{\mathbb{G}}$,

$$\bar{s}(A,\theta) \xrightarrow{j(A,\theta)} sL(A,\theta) \underset{\eta sL(A,\theta)}{\overset{sL\theta}{\rightrightarrows}} srsL(A,\theta)$$

is an equalizer diagram.

Proof: We remark first that for any $A \in \underline{A}$ and $\varphi : IM \longrightarrow A$, we have the diagram

$$\begin{array}{ccc} IM & & \\ {\scriptstyle \theta_M}\downarrow & \searrow^{i_\varphi} & \\ GIM & \xrightarrow{G\varphi} & GA \end{array} \quad .$$

Thus, if $(A,\theta) \in \underline{A}_{\mathbb{G}}$ such a φ is a coalgebra morphism iff $\theta \cdot \varphi = i_\varphi$.

Now suppose we have a natural transformation $\psi : F \longrightarrow sA$ such that in the diagram

$$\begin{array}{ccc} F & & \\ \downarrow{\scriptstyle \psi} & & \\ sA & \underset{\eta sA}{\overset{s\theta}{\rightrightarrows}} & srsA \end{array}$$

$s\theta \cdot \psi = \eta sA \cdot \psi$. That is, for $M \in \underline{M}$ and $x \in FM$

$$\psi M(x) : IM \longrightarrow A$$

satisfies

$$\theta \cdot \psi M(x) \quad = \quad (\eta sA \cdot M)(\psi M(x)) \quad = \quad i_{\psi M(x)} \quad .$$

By the above remark, each $\psi M(x)$ is thus a morphism of coalgebras. But then by definition ψ factors through $j(A,\theta)$ - uniquely, since $j(A,\theta)$ is monic. Of course, also by the above remark, $j(A,\theta)$ itself equalizes.

<u>Proposition 3.2</u>

$$\bar{r} \longrightarrow\!\!\!\!| \; \bar{s}$$

<u>Proof</u>: For $(A,\theta) \in \underline{A}_{\mathbb{G}}$, let $\bar{\varepsilon}(A,\theta) : \bar{r}\bar{s}(A,\theta) \longrightarrow (A,\theta)$ be the composite

$$r\bar{s}(A,\theta) \xrightarrow{rj(A,\theta)} rsA \xrightarrow{\varepsilon A} A \ .$$

This is the unique $\underline{A}$-morphism such that for any coalgebra morphism $\varphi : IM \longrightarrow A$, the diagram

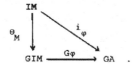

$$\begin{array}{ccc} IM & & \\ {\scriptstyle i_\varphi}\downarrow & \overset{\varphi}{\searrow} & \\ r\bar{s}(A,\theta) & \xrightarrow{\bar{\varepsilon}(A,\theta)} & A \end{array}$$

commutes. Using this, a simple calculation shows that $\bar{\varepsilon}(A,\theta)$ is a morphism of coalgebras. It is clearly natural.

For $F \in (\underline{M}^*,\mathfrak{S})$, naturality of η gives the diagram

$$\begin{array}{ccc} F & \xrightarrow{\eta F} & srF \\ {\scriptstyle \eta F}\downarrow & & \downarrow{\scriptstyle \eta srF} \\ srF & \xrightarrow{sr\eta F} & srsrF \end{array}$$

Then, by 3.1 there is a unique $\bar{\eta}F : F \longrightarrow \bar{s}\bar{r}F$ making

$$\begin{array}{ccccc} & & F & & \\ & \overset{\bar{\eta}F}{\swarrow} & \downarrow{\scriptstyle \eta F} & & \\ \bar{s}\bar{r}F & \xrightarrow{j\bar{r}F} & srF & \overset{sr\eta F}{\underset{\eta srF}{\rightrightarrows}} & srsrF \end{array}$$

commute - i.e. for $M \in \underline{M}$ and $x \in FM$, each $i_x : IM \longrightarrow rF$ is a coalgebra morphism. $\bar{\eta}F$ is trivially natural in F.

Now consider the two composites

$$\bar{s} \xrightarrow{\bar{\eta}\bar{s}} \bar{s}\bar{r}\bar{s} \xrightarrow{\bar{s}\bar{\varepsilon}} \bar{s}$$

and
$$\bar{r} \xrightarrow{\bar{r}\bar{\eta}} \bar{r}\bar{s}\bar{r} \xrightarrow{\bar{\varepsilon}\bar{r}} \bar{r} .$$

Suppose $M \in \underline{M}$, $(A,\theta) \in \underline{A}_{\mathfrak{G}}$, and $\varphi : IM \longrightarrow A$ is a morphism of coalgebras. Then $(\bar{\eta}\bar{s}(A,\theta).M)(\varphi) = i_\varphi$, and $(\bar{s}\bar{\varepsilon}(A,\theta).M)(i_\varphi) = \bar{\varepsilon}(A,\theta)\cdot i_\varphi = \varphi$. Thus, $\bar{s}\bar{\varepsilon}\cdot\bar{\eta}\bar{s} = 1_{\bar{s}}$. Also, if $F \in (\underline{M}^*,\mathfrak{S})$ and $x \in FM$, then $\bar{r}\bar{\eta}F\cdot i_x = i_{i_x}$, and $\bar{\varepsilon}\bar{r}F\cdot i_{i_x} = i_x$. Hence $\bar{\varepsilon}\bar{r}\cdot\bar{r}\bar{\eta} = 1_{\bar{r}}$, and we have $(\bar{\varepsilon},\bar{\eta}) : \bar{r} \longrightarrow\!\!\!\!\!\!\dashv\; \bar{s}$.

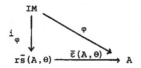

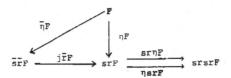

Having $(\bar{\varepsilon},\bar{\eta}) : \bar{r} \dashv \bar{s}$ gives a cotriple $\bar{\mathbb{G}} = (\overline{r}\overline{s},\bar{\varepsilon},\overline{r\eta s})$ on $\underline{A}_{\mathbb{G}}$. We call $\bar{\mathbb{G}}$ the <u>lifted</u> <u>cotriple</u>. In what follows, we shall be interested in those $(A,\theta) \in \underline{A}_{\mathbb{G}}$ for which

$$\bar{\varepsilon}(A,\theta) : \bar{G}(A,\theta) \xrightarrow{\ \sim\ } (A,\theta) .$$

In fact, these (A,θ) will turn out to be precisely the $\underline{M}$-objects of § 2. Thus, we prove a theorem giving necessary and sufficient conditions for this to happen. A more general form of this has been proved by Jon Beck (unpublished).

<u>Theorem 3.3</u>

For $(A,\theta) \in \underline{A}_{\mathbb{G}}$,

$$\bar{\varepsilon}(A,\theta) : \bar{G}(A,\theta) \xrightarrow{\ \sim\ } (A,\theta)$$

iff

$$r\bar{s}(A,\theta) \xrightarrow{rj(A,\theta)} GA \underset{\delta A}{\overset{G\theta}{\rightrightarrows}} G^2A$$

is an equalizer diagram.

<u>Proof</u>: We remark first that for any (A,θ),

$$A \xrightarrow{\ \theta\ } GA \underset{\delta A}{\overset{G\theta}{\rightrightarrows}} G^2A$$

is an equalizer diagram. In fact, θ equalizes by definition, and if we have any $f : A' \longrightarrow GA$ that equalizes, then $\varepsilon A \cdot f : A' \longrightarrow A$ and the following diagram shows that $\theta \cdot (\varepsilon A \cdot f) = f$.

$$\begin{array}{ccccc} A' & \xrightarrow{\ f\ } & GA & \underset{\delta A}{\overset{G\theta}{\rightrightarrows}} & G^2A \\ & & {\scriptstyle \varepsilon A}\downarrow & & \downarrow{\scriptstyle \varepsilon GA} \\ & & A & \xrightarrow{\ \theta\ } & GA \end{array}$$

$\varepsilon A \cdot f$ is unique, since θ is monic. By this remark, and the fact that $L\bar{\varepsilon}(A,\theta) = \varepsilon A \cdot rj(A,\theta)$, we see that we have

$$\begin{array}{ccc} & & r\cdot\bar{s}(A,\theta) \\ & \swarrow{\scriptstyle L\bar{\varepsilon}(A,\theta)} & \downarrow{\scriptstyle rj(A,\theta)} \\ A & \xrightarrow{\ \theta\ } & GA \end{array} \quad .$$

Thus,

$$r\bar{s}(A,\theta) \xrightarrow{rj(A,\theta)} GA \underset{sA}{\overset{G\theta}{\rightrightarrows}} G^2A$$

is an equalizer diagram (in $\underline{A}$) iff $L\bar{\varepsilon}(A,\theta)$ is an isomorphism, iff $\bar{\varepsilon}(A,\theta)$ is an isomorphism, since L clearly reflects isomorphisms.

Note that in 3.3 we have shown that the adjointness $(\bar{\varepsilon},\bar{\eta}) : \bar{r} \longrightarrow\!\!\!\!\!| \ \bar{s}$ exhibits $\underline{A}_{\mathbb{G}}$ as a coretract of $(\underline{M}^*,\mathfrak{S})$ iff for all (A,θ) , r preserves the equalizer

$$\bar{s}(A,\theta) \xrightarrow{j(A,\theta)} sA \underset{\eta sA}{\overset{s\theta}{\rightrightarrows}} srsA \quad .$$

In the next section we shall give an additional condition that is necessary and sufficient for the adjointness to be an equivalence of categories. Note also that to verify that

$$r\bar{s}(A,\theta) \xrightarrow{rj(A,\theta)} GA \underset{sA}{\overset{G\theta}{\rightrightarrows}} G^2A$$

is an equalizer diagram, it suffices to show that $rj(A,\theta)$ is monic, and there is a factorization

$$\begin{array}{ccc} & & A \\ & \swarrow{\scriptstyle \bar{\theta}} & \downarrow{\scriptstyle \theta} \\ r\bar{s}(A,\theta) & \xrightarrow{\ rj(A,\theta)\ } & GA \end{array}$$

of θ in $\underline{A}$. In the presence of a good underlying set functor $U : \underline{A} \longrightarrow \mathfrak{S}$, we shall give a simple sufficient condition for this in § 4 .

Now we proceed to explain the connection between these distinguished coalgebras and the $\underline{M}$-objects of § 2. It is clear that if (A,θ) is a $\mathbb{G}$-coalgebra for which

$$\bar{\varepsilon}(A,\theta) : \bar{G}(A,\theta) \xrightarrow[\sim]{} (A,\theta) \quad ,$$

then A is an $\underline{M}$-object by means of the atlas

$$\bar{s}(A,\theta) \xrightarrow{\;j(A,\theta)\;} sA \; .$$

Conversely, let A be an $\underline{M}$-object with atlas

$$\mathfrak{G} \xrightarrow{\;j\;} sA \; ,$$

and define $\theta : A \longrightarrow GA$ to be the composite

$$A \xrightarrow{\;e^{-1}\;} r\mathfrak{G} \xrightarrow{\;rj\;} rsA \; .$$

Proposition 3.4

θ is a coalgebra structure for A .

Proof: The diagram

$$\begin{array}{ccc} A & \xrightarrow{\;\theta\;} & GA \\ & {\scriptstyle 1_A}\searrow & \downarrow{\scriptstyle \varepsilon A} \\ & & A \end{array}$$

commutes trivially, since the composite

$$r\mathfrak{G} \xrightarrow{\;rj\;} rsA \xrightarrow{\;\varepsilon A\;} A$$

is e . Consider the diagram

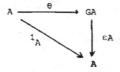

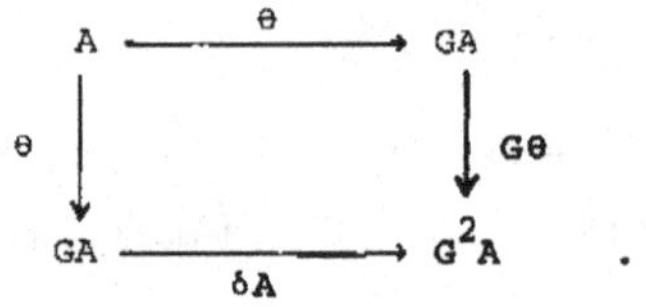

That is, the diagram

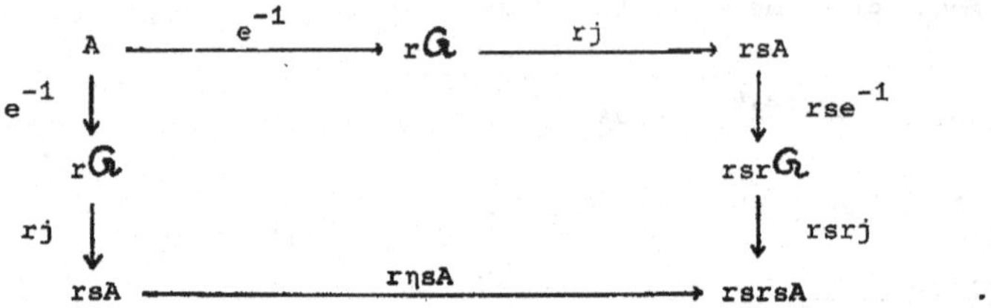

Now for any $M \in \underline{M}$ and $\varphi \in \mathfrak{G}M$, we have the diagram

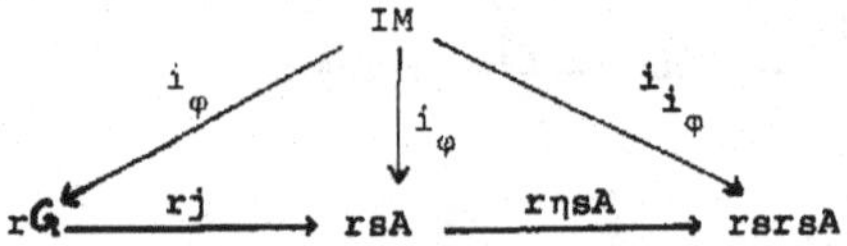

and

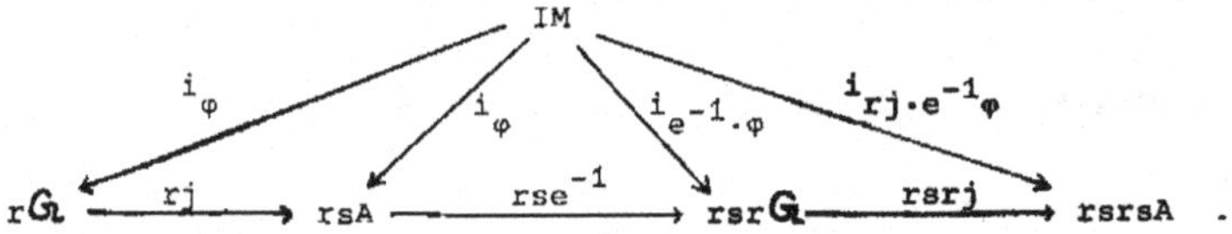

Thus, to complete the proof, it is enough to show that for any $\varphi \in \mathfrak{G}M$

$$i_\varphi = rj \cdot e^{-1} \cdot \varphi = \theta \cdot \varphi \quad .$$

For this, consider the diagram

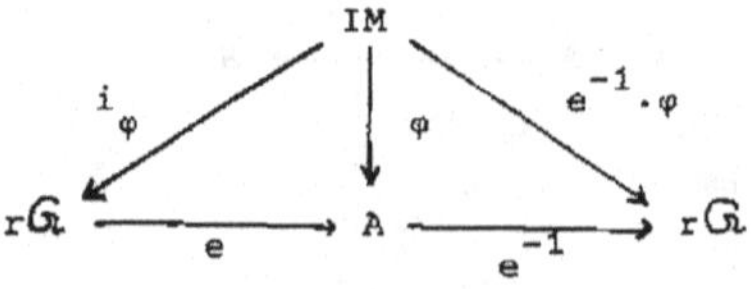

Thus, $i_\varphi = e^{-1}\cdot\varphi$ and hence

$$i_\varphi = rj\cdot i_\varphi = rj\cdot e^{-1}\cdot\varphi = \theta\cdot\varphi \ .$$

Note that in 3.4 we have not only shown that θ is a coalgebra structure, but also that with respect to this θ, each $\varphi \in \mathfrak{G}M$ is a coalgebra morphism. That is, there is a factorization

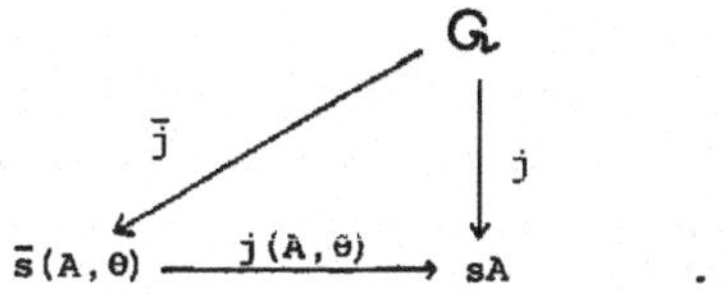

But now, if we apply r to this diagram and compose with e^{-1} we obtain

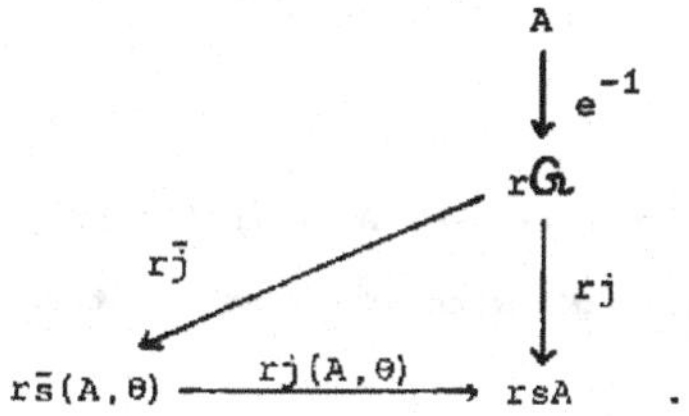

Hence, θ factors through $rj(A,\theta)$, so that if $rj(A,\theta)$ is monic it follows by the remark made after 3.3 that

$$\bar{\varepsilon}(A,\theta) : \bar{G}(A,\theta) \xrightarrow{\sim} (A,\theta) \ .$$

Therefore, under the assumption that $rj(A,\theta)$ is monic for all (A,θ), the $\underline{M}$-objects are precisely these distinguished coalgebras. We shall see in the examples that this is a very mild assumption in general. In fact, we will usually have $\bar{\varepsilon}(A,\theta) : \bar{G}(A,\theta) \xrightarrow{\sim} (A,\theta)$ for <u>all</u> (A,θ), so that the class of $\underline{M}$-objects is the class of $A \in \underline{A}$ admitting a $\mathbf{G}$-coalgebra structure $\theta : A \longrightarrow GA$. When this is the case, we define a <u>morphism</u> of $\underline{M}$-objects to be a coalgebra morphism with respect to the induced coalgebra structures.

<u>Remark</u>. Given an $\underline{M}$-object $A \in \underline{A}$ with atlas $\mathfrak{G}$, $\bar{s}(A,\theta)$ is a <u>maximal</u> atlas for A consisting of all $\underline{A}$-morphisms compatible with the morphisms of $\mathfrak{G}$. To see this intuitively, suppose $\underline{A}$ has a faithful underlying set functor that preserves colimits- i.e. we shall act as if the objects of $\underline{A}$ have elements, and the elements of a colimit are equivalence classes as in § 2. Then for $\psi : IM \longrightarrow A$ to be a coalgebra morphism with respect to the above θ, we must have the diagram

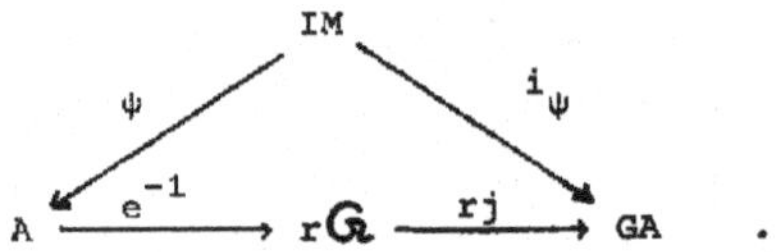

The effect of e^{-1} is the following: for $a \in A$, pick $M' \in \underline{M}$ and $\varphi \in \mathfrak{G}M'$ so that there exists $m' \in IM'$ with $\varphi m' = a$. This can be done since e is epic. Then,

$$e^{-1}a = |\varphi, m'| \in r\mathfrak{G} .$$

Thus, $\bar{s}(A,\theta)M$ consists of all morphisms $\psi : IM \longrightarrow A$ with the property that if $\varphi \in \mathfrak{G}M'$, and there is $m' \in IM'$ and $m \in IM$ with $\psi m = \varphi m'$, then

$$|\psi, m| = |\varphi, m'|$$

in GA. i.e. all $\psi : IM \longrightarrow A$ compatible with the morphisms from $\mathfrak{G}$. (Here, this is the <u>definition</u> of compatibility. For the connection with the usual definition, see 4.1.)

4. The equivalence theorem

In § 3, we gave necessary and sufficient conditions for

$$\bar{\varepsilon}(A,\theta) : \bar{G}(A,\theta) \xrightarrow[\sim]{} (A,\theta)$$

in terms of the preservation of a certain equalizer. Here, we will first investigate this more closely in the presence of an underlying set functor on $\underline{A}$, and then complete the equivalence theorem by giving necessary and sufficient conditions for

$$\bar{\eta} : 1 \xrightarrow[\sim]{} \bar{s}\bar{r} \quad .$$

So, assume we have a functor $U : \underline{A} \longrightarrow \mathbf{S}$, and consider the following condition: given a pair of morphisms

$$\begin{array}{ccc} & & IM_1 \\ & & \Big\downarrow I\alpha_1 \\ IM_2 & \xrightarrow[I\alpha_2]{} & IM \end{array}$$

in $\underline{A}$, and elements $m_i \in UIM_i$, $i = 1,2$, such that

$$UI\alpha_1(m_1) \;=\; UI\alpha_2(m_2)$$

then there are morphisms $\beta_i : M_o \longrightarrow M_i$, $i = 1,2$, in $\underline{M}$ and $m_o \in UIM_o$, satisfying

$$\begin{array}{ccc} M_o & \xrightarrow{\beta_1} & M_1 \\ {\scriptstyle \beta_2}\Big\downarrow & & \Big\downarrow{\scriptstyle \alpha_1} \\ M_2 & \xrightarrow{\alpha_2} & M \end{array}$$

and $UI\beta_i(m_o) = m_i$, $i = 1,2$. We call this condition (a).

<u>Lemma 4.1</u>

Assume $U : \underline{A} \longrightarrow \mathbf{S}$ is colimit preserving, and $I : \underline{M} \longrightarrow \underline{A}$ satisfies (a). Let $F \in (\underline{M}^*, \mathbf{S})$, and suppose that in UrF

$$|M_1, x_1, m_1| \;=\; |M_2, x_2, m_2| \quad .$$

Then there are morphisms $\alpha_i : M \longrightarrow M_i$, $i = 1,2$, in $\underline{M}$ and $m \in UIM$ such that

$$UI\alpha_i(m) = m_i \;, \quad i = 1,2 \quad \text{and} \quad F\alpha_1(x_1) = F\alpha_2(x_2) \; .$$

<u>Proof</u>: Recall from § 2 that UrF can be represented as the set of equivalence classes of triples (M,x,m) for $(M,x) \in (Y,F)$ and $m \in UIM$ under the equivalence relation generated by the relation

$$(M,x,m) \sim (M',x',m')$$

iff there is $\alpha : (M,x) \longrightarrow (M',x')$ in (Y,F) such that $UI\alpha(m) = m'$.

Let $\equiv_o$ be the relation given by the <u>conclusion</u> of the lemma. That is,

$$(M_1,x_1,m_1) \equiv_o (M_2,x_2,m_2)$$

iff there is (M,x,m) such that

$$(M,x,m) \sim (M_1,x_1,m_1)$$

and

$$(M,x,m) \sim (M_2,x_2,m_2) \quad .$$

Now if we show $\equiv_o$ is an equivalence relation containing $\sim$ we are done, since clearly any equivalence relation containing $\sim$ contains $\equiv_o$. Obviously, $\equiv_o$ is reflexive, symmetric, and contains $\sim$. We are left with transitivity, so assume

$$(M_1,x_1,m_1) \equiv_o (M_2,x_2,m_2) \equiv_o (M_3,x_3,m_3) \quad .$$

Then we have

$$(M',x',m') \sim (M_1,x_1,m_1)$$

$$(M',x',m') \sim (M_2,x_2,m_2)$$

and

$$(M'',x'',m'') \sim (M_2,x_2,m_2)$$

$$(M'',x'',m'') \sim (M_3,x_3,m_3) \quad ,$$

so there is a string of morphisms

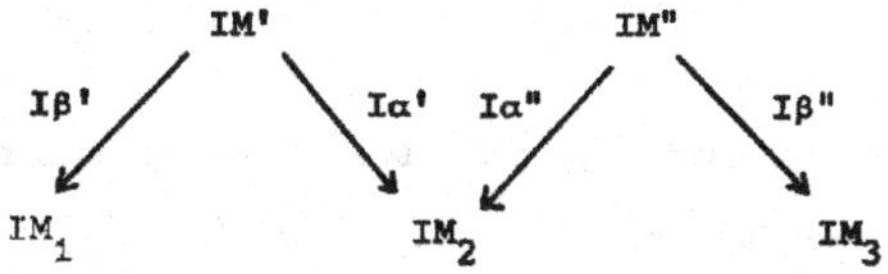

such that

$$F\beta'(x_1) = x' = F\alpha'(x_2)$$

$$F\alpha''(x_2) = x'' = F\beta''(x_3)$$

and
$$UI\beta'(m') = m_1$$
$$UI\alpha'(m') = m_2$$
$$UI\alpha''(m'') = m_2$$
$$UI\beta''(m'') = m_3 \quad .$$

By (a) we can find a diagram

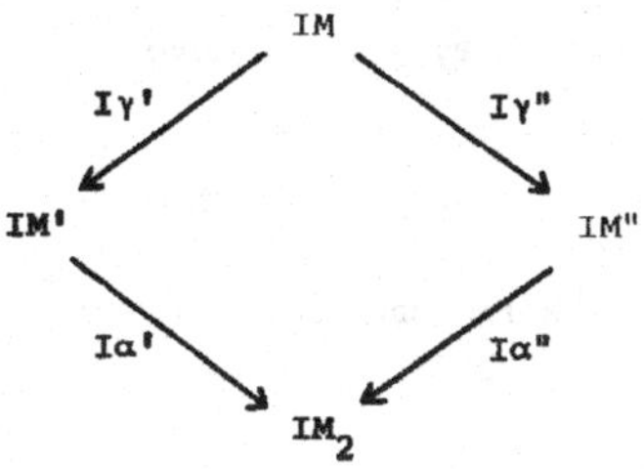

and an $m \in UIM$ such that $UI\gamma'(m) = m'$, $UI\gamma''(m) = m''$.

Let $x = F\gamma'(x') = F\gamma''(x'')$. Then

$$(M,x,m) \sim (M_1,x_1,m_1)$$

$$(M,x,m) \sim (M_3,x_3,m_3)$$

so $(M_1,x_1,m_1) \equiv_o (M_3,x_3,m_3)$.

Corollary 4.2

If U is faithful and colimit preserving, and $I : \underline{M} \longrightarrow \underline{A}$ satisfies (a), then $r : (\underline{M}^*, \mathfrak{S}) \longrightarrow \underline{A}$ preserves monomorphisms.

Proof: Let $F', F \in (\underline{M}^*, \mathfrak{S})$ and let

$$j : F' \longrightarrow F$$

be a monomorphism. Since U is faithful, it reflects monomorphisms, so it is only necessary to check that

$$Urj : UrF' \longrightarrow UrF$$

is monic. So suppose $|x,m|$ and $|x',m'|$ are elements of UrF' such that

$$Urj(|x,m|) \;=\; Urj(|x',m'|)$$

i.e. $$|jM(x),m| \;=\; |jM'(x'),m'|$$

(Note we have dropped the model in the notation.) By 4.1, we have

$$IM \xleftarrow{\;I\alpha\;} IM_o \xrightarrow{\;I\beta\;} IM'$$

with $m_o \in UIM_o$ such that $F\alpha(jM(x)) = F\beta(jM'(x'))$ and $UI\alpha(m_o) = m$, $UI\beta(m_o) = m'$. By naturality we have the diagram

$$\begin{array}{ccccc} F'M & \xrightarrow{F'\alpha} & F'M_o & \xleftarrow{F'\beta} & F'M' \\ \downarrow{\scriptstyle jM} & & \downarrow{\scriptstyle jM_o} & & \downarrow{\scriptstyle jM'} \\ FM & \xrightarrow{F\alpha} & FM_o & \xleftarrow{F\beta} & FM' \end{array} ,$$

and jM_o monic gives $F'\alpha(x) = F'\beta(x')$. But then $|x,m| = |x',m'|$ in UrF' , so Urj is monic.

Corollary 4.3

If U reflects equalizers and preserves colimits, and $I : \underline{M} \longrightarrow \underline{A}$ satisfies (a), then

$$\bar{\varepsilon}(A,\theta) : \bar{G}(A,\theta) \xrightarrow[\sim]{} (A,\theta)$$

for all $(A,\theta) \in \underline{A}_{\mathbb{G}}$.

Proof: We have to show, by 3.3, that for each (A,θ) , the diagram

$$r\bar{s}(A,\theta) \xrightarrow{rj(A,\theta)} rsA \underset{r\eta sA}{\overset{rs\theta}{\rightrightarrows}} (rs)^2A$$

is an equalizer diagram. Since U reflects equalizers, it is enough to show that after applying U we obtain an equalizer diagram in $\mathcal{S}$. Using (a) as in 4.2, $Urj(A,\theta)$ is monic. Therefore, we must show that its image - the set of all $|\varphi,m|$ such that $(\varphi,m) \equiv (\psi,m')$ for some coalgebra morphism ψ - is exactly the set on which $rs\theta$ and $r\eta sA$ agree. The image is clearly contained in this set, so suppose $|\varphi,m|$ is an element of rsA such that

$$rs\theta(|\varphi,m|) = r\eta sA(|\varphi,m|)$$

i.e. $$|\theta\cdot\varphi,m| = |i_{\varphi},m|$$

where $\varphi : IM \longrightarrow A$ and $m \in UIM$. Then by 4.1 we have a diagram

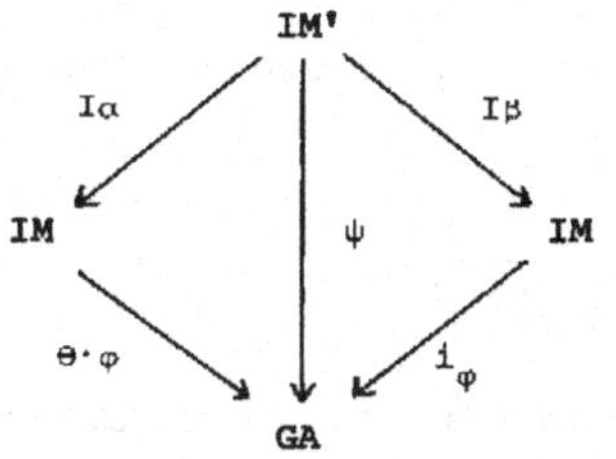

and an element $m' \in UIM'$ such that $UI\alpha(m') = m = UI\beta(m')$. Composing with $\varepsilon A : GA \longrightarrow A$, we obtain $\varphi\cdot I\alpha = \varphi\cdot I\beta$. Call the common value $\varphi' : IM' \longrightarrow A$

Recalling that $i_\varphi \cdot I\beta = i_{\varphi \cdot I\beta}$, the above diagram gives

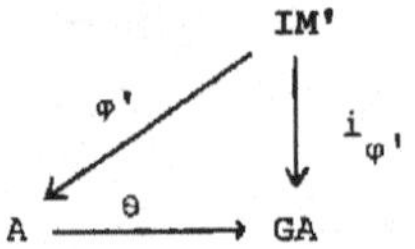

so that φ' is a morphism of coalgebras. Furthermore,

$$|\varphi',m'| = |\varphi \cdot I\alpha,m'| = |\varphi,UI\alpha(m')| = |\varphi,m| \quad ,$$

and hence $|\varphi,m|$ is in the image of $Urj(A,\theta)$.

<u>Remark</u>: There is a condition (b) on $I : \underline{M} \longrightarrow \underline{A}$, similar in nature to (a), which together with the assumption of 4.3 implies that r preserves <u>all</u> equalizers. We do not need this stronger result in what follows, however, and hence we omit a discussion of it here.

Also, some remarks are in order concerning the use of 4.2 and 4.3. In practice, 4.2 is useful, and 4.3 is not. That is, our underlying functors are often faithful and colimit preserving, but then rarely satisfy the stronger property of reflecting equalizers. If they reflect equalizers, they usually do not preserve colimits. For example, the category $\underline{A}$ occuring most often in the examples is <u>Top</u> - the category of topological spaces and continuous maps. Here the obvious underlying set functor preserves colimits since it has an adjoint (the indiscrete topology) and is certainly faithful. It does <u>not</u>, however, reflect equalizers. What one gets from (a) here is that in the diagram

$$r\bar{s}(A,\theta) \xrightarrow{rj(A,\theta)} rsA \underset{r\eta sA}{\overset{rs\theta}{\rightrightarrows}} (rs)^2A$$

$rj(A,\theta)$ is monic, and the underlying set of $r\bar{s}(A,\theta)$ is that of the equalizer of $rs\theta$ and $r\eta sA$. One must still check that $r\bar{s}(A,\theta)$ has the subspace topology from rsA . Even though some such modification is generally necessary in practice, it seemed worthwhile to present the result in a form that would not overly obscure the basic idea involved - hence the assumption of reflecting equalizers in 4.3.

We return now to the general situation and complete the picture by giving necessary and sufficient conditions for

$$\underline{A}_{\mathfrak{G}} \underset{\bar{r}}{\overset{\bar{s}}{\rightrightarrows}} (\underline{M}^*, \mathfrak{S})$$

to be an equivalence of categories.

<u>Theorem 4.4</u>

Suppose $\bar{\varepsilon} : \bar{r}\bar{s} \xrightarrow{\sim} 1$. Then

$$\bar{\eta} : 1 \xrightarrow{\sim} \bar{s}\bar{r}$$

iff r reflects isomorphisms.

<u>Proof</u>: Consider the diagram

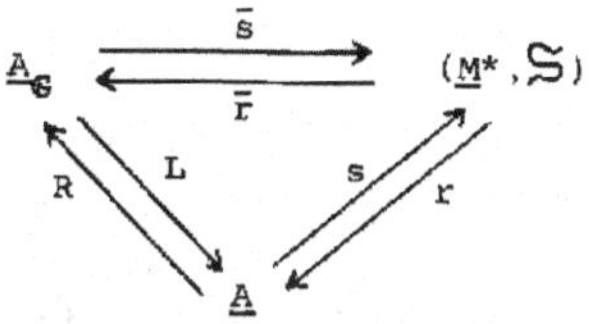

Here $r \dashv s$, $\bar{r} \dashv \bar{s}$, and $L \dashv R$. Furthermore, $L\bar{r} = r$. (Note that this makes $\bar{s}R$ naturally equivalent to s .) So, if $\bar{\varepsilon} : \bar{r}\bar{s} \xrightarrow{\sim} 1$ <u>and</u> $\bar{\eta} : 1 \xrightarrow{\sim} \bar{s}\bar{r}$, the top two categories are equivalent. Therefore, since L reflects isomorphisms, so does r .

On the other hand, suppose r reflects isomorphisms. Let $F \in (\underline{M}^*, \mathfrak{S})$ and consider the diagram

$$\begin{array}{ccccc} \bar{s}\bar{r}F & \xrightarrow{\;j\bar{r}F\;} & srF & \overset{sr\eta F}{\underset{\eta srF}{\rightrightarrows}} & srsrF \\ & {\scriptstyle \bar{\eta}F}\nwarrow & \uparrow{\scriptstyle \eta F} & & \\ & & F & & \end{array}$$

The top row is an equalizer by 3.1, and since $\bar{\varepsilon} : \bar{r}\bar{s} \xrightarrow{\sim} 1$, r preserves it by 3.3. But $r\eta F$ is a coalgebra structure, so that the diagram

$$F \xrightarrow{\;\eta F\;} srF \overset{sr\eta F}{\underset{\eta srF}{\rightrightarrows}} srsrF$$

also becomes an equalizer upon application of r . But then $r\bar{\eta}F$ is an isomorphism, and hence so is $\bar{\eta}F$ since r reflects isomorphisms.

<u>Proposition 4.5</u>

If $\bar{\varepsilon} : \bar{r}\bar{s} \xrightarrow{\sim} 1$, then r reflects isomorphisms iff r is faithful.

<u>Proof</u>: If r reflects isomorphisms, then

$$\underline{A}_{\mathfrak{C}} \underset{\bar{r}}{\overset{\bar{s}}{\rightleftarrows}} (\underline{M}^*, \mathfrak{S})$$

is an equivalence of categories by 4.4, so r is faithful since L is. On the other hand, suppose r is faithful and $\gamma : F' \longrightarrow F$ is a natural transformation such that $r\gamma : rF' \xrightarrow{\sim} rF$. Then $r\gamma$ is epic and monic, and hence so is γ since r is faithful. But then γ is an isomorphism. This direction, of course, is independent of $\bar{\varepsilon} : \bar{r}\bar{s} \xrightarrow{\sim} 1$, and uses only the fact that the domain of r is a category of set valued functors.

We shall apply these theorems now to some particular examples of categories with models.

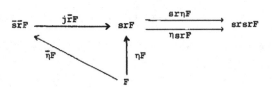

5. Some examples and applications

(1.) Simplicial spaces

Let $\underline{\Delta}$ be the simplicial category. That is, the objects of $\underline{\Delta}$ are sequences $[n] = (0,\dots,n)$ for $n > 0$ an integer, and a morphism $\alpha : [m] \longrightarrow [n]$ is a monotone map. Define

$$I : \underline{\Delta} \longrightarrow \underline{\text{Top}}$$

as follows: $I[n] = \Delta_n$, the standard n-simplex, and if $\alpha : [m] \longrightarrow [n]$, then $I\alpha = \Delta_\alpha : \Delta_m \longrightarrow \Delta_n$ is the affine map determined by $\Delta_\alpha(e_i) = e_{\alpha(i)}$ where the e_i are the vertices of Δ_m. Then in the standard, by now, diagram

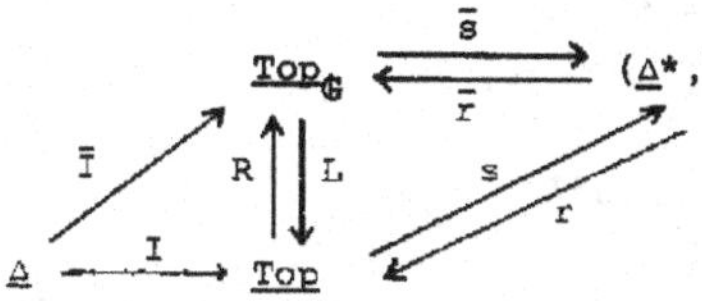

$(\underline{\Delta}^*, \mathfrak{S})$ is the category of simplicial sets, s is the usual singular functor, and r is the geometric realization of Milnor [6]. The underlying set functor of $\underline{\text{Top}}$ is the usual one, and we omit it from the notation.

We will call a $\underline{\Delta}$-object a simplicial space, and we will show that the category of simplicial spaces is equivalent to the category of simplicial sets. We first verify condition (a) of § 4.

Let, for $0 < i < n$

$$\varepsilon^i : [n-1] \longrightarrow [n]$$

be the morphism defined by

$$\varepsilon^i(j) = j \qquad \text{for } j < i$$

$$\varepsilon^i(j) = j+1 \qquad \text{for } j > i \ ,$$

and

$$\eta^i : [n+1] \longrightarrow [n]$$

be the morphism defined by

$$\eta^i(j) = j \qquad \text{for } j \leq i$$

$$\eta^i(j) = j-1 \qquad \text{for } j > i \quad .$$

These morphisms satisfy the following well-known system of identities:

$$\varepsilon^j\varepsilon^i = \varepsilon^i\varepsilon^{j-1} \qquad i < j$$

$$\eta^j\eta^i = \eta^i\eta^{j+1} \qquad i \leq j$$

$$\eta^j\varepsilon^i = \begin{cases} \varepsilon^i\eta^{j-1} & i < j \\ 1 & i = j,\ j+1 \\ \varepsilon^{i-1}\eta^j & i > j+1 \end{cases} .$$

As a result of these, any morphism $\alpha : [m] \longrightarrow [n]$ in $\underline{\Delta}$ may be written uniquely in the form

$$\alpha = \varepsilon^{i_s}\varepsilon^{i_{s-1}} \ldots \varepsilon^{i_1}\eta^{j_t}\eta^{j_{t-1}} \ldots \eta^{j_1}$$

where $\quad n \geq i_s > i_{s-1} > \ldots > i_1 \geq 0$

and $\quad m > j_1 > j_2 > \ldots > j_t \geq 0 \quad .$

The j's are those $j \in [m]$ such that $\alpha(j) = \alpha(j+1)$, and the i's are those $i \in [n]$ such that $i \notin \text{image } \alpha$.

To establish (a), we first settle various special cases involving the ε^i and η^j , and then use the above factorization. We express points $s \in \Delta_n$ by their barycentric coordinates - i.e. $s = (s_o,\ldots,s_n)$ where $0 \leq s_i \leq 1$ and $\sum s_i = 1$.

<u>Case</u> (i.) Consider

$$\begin{array}{ccc} & & \Delta_{n-1} \\ & & \downarrow{\scriptstyle \Delta_{\varepsilon^i}} \\ \Delta_{n-1} & \xrightarrow[\Delta_{\varepsilon^j}]{} & \Delta_n \end{array}$$

together with $s,t \in \Delta_{n-1}$ such that $\Delta_{\varepsilon^i}(s) = \Delta_{\varepsilon^j}(t)$. $i = j$ is trivial, since Δ_{ε^i} is monic, so suppose, say, $i < j$. Then, if $s = (s_o,\ldots,s_n)$ and $t = (t_o,\ldots,t_n)$ we have

$$\Delta_{\varepsilon^i}(s) = (s_o,\ldots,s_{i-1},0,s_i,\ldots,s_{n-1})$$

$$\Delta_{\varepsilon^j}(t) = (t_o,\ldots,t_{j-1},0,t_j,\ldots,t_{n-1}) .$$

Since $i < j$, $t_i = 0$, $s_{j-1} = 0$, and the remaining coordinates are equal - with the appropriate shift in indexing unless $i = j-1$. We have $\varepsilon^j\varepsilon^i = \varepsilon^i\varepsilon^{j-1}$, hence

$$\begin{array}{ccc} \Delta_{n-2} & \xrightarrow{\Delta_{\varepsilon^{j-1}}} & \Delta_{n-1} \\ {\scriptstyle \Delta_{\varepsilon^i}}\downarrow & & \downarrow{\scriptstyle \Delta_{\varepsilon^i}} \\ \Delta_{n-1} & \xrightarrow[\Delta_{\varepsilon^j}]{} & \Delta_n \end{array}$$

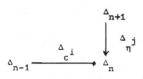

and the point of Δ_{n-2} with the common coordinates hits both s and t .

<u>Case</u> (ii.) Consider

$$\begin{array}{ccc} & & \Delta_{n+1} \\ & & \downarrow{\scriptstyle \Delta_{\eta^j}} \\ \Delta_{n-1} & \xrightarrow{\Delta_{\varepsilon^i}} & \Delta_n \end{array}$$

together with $t \in \Delta_{n+1}$ and $s \in \Delta_{n-1}$ such that $\Delta_{\eta^j}(t) = \Delta_{\varepsilon^i}(s)$. If $t = (t_o,\ldots,t_{n+1})$ and $s = (s_o,\ldots,s_{n-1})$ then

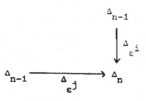

$$\Delta_{\eta^j}(t) = (t_o,\dots,t_{j-1},t_j+t_{j+1},t_{j+2},\dots,t_{n+1})$$

$$\Delta_{\varepsilon^i}(s) = (s_o,\dots,s_{i-1}\ 0,s_i,\dots,s_{n-1}) \ .$$

If $i < j$, we have $t_i = 0$, $s_{j-1} = t_j+t_{j+1}$, $\eta^j\varepsilon^i = \varepsilon^i\eta^{j-1}$, and in

$$\begin{array}{ccc} \Delta_n & \xrightarrow{\Delta_{\varepsilon^i}} & \Delta_{n+1} \\ {\scriptstyle \Delta_{\eta^{j-1}}}\downarrow & & \downarrow{\scriptstyle \Delta_{\eta^j}} \\ \Delta_{n-1} & \xrightarrow[\Delta_{\varepsilon^i}]{} & \Delta_n \end{array}$$

the point $(s_o,\dots,s_{j-2},t_j,t_{j+1},s_j,\dots,s_{n-1})$ of Δ_n hits both s and t . If $i = j$, we have $t_j = t_{j+1} = 0$ and

$$\begin{array}{ccc} \Delta_{n-1} & \xrightarrow{\Delta_{\varepsilon^{i+1}\varepsilon^i}} & \Delta_{n+1} \\ {\scriptstyle 1}\downarrow & & \downarrow{\scriptstyle \Delta_{\eta^i}} \\ \Delta_{n-1} & \xrightarrow[\Delta_{\varepsilon^i}]{} & \Delta_n \end{array}$$

works (since $\eta^i\varepsilon^{i+1} = 1$). If $i > j+1$, we use the relation $\eta^j\varepsilon^{i+1} = \varepsilon^i\eta^j$.

<u>Case</u> (iii.) Consider

$$\begin{array}{ccc} & & \Delta_{n+1} \\ & & \downarrow{\scriptstyle \Delta_{\eta^i}} \\ \Delta_{n+1} & \xrightarrow{\Delta_{\eta^j}} & \Delta_n \end{array}$$

together with $s,t \in \Delta_{n+1}$ such that $\Delta_{\eta^i}(s) = \Delta_{\eta^j}(t)$. Thus

$$\Delta_{\eta^i}(s) = (s_o,\dots,s_{i-1},s_i+s_{i+1},s_{i+2},\dots,s_{n+1})$$

$$\Delta_{\eta^i}(t) = (t_o,\dots,t_{j-1},t_j+t_{j+1},t_{j+2},\dots,t_{n+1}) \ .$$

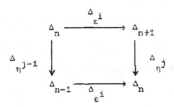

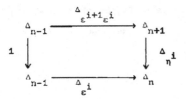

If $i < j$, use $\eta^j\eta^i = \eta^i\eta^{j+1}$. If $i = j$, we have $t_i + t_{i+1} = s_i + s_{i+1}$ and all other coordinates are equal. Suppose, say, $t_i < s_i$. Thus $\eta^i\eta^i = \eta^i\eta^{i+1}$ so

$$\begin{array}{ccc} \Delta_{n+2} & \xrightarrow{\Delta_{\eta^i}} & \Delta_{n+1} \\ {\scriptstyle \Delta_{\eta^{i+1}}}\downarrow & & \downarrow{\scriptstyle \Delta_{\eta^i}} \\ \Delta_{n+1} & \xrightarrow[\Delta_{\eta^i}]{} & \Delta_n \end{array}$$

works with $(s_o,\ldots,s_{i-1},t_i,s_i-t_i,s_{i+1},\ldots,s_{n+1})$ hitting s and t.

For the general case

$$\begin{array}{ccc} & & \Delta_{m_1} \\ & & \downarrow{\scriptstyle \Delta_{\alpha_1}} \\ \Delta_{m_2} & \xrightarrow{\Delta_{\alpha_2}} & \Delta_m \end{array}$$

with $s \in \Delta_{m_1}$, $t \in \Delta_{m_2}$ such that $\Delta_{\alpha_1}(s) = \Delta_{\alpha_2}(t)$, simply write α_1 and α_2 as composites of ε^i's and η^j's, and use (i)-(iii) repeatedly. Except for the case $i = j$ of (ii.), one obtains only blocks involving a single ε^i or η^j. This case cannot cause any trouble however, since the factorizations of α_1 and α_2 are finite. Thus $I : \underline{\Delta} \longrightarrow \underline{\mathrm{Top}}$ satisfies (a).

<u>Lemma 5.1</u>

Let $j : K \longrightarrow K'$ be a monomorphism of simplicial sets. If $\sigma \in K_n$ is non-degenerate, then so is $j\sigma \in K'_n$.

<u>Proof</u>: Suppose $j\sigma = s\tau$, where s is an iterated degeneracy operator and $\tau \in K_m$ for $m < n$. Then there is an iterated face operator $d : L_n \longrightarrow L_m$ such that ds = identity. Therefore,

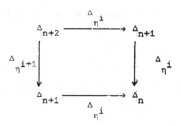

$$jd\sigma = dj\sigma = ds\tau = \tau ,$$

and hence $jsd\sigma = j\sigma$, so that $\sigma = sd\sigma$ is degenerate. In the next lemma we reprove, since condition (a) makes it so easy, a basic lemma of Milnor [6].

Lemma 5.2

If L is a simplicial set, then every element $x \in rL$ has a unique representation of the form

$$x = |\sigma,t|$$

where $\sigma \in L_n$ is non-degenerate and $t \in \text{int}\Delta_n$.

Proof: For existence, let $x \in rL$. Then $x = |\sigma,t|$ for some $\sigma \in L_m$ and $t \in \Delta_m$. t lies in the interior of some face of Δ_m , so there is an injection $\varepsilon : [m'] \longrightarrow [m]$ such that $t = \Delta_\varepsilon(t')$ with $t' \in \text{int}\Delta_{m'}$. Hence

$$|\sigma,t| = |\sigma,\Delta_\varepsilon(t')| = |L_\varepsilon\sigma,t'| .$$

As is well known, Eilenberg-Zilber [2], any $\tau \in L$ can be represented uniquely in the form $\tau = L_\eta\tau'$ where η is a surjection in $\underline{\Delta}$ and τ' is non-degenerate. Hence we can write

$$L_\varepsilon\sigma = L_\eta\sigma" ,$$

where $\eta : [m'] \longrightarrow [m"]$ is a surjection and $\sigma" \in L_{m"}$ is non-degenerate. Then,

$$|L_\varepsilon\sigma,t'| = |L_\eta\sigma",t'| = |\sigma",\Delta_\eta(t')| .$$

Since $t' \in \text{int}\Delta_{m'}$, $t" = \Delta_\eta(t') \in \text{int}\Delta_{m"}$. Thus,

$$x = |\sigma",t"|$$

provides such a representation.

For uniqueness, suppose

$$|\sigma,t| = |\sigma',t'|$$

in rL, where $\sigma \in L_n$ and $\sigma' \in L_{n'}$ are non-degenerate, and $t \in \text{int}\Delta_n$, $t' \in \text{int}\Delta_{n'}$. By 4.1 there are morphisms $\alpha : [m] \longrightarrow [n]$ and $\alpha' : [m] \longrightarrow [n']$ in $\underline{\Delta}$ together with a point $t_o \in \Delta_m$ such that

$$L_\alpha \sigma = L_{\alpha'} \sigma' \quad \text{and} \quad t = \Delta_\alpha(t_o) , \quad t' = \Delta_{\alpha'}(t_o) .$$

Since t and t' are interior points, α and α' must be surjections. But then by the above mentioned uniqueness of the representation of $L_\alpha \sigma = L_{\alpha'} \sigma'$ we must have $\alpha = \alpha'$ and $\sigma = \sigma'$ and hence the result.

<u>Lemma 5.3</u>

If $j : K \longrightarrow K'$ is a monomorphism of simplicial sets, then

$$rj : rK \longrightarrow rK'$$

is closed.

<u>Proof</u>: Let L be an arbitrary simplicial set. By 5.2, each $x \in rL$ can be written uniquely in the form

$$x = |\sigma,t|$$

for $\sigma \in L_n$ non-degenerate and $t \in \text{int}\Delta_n$. For non-degenerate σ, let

$$\mathring{e}_\sigma = \{|\sigma,t| \; ; \; t \in \text{int}\Delta_n\}$$

and $$e_\sigma = \{|\sigma,t| \; ; \; t \in \Delta_n\} .$$

Then the $e_\sigma (\mathring{e}_\sigma)$ are the closed (open) cells for a CW-decomposition of rL. In particular, $C \subset rL$ is closed iff $C \cap e_\sigma$ is closed in e_σ for all non-degenerate $\sigma \in L$.

Now consider

$$rj : rK \longrightarrow rK' .$$

and let $C \subset rK$ be closed (assume $C \neq \emptyset$). Put $C' = rj(C)$. If $\sigma' \in K'$ is non-degenerate and

$$C' \cap \mathring{e}_{\sigma'} \neq \emptyset ,$$

we claim $\sigma' = j\sigma$ for $\sigma \in K$. In fact, let $x \in C' \cap \mathring{e}_{\sigma'}$. Then $x = |j\sigma, s|$ for $\sigma \in K_m$ non-degenerate and $s \in \mathrm{int}\Delta_m$. Also, $x = |\sigma', t|$ for $t \in \mathrm{int}\Delta_n$. By 5.1 $j\sigma$ is non-degenerate, so by uniqueness, $\sigma' = j\sigma$. In this case, we have

$$C' \cap e_{\sigma'} = C' \cap e_{j\sigma} = rj(C) \cap rj(e_\sigma) = rj(C \cap e_\sigma) .$$

(The last equality since rj is monic, which follows from 4.2, or easily directly from 5.1.) But C is closed in rK, so $C \cap e_\sigma$ is closed, and hence compact, in e_σ. Thus, $C' \cap e_{\sigma'}$ is a compact subset of $e_{\sigma'}$, and therefore closed. Let σ' be an arbitrary non-degenerate element of K', and let $\mathcal{T}$ be the set of faces τ' of σ' such that $C' \cap \mathring{e}_{\tau'} \neq \emptyset$. Then

$$C' \cap e_{\sigma'} = C' \cap (\bigcup_{\tau' \in \mathcal{T}} e_{\tau'}) = \bigcup_{\tau' \in \mathcal{T}} (C' \cap e_{\tau'}) .$$

By the above, $C' \cap e_{\tau'}$ is closed in $e_{\tau'}$, which is closed in e_σ. But $\mathcal{T}$ is finite, so $C' \cap e_{\sigma'}$ is closed in $e_{\sigma'}$, and C' is closed in rK'.

Summing up, we have the following for any $(X,\theta) \in \underline{\mathrm{Top}}_G$. By (a), the underlying set of

$$r\bar{s}(X,\theta) \xrightarrow{rj(X,\theta)} rsX \underset{r\eta sX}{\overset{rs\theta}{\rightrightarrows}} (rs)^2X$$

is that of the equalizer of $rs\theta$ and $r\eta sX$. That is, $rj(X,\theta)$ is monic, and its image is the set of points of rsX on which $rs\theta$ and $r\eta sX$ agree. Furthermore, by 5.3 $rj(X,\theta)$ is closed. Thus, if we identify $r\bar{s}(X,\theta)$ with a subset of rsX by means of $rj(X,\theta)$, then the given topology of $r\bar{s}(X,\theta)$ is the induced topology as a closed subspace of rsX. Thus the above is an equalizer diagram in $\underline{\mathrm{Top}}$, so by 3.3

$$\bar{\varepsilon}(X,\theta) : \bar{G}(X,\theta) \xrightarrow{\approx} (X,\theta) .$$

The desired equivalence of categories follows now from 4.4, 4.5 and the following proposition.

Proposition 5.4

r is faithful.

Proof: Suppose

$$K \underset{\gamma_2}{\overset{\gamma_1}{\rightrightarrows}} L$$

are morphisms of simplicial sets, and

$$r\gamma_1 = r\gamma_2 : rK \longrightarrow rL .$$

Let $\sigma \in K_n$, and write

$$\gamma_1\sigma = s_1\tau_1$$

$$\gamma_2\sigma = s_2\tau_2$$

where $\tau_i \in L_{m_i}$, $m_i < n$, is non-degenerate for $i = 1,2$, and $s_i = L_{\eta_i}$ for $\eta_i : [n] \longrightarrow [m_i]$ an epimorphism in $\underline{\Delta}$, $i = 1,2$. Since $r\gamma_1 = r\gamma_2$, if $t \in \mathrm{int}\Delta_n$ we have

$$|\gamma_1\sigma, t| = |\gamma_2\sigma, t|$$

or

$$|L_{\eta_1}(\tau_1), t| = |L_{\eta_2}(\tau_2), t|$$

or

$$|\tau_1, \Delta_{\eta_1}(t)| = |\tau_2, \Delta_{\eta}\quad)| .$$

But then, since $\Delta_{\eta_1}(t)$ and $\Delta_{\eta_2}(t)$ e interior points,

$$\tau_1 = \tau_2 \quad \text{and} \quad \Delta_{\eta_1}(t) = \Delta_{\eta_2}(t) .$$

Δ_{η_1} and Δ_{η_2} are simplicial, and hence agree on the carrier of t , which is Δ_n .

Thus $\Delta_{\eta_1} = \Delta_{\eta_2}$, and hence $\eta_1 = \eta_2$ (I is faithful), which gives $\gamma_1\sigma = \gamma_2\sigma$.

We describe now in more detail what it means to be a simplicial space. Namely, we claim that an X in Top is a simplicial space iff there exists a familiy $\mathcal{F}$ of continuous maps $\varphi : \Delta_n \longrightarrow X$ (n variable) with the following properties:

(i) $\mathcal{F}$ covers X . That is, for each $x \in X$ there exists $\varphi : \Delta_n \longrightarrow X$ in $\mathcal{F}$ and $t \in \Delta_n$ such that $\varphi t = x$.

(ii) The φ's in $\mathcal{F}$ are compatible. That is, if (φ,ψ) is a pair of morphisms $\varphi : \Delta_n \longrightarrow X$ and $\psi : \Delta_m \longrightarrow X$ in $\mathcal{F}$, and $\varphi t = \psi t'$ for $t \in \Delta_n$ and $t' \in \Delta_m$, then there is a commutative diagram

$$\begin{array}{ccccc} & & \Delta_q & & \\ & \swarrow^{\Delta_\alpha} & & \searrow^{\Delta_\beta} & \\ \Delta_n & & & & \Delta_m \\ & \searrow_{\varphi} & & \swarrow_{\psi} & \\ & & X & & \end{array}$$

together with $t'' \in \Delta_q$ such that $t = \Delta_\alpha(t'')$, $t' = \Delta_\beta(t'')$.

(iii) X has the weak topology with respect to $\mathcal{F}$. That is, $U \subset X$ is open iff for each $\varphi : \Delta_n \longrightarrow X$ in $\mathcal{F}$, $\varphi^{-1}U$ is open in Δ_n .

Well, if X is a simplicial space with atlas

$$\mathfrak{G} \hookrightarrow sX$$

then we claim that any generating set $\mathcal{F}$ of $\mathfrak{G}$ provides a family satisfying (i) - (iii). In fact,

$$e : r\mathfrak{G} \xrightarrow{\sim} X$$

and is given by $e|\varphi,t| = \varphi t$ for $\varphi : \Delta_n \longrightarrow X$ in $\mathfrak{G}_n$ and $t \in \Delta_n$. Let $\mathcal{F}$ be a generating set for $\mathfrak{G}$. Since e is surjective, if $x \in X$ there is

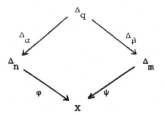

$\varphi : \Delta_n \longrightarrow X$ in $\mathcal{G}$ and $t \in \Delta_n$ such that $x = e|\varphi,t| = \varphi t$. But $\varphi = \psi \cdot \Delta_\alpha$ for some $\alpha : [n] \longrightarrow [m]$ in $\underline{\Delta}$ and ψ in $\mathcal{F}$. Thus, $x = \psi(\Delta_\alpha(t))$ and $\mathcal{F}$ satisfies (i). (ii) follows from the injectivity of e and condition (a). For (iii), e is a homeomorphism, $r\mathcal{G}$ has the weak topology with respect to the canonical maps

$$i_\varphi : \Delta_n \longrightarrow r\mathcal{G}$$

for $\varphi \in \mathcal{G}_n$, and $ei_\varphi = \varphi$. Thus, X has the weak topology with respect to the family of __all__ maps φ in $\mathcal{G}$. Now suppose $U \subset X$ has the property that for all $\psi : \Delta_m \longrightarrow X$ in $\mathcal{F}$, $\psi^{-1}U$ is open in Δ_m . Then if φ is any map in $\mathcal{G}$, we can write $\varphi = \psi\Delta_\alpha$ as above, so that $\varphi^{-1}U = \Delta_\alpha^{-1}(\psi^{-1}U)$ is open in Δ_n . Thus, U is open in X , and $\mathcal{F}$ satisfies (iii). The most interesting generating family in $\mathcal{G}$ consists of the non-degenerate elements of $\mathcal{G}$. Namely, we know then that every point in $r\mathcal{G}$ has a __unique__ representation of the form $|\varphi,t|$ for φ a non-degenerate element of $\mathcal{G}_n$, and $t \in \text{int}\Delta_n$. Thus, this $\mathcal{F}$ satisfies the stronger condition:

(i') For each $x \in X$ there is a __unique__ $\varphi : \Delta_n \longrightarrow X$ and a __unique__ $t \in \text{int}\Delta_n$ such that $x = \varphi t$.

Therefore, $\mathcal{F}$ provides a family of characteristic maps for a CW-decomposition of X .

On the other hand, suppose for $X \in \underline{\text{Top}}$ that there exists a family $\mathcal{F}$ satisfying (i) - (iii). Let $\mathcal{F}$ generate a pre-atlas, and consider

$$e : r\mathcal{G} \longrightarrow X .$$

By (i), e is surjective. Suppose

$$\psi_1 : \Delta_{n_1} \longrightarrow X$$

and
$$\psi_2 : \Delta_{n_2} \longrightarrow X$$

are in $\mathcal{G}$, $t_i \in \Delta_{n_i}$, $i = 1,2$, and

$$e|\psi_1,t_1| = e|\psi_2,t_2|$$

i.e. $\psi_1 t_1 = \psi_2 t_2$. Then,

$$\psi_1 = \varphi_1 \cdot \Delta_{\alpha_1}$$

$$\psi_2 = \varphi_2 \cdot \Delta_{\alpha_2}$$

for φ_1, φ_2 in $\mathcal{F}$, and

$$|\psi_1, t_1| = |\varphi_1, \Delta_{\alpha_1}(t_1)|$$

$$|\psi_2, t_2| = |\varphi_2, \Delta_{\alpha_2}(t_2)| \quad .$$

By (ii), however,

$$|\varphi_1, \Delta_{\alpha_1}(t_1)| = |\varphi_2, \Delta_{\alpha_2}(t_2)|$$

so e is injective. Let $U \subset r\mathcal{G}$ be open, and consider $eU \subset X$. Let $\varphi \in \mathcal{F}$. Then

$$\begin{array}{ccc} r\mathcal{G} & \xrightarrow{\ e\ } & X \\ {\scriptstyle i_\varphi}\uparrow & \nearrow_{\varphi} & \\ \Delta_n & & \end{array}$$

commutes, so $\varphi^{-1}(eU) = (i_\varphi^{-1} e^{-1})(eU) = i_\varphi^{-1} U$, which is open in Δ_n . Thus, by (iii), eU is open in X , and e is a homeomorphism. By taking the non-degenerate elements of $\mathcal{G}$, which are composites of φ's in $\mathcal{F}$ with injections $\Delta_\varepsilon : \Delta_m \longrightarrow \Delta_n$, we can again modify $\mathcal{F}$ to obtain a family $\mathcal{F}'$ satisfying the stronger condition (i').

We determine now the regular $\underline{\Delta}$-objects. Recall from § 2, that X is a regular $\underline{\Delta}$-object iff X has an atlas with a regular generating set $\mathcal{F}$, where regularity for $\mathcal{F}$ is the condition:

(ii') If for $\varphi : \Delta_n \longrightarrow X$ and $\psi : \Delta_m \longrightarrow X$ in $\mathcal{F}$ we have $\varphi t = \psi t'$ for $t \in \Delta_n$ and $t' \in \Delta_m$, then there is a pullback diagram in Top of the form:

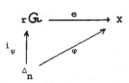

$$\begin{array}{ccccc} & & \Delta_q & & \\ & \swarrow^{\Delta_\alpha} & & \searrow^{\Delta_\beta} & \\ \Delta_n & & & & \Delta_m \\ & \searrow_{\varphi} & & \swarrow_{\psi} & \\ & & X & & \end{array} .$$

In particular, this is true for the pair (φ,φ) , where $\varphi : \Delta_n \longrightarrow X$ is any element of $\mathcal{F}$. i.e. there is a pullback diagram of the form:

$$\begin{array}{ccccc} & & \Delta_m & & \\ & \swarrow^{\Delta_\alpha} & & \searrow^{\Delta_\beta} & \\ \Delta_n & & & & \Delta_n \\ & \searrow_{\varphi} & & \swarrow_{\varphi} & \\ & & X & & \end{array} .$$

We will show that $\alpha = \beta$, which implies φ is a monomorphism. From this it follows that X is a classical (ordered) simplicial complex, since if each $\varphi \in \mathcal{F}$ is a monomorphism, then the α's and β's appearing in pullbacks of pairs (φ,ψ) in must all be injections. On the other hand, any simplicial complex clearly has an atlas with such a regular generating set.

To prove that $\alpha = \beta$, consider the above pullback diagram for the pair (φ,φ) . Δ_α and Δ_β give a map

$$\Delta_m \xrightarrow{\Delta_\alpha \times \Delta_\beta} \Delta_n \times \Delta_n ,$$

and the image of $\Delta_\alpha \times \Delta_\beta$ is the set of pairs (t,t') in $\Delta_n \times \Delta_n$ such that $\varphi t = \varphi t'$. In particular, it is symmetric i.e. if (t,t') is a member, so is (t',t). Now if we give $\Delta_n \times \Delta_n$ the standard triangulation as the realization of the product of two standard n-simplices in the category of simplicial sets, then $\Delta_\alpha \times \Delta_\beta$ is simplicial, and the vertices of the image m-simplex are

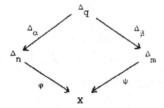

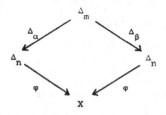

$$(e_{\alpha(o)}, e_{\beta(o)}),\ (e_{\alpha(1)}, e_{\beta(1)}),\ \ldots\ ,\ (e_{\alpha(m)}, e_{\beta(m)})\ .$$

Suppose $\alpha \neq \beta$, and let $0 < 1 < m$ be the first integer for which $\alpha(1) \neq \beta(1)$. By symmetry, the vertex $(e_{\beta(1)}, e_{\alpha(1)})$ must also occur in the image, and must be the image of a vertex of Δ_m , since $\Delta_\alpha \times \Delta_\beta$ is simplicial. Thus, there is $q > 1$ such that

$$\alpha(q) = \beta(1)$$

and $$\beta(q) = \alpha(1)\ .$$

However, if $\alpha(1) < \beta(1)$ then $\beta(q) < \beta(1)$ contradicting the monotonicity of β , and if $\alpha(1) > \beta(1)$, thus $\alpha(q) < \alpha(1)$ contradicting the monotonicity of α . Thus, $\alpha = \beta$. In fact, although we do not need this, an easy further argument shows that $m = n$ and both α and β are the identity.

Thus, the classical objects of study are the regular $\underline{\Delta}$-objects, and the others are generalizations of these. This will be a feature of most of the examples. The generalized objects are of interest,among other reasons, because they will almost always be at least a coreflective subcategory of the functor category, even if not fully equivalent to it as in this case. Thus, they can be treated as functors or objects of the ambient category $\underline{A}$, and automatically have good limit properties, etc.

(2) Simplicial modules.

As in (1), let $\underline{\Delta}$ be the simplicial category. Let Λ be a commutative ring with unit, and denote the category of Λ-modules by $\underline{\mathrm{Mod}}(\Lambda)$. Define

$$I : \underline{\Delta} \longrightarrow \underline{\mathrm{Mod}}(\Lambda)$$

by setting $I[n]$ = free Λ-module on the injections $\varepsilon : [q] \longrightarrow [n]$. If $\alpha : [n] \longrightarrow [m]$ define $I\alpha : I[n] \longrightarrow I[m]$ by putting

$$I\alpha(\varepsilon) = \begin{cases} \alpha \cdot \varepsilon & \text{if this is monic} \\ 0 & \text{otherwise} \end{cases}$$

Clearly, $I1_{[n]} = 1_{I[n]}$. Suppose $\beta : [m] \longrightarrow [l]$. If $\varepsilon : [q] \longrightarrow [n]$ is an injection, let $\alpha \cdot \varepsilon = \varepsilon' \cdot \eta'$ and $\beta \cdot \varepsilon' = \varepsilon'' \cdot \eta''$. Then, $(\beta \cdot \alpha) \cdot \varepsilon = \varepsilon'' \cdot \eta'' \cdot \eta'$, from which it follows that $I(\beta \cdot \alpha) = I\beta \cdot I\alpha$. Thus, I is a functor and we have the, by now, standard diagram:

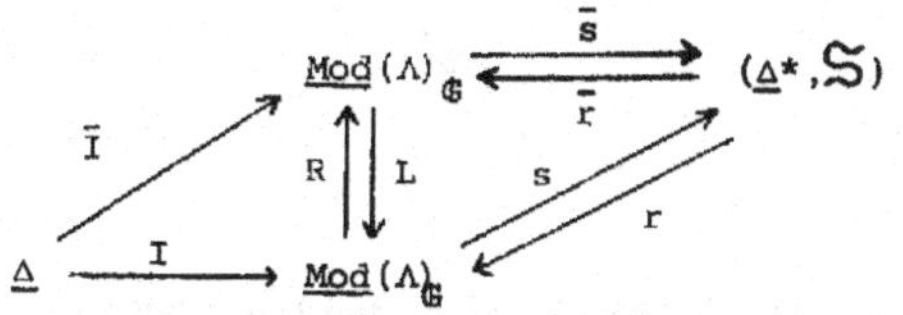

where $(\underline{\Delta}^*, \mathfrak{S})$ is again the category of simplicial sets, and s is given by

$$(sM)_n = \underline{\mathrm{Mod}}(\Lambda)(I[n], M)$$

for $M \in \underline{\mathrm{Mod}}(\Lambda)$ and $n \geq 0$.

Here the usual underlying set functor of $\mathrm{Mod}(\Lambda)$ is not colimit preserving, so we must identify r directly. For this we have

<u>Proposition 5.5</u>

Let $K \in (\underline{\Delta}^*, \mathfrak{S})$. Then

$$rK \approx \text{free } \Lambda\text{-module on the non-degenerate elements of } K .$$

<u>Proof</u>: Let $r'K$ denote the above free Λ-module, and if $\sigma \in K_n$ denote by $\bar{\sigma}$ the element of $r'K$ that is σ if σ is non-degenerate and 0 otherwise. If $\gamma : K \longrightarrow L$ is a morphism of simplicial sets, define

$$r'\gamma : r'K \longrightarrow r'L$$

by setting, for σ a non-degenerate element of K ,

$$r'\gamma(\sigma) = \overline{\gamma\sigma} .$$

It is easy to check that this makes r' a functor.

If $K \in (\underline{\Delta}^*, \mathfrak{S})$, let

$$\eta K : K \longrightarrow sr'K$$

be given as follows: for $\sigma \in K_n$

$$\eta K.\sigma : I[n] \longrightarrow r'K$$

is the Λ-morphism determined by

$$(\eta K.\sigma)(\varepsilon) = \overline{K_\varepsilon \sigma}$$

for $\varepsilon : [q] \longrightarrow [n]$ an injection. It is then easy to verify that ηK is a morphism of simplicial sets, which is natural in K .

For $M \in \underline{Mod}(\Lambda)$, define

$$\varepsilon M : r'sM \longrightarrow M$$

by $\varepsilon\varphi = \varphi(1_{[n]})$, where $\varphi : I[n] \longrightarrow M$ is a non-degenerate element of $(sM)_n$. Again, it is easy to see that ε is natural in M . Now, a simple computation shows that both composites

$$sM \xrightarrow{\eta sM} sr'sM \xrightarrow{s\varepsilon M} sM$$

and $$r'K \xrightarrow{r'\eta K} r'sr'K \xrightarrow{\varepsilon r'K} r'K$$

are the respective identities. In all of these verfifications, one should note that a necessary condition for an element

$$\varphi : I[n] \longrightarrow M$$

of $(sM)_n$ to be degenerate is that $\varphi(1_{[n]}) = 0$. Thus, for example, ηK takes non-degenerate elements to non-degenerate elements etc. In any case, the above shows that

$$(\varepsilon,\eta) : r' \longrightarrow\!\!\!\!\!\dashv\ s ,$$

and hence, for $K \in (\underline{\Delta}^*, \mathfrak{S})$, there is a natural isomorphism

$$rK \approx r'K .$$

We will drop the prime notation, and identify rK and $r'K$ by means of this isomorphism.

In this setting, we shall call an $\underline{\Delta}$-object a <u>simplicial module</u>, and we will show that these are again equivalent to the category of simplicial sets. First of all, if $j : K \longrightarrow L$ is a monomorphism of simplicial sets, then j takes non-degenerate elements to non-degenerate elements by Lemma 5.1, so that $rj : rK \longrightarrow rL$ is also monic. Let $(M,\theta) \in \underline{\mathrm{Mod}}(\Lambda)_{\mathbb{G}}$ be a coalgebra for the model induced cotriple $\mathbb{G}$, and consider the diagram

$$(*) \qquad r\bar{s}(M,\theta) \xrightarrow{rj(M,\theta)} rsM \underset{r\eta sM}{\overset{rs\theta}{\rightrightarrows}} (rs)^2M .$$

Both $s\theta$ and ηsM are monic, so that the same is true of $rs\theta$ and $r\eta sM$. Let $\lambda_1\varphi_1 + \ldots + \lambda_n\varphi_n$ $(\lambda_i \neq 0)$ be an element of rsM for which

$$rs\theta(\lambda_1\varphi_1 + \ldots + \lambda_n\varphi_n) = r\eta sM(\lambda_1\varphi_1 + \ldots + \lambda_n\varphi_n)$$

i.e.

$$\lambda_1\theta\cdot\varphi_1 + \ldots + \lambda_n\theta\cdot\varphi_n = \lambda_1\eta sM(\varphi_1) + \ldots + \lambda_n\eta sM(\varphi_n) .$$

Thus, there is a permutation π of $\{1,\ldots,n\}$ such that

$$\lambda_i = \lambda_{\pi(i)} \quad \text{and} \quad \theta\cdot\varphi_i = \eta sM(\varphi_{\pi(i)}) .$$

But then

$$s\varepsilon M(\theta\cdot\varphi_i) = s\varepsilon M(\eta sM(\varphi_{\pi(i)}))$$

or

$$\varphi_i = \varphi_{\pi(i)} .$$

Thus, for each i, we have

$$\theta\cdot\varphi_i = \eta sM(\varphi_i) .$$

Therefore, each φ_i is a morphism of coalgebras, and is in the image of $rj(M,\theta)$. Hence, $(*)$ is an equalizer diagram for each (M,θ), so by 3.2 we have

$$\bar{\varepsilon} : \bar{r}\bar{s} \xrightarrow{\approx} 1 .$$

The equivalence of categories is established now by 4.3 and the following proposition.

Proposition 5.6

r reflects isomorphisms.

Proof: Suppose $\gamma : K \longrightarrow L$ is a morphism of simplicial sets such that $r\gamma : rK \longrightarrow rL$ is an isomorphism. Then, γ takes non-degenerate elements to non-degenerate elements, and must be monic on these. Moreover, it must be epic on these, for if $\tau \in L$ is non-degenerate, then since $r\gamma$ is epic we have

$$r\gamma(\lambda_1\sigma_1 + \ldots + \lambda_n\sigma_n) = \tau$$

for some element $\lambda_1\sigma_1 + \ldots + \lambda_n\sigma_n$ in rK . But then

$$\lambda_1\gamma\sigma_1 + \ldots + \lambda_n\gamma\sigma_n = \tau$$

so that, for some i , $\tau = \lambda_i\gamma\sigma_i$ with $\lambda_i = 1$, and the other λ's are 0 . Thus, γ maps the non-degenerate elements of K bijectively on the non-degenerate elements of L . But this implies γ is bijective everywhere. In fact, let $\sigma_1,\sigma_2 \in K$ such that $\gamma\sigma_1 = \gamma\sigma_2$. Write

$$\sigma_i = K_{\eta_i}\sigma_i' \qquad i = 1,2$$

where η_i is a surjection, and σ_i' is non-degenerate for $i = 1,2$. Then

$$L_{\eta_1}\gamma\sigma_1' = L_{\eta_2}\gamma\sigma_2' .$$

By uniqueness, since $\gamma\sigma_1'$ and $\gamma\sigma_2'$ are non-degenerate,

$$\eta_1 = \eta_2 \text{ and } \gamma\sigma_1' = \gamma\sigma_2' .$$

But then $\sigma_1' = \sigma_2'$, so $\sigma_1 = \sigma_2$, and γ is monic. Also, let $\tau \in L$, and write

$$\tau = L_\eta\tau'$$

for η a surjection, and τ' non-degenerate. Then $\tau' = \gamma\sigma'$ for $\sigma' \in K$ non-degenerate. Thus,

$$\gamma(K_\eta \sigma') = L_\eta \gamma\sigma' = L_\eta \tau' = \tau ,$$

and γ is epic, which proves the proposition.

We will now characterize those modules in Mod(Λ) that are simplicial, but before doing this we need a lemma. Let $n > 1$, $0 < j < n-1$ and consider the morphism

$$I\eta^j : I[n] \longrightarrow I[n-1] .$$

Lemma 5.7

A basis for $\ker I\eta^j$ is given by those $\varepsilon : [q] \longrightarrow [n]$ that hit both j and $j+1$, together with all elements of the form $\varepsilon^{j+1}\varepsilon' - \varepsilon^j\varepsilon'$, where $\varepsilon' : [q] \longrightarrow [n-1]$ is any injection that hits j .

Proof: Since $\eta^j\varepsilon^j = 1$, we have a split exact sequence

$$0 \longrightarrow \ker I\eta^j \longrightarrow I[n] \underset{I\varepsilon^j}{\overset{I\eta^j}{\rightleftarrows}} I[n-1] \longrightarrow 0 ,$$

and $f = 1 - I(\varepsilon^j\eta^j)$ gives an isomorphism

$$f : I[n]/I\varepsilon^j(I[n-1]) \longrightarrow \ker I\eta^j .$$

A basis for $I\varepsilon^j(I[n-1])$ consists of injections $\varepsilon : [q] \longrightarrow [n]$ that can be factored in the form

$$[q] \xrightarrow{\ \varepsilon'\ } [n-1] \xrightarrow{\ \varepsilon^j\ } [n]$$

for arbitrary ε' , and using the first simplicial identity, it is easy to see that these are exactly those ε that miss j . Thus, $I[n]/I\varepsilon^j(I[n-1])$ is isomorphic to the free Λ-module on the injections $\varepsilon : [q] \longrightarrow [n]$ that hit j , and f applied to these gives a basic for $\ker I\eta^j$. For such an ε ,

$$f\varepsilon = \begin{cases} \varepsilon - \varepsilon^j\eta^j\varepsilon & \text{if } \eta^j\varepsilon \text{ is monic,} \\ \varepsilon & \text{otherwise.} \end{cases}$$

Among the ε hitting j, the ones for which $\eta^j\varepsilon$ is <u>not</u> monic are precisely those that also hit $j+1$, so these form part of a basis for $\ker I\eta^j$. Those ε that hit j, but miss $j+1$, are of the form

$$\varepsilon = \varepsilon^{j+1}.\varepsilon' ,$$

where $\varepsilon' : [q] \longrightarrow [n-1]$ is any injection hitting j. For these,

$$f\varepsilon = \varepsilon^{j+1}.\varepsilon' - \varepsilon^j.\varepsilon' ,$$

which proves the lemma.

We claim now that if $M \in \underline{\text{Mod}}(\Lambda)$, then M is simplicial iff M is positively graded and has a homogeneous basis B with the following structure: if $\varepsilon : [q] \longrightarrow [n]$ is an injection, then there is a function

$$B_n \longrightarrow B_q \cup \{0\} ,$$

which we write as $b \longrightarrow b_\varepsilon$. For this operation we have $b_{1_{[n]}} = b$, and if $b_\varepsilon \neq 0$, then $(b_\varepsilon)_{\varepsilon'} = b_{\varepsilon\varepsilon'}$. If $b_\varepsilon = 0$, then there is $0 < j < q-1$ such that for any $\varepsilon' : [m] \longrightarrow [q]$ hitting both j and $j+1$, $b_{\varepsilon\varepsilon'} = 0$, and for any $\varepsilon'' : [m] \longrightarrow [q-1]$ hitting j,

$$b_{\varepsilon(\varepsilon^{j+1}\varepsilon'')} = b_{\varepsilon(\varepsilon^j\varepsilon'')} .$$

Well, suppose M has an atlas $\mathcal{G} \xrightarrow{\sim} sM$. Thus

$$e : r\mathcal{G} \xrightarrow{\sim} M ,$$

where $r\mathcal{G}$ is the free Λ-module on the non-degenerate $\varphi : I[n] \longrightarrow M$ in $\mathcal{G}_n$, and $e\varphi = \varphi 1_{[n]}$. Thus M is graded in the obvious way, and these $\varphi 1_{[n]}$ provide such a basis B for M by setting, for $b = \varphi 1_{[n]}$ and $\varepsilon : [q] \longrightarrow [n]$,

$b_\varepsilon = \varphi\varepsilon = (\varphi\cdot I\varepsilon)(1_{[q]})$. Then b_ε is either 0 or a basis element of dimension q , depending on whether $\varphi\cdot I\varepsilon$ is degenerate or not. If $\varphi\cdot I\varepsilon$ is non-degenerate and $\varepsilon' : [m] \longrightarrow [q]$, then

$$(b_\varepsilon)_{\varepsilon'} = (\varphi\cdot I\varepsilon)(\varepsilon') = \varphi(\varepsilon\varepsilon') = b_{\varepsilon\varepsilon'} .$$

If $b_\varepsilon = 0$, then $\varphi\cdot I\varepsilon$ is degenerate, so there is a factorization

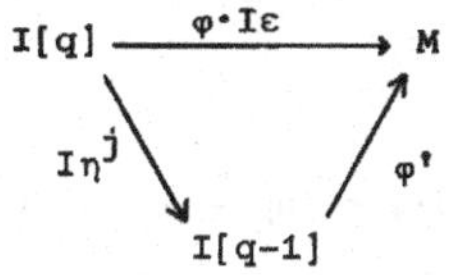

for some $0 < j < q-1$. Thus, $\varphi\cdot I\varepsilon$ vanishes on $\ker I\eta^j$, so by Lemma 5.7 we have

$$(\varphi\cdot I\varepsilon)(\varepsilon') = b_{\varepsilon\varepsilon'} = 0$$

for any $\varepsilon' : [m] \longrightarrow [q]$ hitting j and $j+1$, and

$$\varphi\cdot I\varepsilon(\varepsilon^{j+1}\cdot\varepsilon'' - \varepsilon^{j}\cdot\varepsilon'') = b_{\varepsilon(\varepsilon^{j+1}\varepsilon'')} - b_{\varepsilon(\varepsilon^{j}\varepsilon'')} = 0$$

for any $\varepsilon'' : [m] \longrightarrow [q-1]$ hitting j .

On the other hand, suppose M has such a basis B . For $b \in B_n$, define

$$\varphi_b : I[n] \longrightarrow M$$

by $\varphi_b\varepsilon = b_\varepsilon$, and let $\mathcal{A}$ be the atlas generated by the φ_b for $b \in B$. Let Φ denote the set of non-degenerate elements of $\mathcal{A}$, and consider

$$e : r\mathcal{A} \longrightarrow M .$$

We claim, $e : \Phi \longrightarrow B$. Well, a $\varphi \in \Phi$ is of the form

$$\varphi = \varphi_b\cdot I\alpha$$

for some $b \in B$ and α in $\underline{\Delta}$. Since φ is non-degenerate, $\alpha = \varepsilon$, an injection, and

$$\varphi = \varphi_b \cdot I\varepsilon : I[q] \longrightarrow M$$

has the property that $e\varphi = \varphi 1_{[q]} = b_\varepsilon \neq 0$. (If $b_\varepsilon = 0$, then φ vanishes on ker $I\eta^j$ for some j , making η degenerate in sM , and hence in $\mathfrak{G}$, since $j : \mathfrak{G} \longrightarrow sM$ takes non-degenerate elements to non-degenerate elements.) Thus, $e\varphi \in B$.

Define

$$f : B \longrightarrow \Phi$$

by $fb = \varphi_b$. Clearly, $ef \cdot b = b$, so $ef = 1$. $fe \cdot \varphi = fb_\varepsilon = \varphi_{b_\varepsilon}$, where $\varphi = \varphi_b \cdot I\varepsilon$ is as above. However, if $\varepsilon' : [m] \longrightarrow [q]$ is an injection, then

$$\varphi_{b_\varepsilon}(\varepsilon') = (b_\varepsilon)_{\varepsilon'} = b_{\varepsilon\varepsilon'} = (\varphi_b \cdot I\varepsilon)(\varepsilon') .$$

Thus, $\varphi_{b_\varepsilon} = \varphi$, so $fe = 1$, and

$$e : r\mathfrak{G} \longrightarrow M$$

is an isomorphism.

<u>Remarks</u>.

(i) If we combine the equivalences of (1) and (2), we find that the category of simplicial spaces is equivalent to the category of simplicial Λ-modules for any Λ . Furthermore, it is easy to see that the composite equivalence is simply the functor that assigns to a simplicial space its cellular chain complex over Λ .

(ii) Since there are no well-known classical objects among the simplicial modules, we omit the calculation of the regular objects. One can show, however, that if M is a regular $\underline{\Delta}$-object, then the elements $\varphi : I[n] \longrightarrow M$ in the regular generating set are monic. This in turn, gives a graded basis B for M with the property that if $b \in B_n$ and $\varepsilon : [q] \longrightarrow [n]$, then $b_\varepsilon \in B_q$ (i.e. $b_\varepsilon \neq 0$). If X is a classical ordered simplicial complex, then, of course, its chain complex is of this form.

(iii) Consider the functor

$$I : \underline{\Delta} \longrightarrow \underline{\text{Mod}}(\Lambda) \ ,$$

and the resulting singular functor

$$s : \underline{\text{Mod}}(\Lambda) \longrightarrow (\underline{\Delta}^*, \mathfrak{S}) \ .$$

For $M \in \underline{\text{Mod}}(\Lambda)$, we have

$$(sM)_n = \underline{\text{Mod}}(\Lambda)(I[n], M) \ .$$

Since Λ is commutative, this set has a canonical Λ-module structure, and we can consider s as a functor

$$s : \underline{\text{Mod}}(\Lambda) \longrightarrow (\underline{\Delta}^*, \underline{\text{Mod}}(\Lambda)) \ .$$

A coadjoint to s still exists in this situation, and the equivalence theorem applied here gives the theorem of Dold and Kan, which asserts that the category of FD-Modules over Λ is equivalent to the category of positive Λ-chain complexes. All of this results from the fact that $\underline{\text{Mod}}(\Lambda)$ is a "closed" category in, say, the sense of Eilenberg and Kelley. We will discuss this situation in detail in a later paper.

(3) Manifolds.

Let Γ be a pseudogroup of transformations defined on open subsets of n-dimensional Euclidean space E^n for some fixed n . That is, elements $g \in \Gamma$ are homeomorphisms into

$$g : U \longrightarrow V$$

where U and V are open in E^n , such that

(i) If $g_1, g_2 \in \Gamma$ and $g_1 g_2$ is defined, then $g_1 g_2 \in \Gamma$.

(ii) If $g \in \Gamma$, then $g^{-1} \in \Gamma$.

(iii) If $i : U \longrightarrow V$ is an inclusion, then $i \in \Gamma$.

(iv) Γ is local. That is, if $g : U \longrightarrow V$ is a homeomorphism into, and each $x \in U$ has a neighborhood $U(x)$ such that $g|U(x) \in \Gamma$, then $g \in \Gamma$.

The kinds of examples of Γ that we have in mind are the following (there are, of course, others).

$$\Gamma = \left\{ \begin{array}{l} \text{all homeomorphisms into,} \\ \text{orientation preserving homeomorphisms into defined on oriented open} \\ \text{subsets of } E^n \text{ ,} \\ \text{PL homeomorphisms into,} \\ \text{diffeomorphisms into } g : U \longrightarrow V \text{ whose Jacobian } Jg \text{ is an element} \\ \text{of a subgroup } G \subset GL(n,R) \text{ ,} \\ \text{real or complex } (n = 2m) \text{ analytic isomorphisms into.} \end{array} \right.$$

Let $\underline{E}_\Gamma$ be the category whose objects are domains of elements of Γ , and whose morphisms are the elements of Γ . Let

$$I : \underline{E}_\Gamma \longrightarrow \underline{Top}$$

be the obvious embedding, which we will henceforth omit from the notation . Top has again its standard underlying set functor, which we also omit.

In this example, condition (a) is trivially satisfied. Namely, consider a pair of morphisms

$$\begin{array}{ccc} & & U_1 \\ & & \big\downarrow g_1 \\ U_2 & \xrightarrow[g_2]{} & U \end{array}$$

in Γ , together with $x_i \in U_i$, $i = 1,2$ such that $g_1(x_1) = g_2(x_2)$. Then the diagram

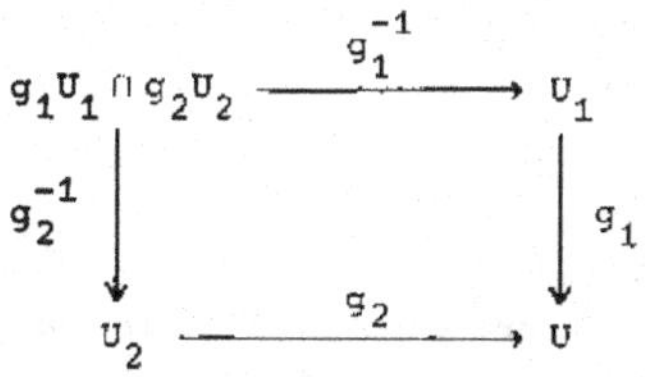

is a pullback diagram in Top , and we have (a). Thus, by 4.2, in the diagram

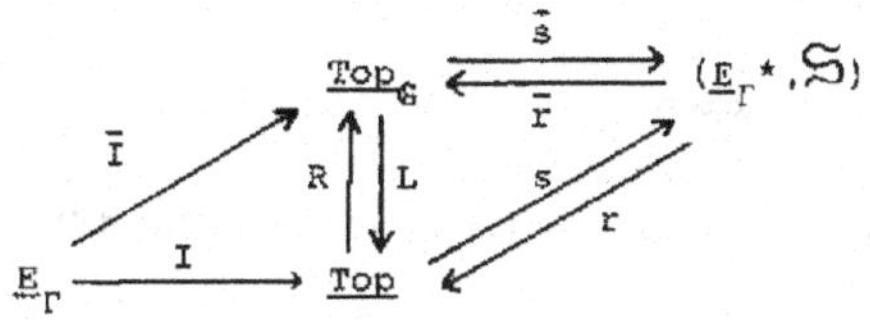

r preserves monomorphisms.

Now, for $F \in (\underline{E}_\Gamma^*, \mathfrak{S})$ we investigate in detail the structure of rF . As a set, rF consists of equivalence classes $|U,x,u|$ where $U \in \underline{E}_\Gamma$, $x \in FU$, and $u \in U$. We abbreviate these as $|x,u|$. Since (a) is satisfied, $|x_1,u_1| = |x_2,u_2|$ iff there is a pair of morphisms

$$U_1 \xleftarrow{\;g_1\;} U \xrightarrow{\;g_2\;} U_2$$

in $\underline{E}_\Gamma$, and a $u \in U$ such that $Fg_1(x_1) = Fg_2(x_2)$ and $g_1 u = u_1$, $g_2 u = u_2$. As a topological space, rF has the quotient topology with respect to the universal morphisms

$$i_x : U \longrightarrow rF$$

given by $i_x(u) = |x,u|$ for $U \in \underline{E}_\Gamma$, $x \in FU$, and $u \in U$. In fact, we have the general result:

Proposition 5.8

Let $\underline{M}$ be small, and suppose

$$I : \underline{M} \longrightarrow \underline{\text{Top}}$$

satisfies condition (a). Then if $I\alpha : IM_1 \longrightarrow IM_2$ is open for each $\alpha : M_1 \longrightarrow M_2$ in $\underline{M}$, we have

$$i(M,x) : IM \longrightarrow rF$$

open for all $F \in (M^*, \mathfrak{S})$ and $(M,x) \in (Y,F)$. Furthermore, a basis for the topology of rF is given by the collection of all open sets of the form

$$i(M,x)(U) ,$$

where $(M,x) \in (Y,F)$, and $U \subset IM$ is open.

<u>Proof:</u> Let $(M,x) \in (Y,F)$, and let $U \subset IM$ be open. Then $i(M,x)(U)$ is open in rF iff for each $(M',x') \in (Y,F)$ we have

$$i(M',x')^{-1}i(M,x)(U)$$

open in IM'. If this set is empty we are done, and if not let

$$m' \in i(M',x')^{-1}i(M,x)(U) .$$

Then there is $m \in U$ such that

$$|x',m'| = |x,m|$$

in rF.

Since (a) is satisfied, we obtain a pair of maps

$$M' \xleftarrow{\alpha'} M_o \xrightarrow{\alpha} M$$

in $\underline{M}$, together with an $m_o \in IM_o$, such that $F\alpha(x) = F\alpha'(x')$ and $I\alpha(m_o) = m$, $I\alpha'(m_o) = m'$. Let $U_o = I\alpha^{-1}(U)$, and $U' = I\alpha'(U_o)$. U' is open in IM' by assumption. Let $\bar{m}' \in U'$, say $\bar{m}' = I\alpha'(\bar{m}_o)$. Then,

$$|x',\bar{m}'| = |F\alpha'(x'),\bar{m}_o| = |F\alpha(x),\bar{m}_o| = |x,I\alpha(\bar{m}_o)| .$$

Thus, $m' \in U' \subset i(M',x')^{-1}i(M,x)(U)$ and the latter set is open in IM'. To conclude the proof, let $\bar{V} \subset rF$ be any open set. By the preceding,

$$i(M,x)(i(M,x)^{-1}\bar{V}) \subset \bar{V}$$

is open for any $(M,x) \in (Y,F)$. Thus,

$$\bar{V} = \bigcup_{(M,x) \in (Y,F)} i(M,x)(i(M,x)^{-1}\bar{V}) \ .$$

Still in the situation of 5.8, let $\gamma : F \longrightarrow F'$ be a morphism in $(\underline{M}^*, \mathfrak{S})$. Then we claim $r\gamma : rF \longrightarrow rF'$ is open. It is enough to show this on the basis given by 5.8, but if $x \in FM$ for $M \in \underline{M}$ then we have

$$\begin{array}{ccc} IM & & \\ {\scriptstyle i(M,x)}\downarrow & \searrow {\scriptstyle i_{\gamma M(x)}} & \\ rF & \xrightarrow{\ r\gamma\ } & rF' \end{array} ,$$

so if $U \subset IM$ is open,

$$r\gamma\{i(M,x)(U)\} = i_{\gamma M(x)}(U) \ ,$$

which is open in rF' .

Now let $\mathbb{G} = (G,\varepsilon,\delta)$ be the cotriple in $\underline{\text{Top}}$ induced by $I : \underline{M} \longrightarrow \underline{\text{Top}}$, and consider $(X,\theta) \in \underline{\text{Top}}_{\mathbb{G}}$. By condition (a), the underlying set of

$$r\bar{s}(X,\theta) \xrightarrow{\ rj(X,\theta)\ } GX \underset{\delta X}{\overset{G\theta}{\rightrightarrows}} G^2X$$

is that of the equalizer of $G\theta$ and δX . Furthermore, by the above remark $rj(X,\theta)$ is open. Therefore, the above is an equalizer at the level of spaces, and hence

$$\bar{\varepsilon} : \bar{r}\bar{s} \xrightarrow{\ \approx\ } 1$$

in the adjoint pair

$$\text{Top}_{\mathbb{G}} \underset{\bar{r}}{\overset{\bar{s}}{\rightleftarrows}} (\underline{M}^*, \mathfrak{S}) \ .$$

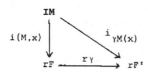

From this it follows that the $\underline{M}$-objects in $\underline{Top}$ are spaces X admitting a $\mathbb{G}$-coalgebra structure $\theta : X \longrightarrow GX$. In particular, this is all true for $\underline{M} = \underline{E}_\Gamma$.

In terms of atlases, we can say the following. $X \in \underline{Top}$ is an $\underline{E}_\Gamma$-object iff there is a family $\mathcal{F}$ of morphisms $\varphi : U \longrightarrow X$ with the following properties.

(i) $\mathcal{F}$ covers X . That is, for $x \in X$ there exists $\varphi : U \longrightarrow X$ in $\mathcal{F}$ and $u \in U$, such that $\varphi u = x$.

(ii) $\mathcal{F}$ is a compatible family. That is, if $\varphi : U \longrightarrow X$ and $\psi : V \longrightarrow X$ are a pair of morphisms from $\mathcal{F}$, and $\varphi u = \psi v$ for $u \in U$, $v \in V$, then there is a diagram

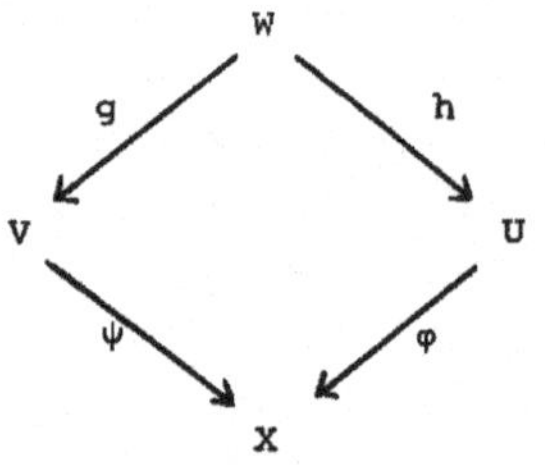

with $g,h \in \Gamma$, and $w \in W$ such that $gw = v$, $hw = u$.

(iii) $\mathcal{F}$ is open, i.e. each $\varphi : U \longrightarrow X$ in $\mathcal{F}$ is open.

The proof of this is very much like the corresponding statement for simplicial spaces, so we only give a sketch. If X has an $\underline{E}_\Gamma$ -atlas

$$\mathcal{G} \hookrightarrow sX$$

then $e : r\mathcal{G} \xrightarrow{\approx} X$, and any generating family $\mathcal{F}$ of $\mathcal{G}$ satisfies (i)-(iii). (i) since e is surjective, (ii) since e is injective and condition (a) is satisfied, and (iii) since e is open, as is each $i_\varphi : U \longrightarrow r\mathcal{G}$. On the other hand, if $\mathcal{F}$ is a family satisfying (i)-(iii) and $\mathcal{G}$ is the preatlas generated by $\mathcal{F}$, then it is easy to see that

$$e : r\mathcal{G} \xrightarrow{\approx} X .$$

We claim that the classical $\underline{E}_\Gamma$-objects, namely the Γ-manifolds, are precisely the regular $\underline{E}_\Gamma$-objects defined in § 2. To see this, recall that a regular E_Γ-object is an $\underline{E}_\Gamma$-object X with an atlas $\mathfrak{A}$ having a generating set $\mathfrak{F}$ of morphisms $\varphi : U \longrightarrow X$ with the property that for any pair (φ, ψ) of morphisms in $\mathfrak{F}$ the pullback of

$$\begin{array}{ccc} & & U \\ & & \downarrow \varphi \\ V & \xrightarrow{\psi} & X \end{array}$$

in $\underline{\text{Top}}$ is of the form

$$\begin{array}{ccc} W & \xrightarrow{g} & U \\ h \downarrow & & \downarrow \varphi \\ V & \xrightarrow{\psi} & X \end{array}$$

where $g, h \in \Gamma$. Since elements of Γ are monomorphisms in $\underline{\text{Top}}$, if we apply this to the pair (φ, φ) for $\varphi \in \mathfrak{F}$, we see by 2.2 that each φ is monic, and hence a homeomorphism into. Thus, X is a manifold of the appropriate type. Conversely, if X is a Γ-manifold its charts generate a regular $\underline{E}_\Gamma$-atlas for X. Note that we do not require a Γ-manifold to be Hausdorff.

We give briefly some examples of the kind of objects that can appear as non-regular $\underline{E}_\Gamma$-objects. For simplicity of statement, we restrict to the topological case - i.e. Γ = all homeomorphisms into. The necessary modifications for other Γ will be obvious.

(i) Let X be an m-dimensional manifold where $m < n$. Then X has a system of charts

$$\varphi_i : U_i^m \longrightarrow X ,$$

for U_i^m open in E^m which is compatible, open, and covers X. Let $\varphi_i^!$ be the composite of

$$U_i^m \times E^{n-m} \longrightarrow U_i^m \xrightarrow{\varphi_i} X$$

where the first morphism is projection on the first factor. Then each $\varphi_i^!$ is open, and they cover X. Furthermore, they are compatible. For, suppose

$$\varphi_i^!(u_i,t_i) = \varphi_j^!(u_j,t_j)$$

where $t_i,t_j \in E^{n-m}$, $u_i \in U_i^m$, $u_j \in U_j^m$. Then $\varphi_i(u_i) = \varphi_j(u_j)$ so we have a diagram

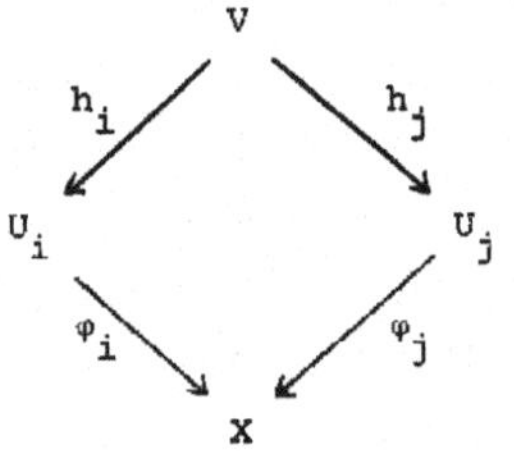

and $v \in V$ with $h_i v = u_i$, $h_j v = u_j$. Crossing with E^{n-m} gives a diagram

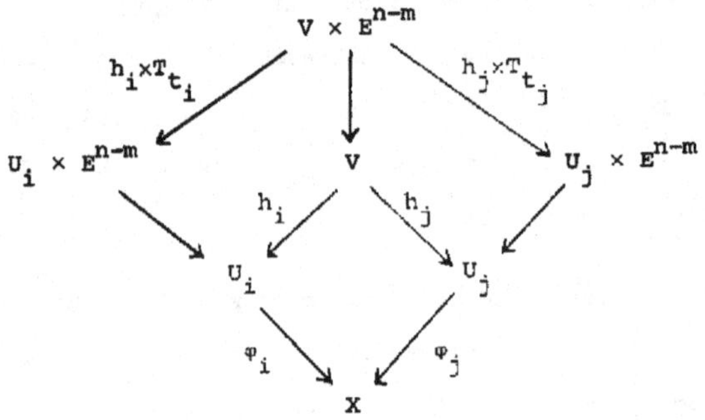

where T_t denotes translation by $t \in E^{n-m}$. Also,

$$(h_i \times T_{t_i})(v,0) = (u_i,t_i)$$

$$(h_j \times T_{t_j})(v,0) = (u_j,t_j) .$$

Thus, the $\varphi_i^!$ generate an atlas making X an n-dimensional $\underline{E}_\Gamma$-object. Of course, the same argument shows that any m-dimensional object for $m < n$ appear also as an n-dimensional object.

(ii) Let X be an n-dimensional manifold with boundary. Then every point of X has an open neighborhood homeomorphic to either an open disc

$$D^n(a,\varepsilon) = \{t \in E^n : \|t-a\| < \varepsilon\}$$

or to a $\frac{1}{2}$ -open disc

$$D^n_+(a,\varepsilon) = \{t \in E^n : \|t-a\| < \varepsilon, \quad t_n \geq 0\}$$

where for $a = (a_1,\dots,a_{n-1},a_n)$ in $D^n_+(a,\varepsilon)$ we have $a_n = 0$. Then the folding map

$$f : D^n(a,\varepsilon) \longrightarrow D^n_+(a,\varepsilon)$$

given by $f(t_1,\dots,t_{n-1},t_n) = (t_1,\dots,t_{n-1},|t_n|)$ is obviously open and surjective. Consider the following system of morphisms. For points of X having neighborhoods homeomorphic to $D^n(a,\varepsilon)$'s, take the given morphism

$$\varphi' = \varphi : D^n(a,\varepsilon) \longrightarrow X .$$

For points of X having neighborhoods homeomorphic to $D^n_+(a,\varepsilon)$'s, take the composites

$$\varphi' = \varphi \cdot f : D^n(a,\varepsilon) \longrightarrow D^n_+(a,\varepsilon) \longrightarrow X$$

where φ is the given homeomorphism. This is an open system, and it covers X. For compatibility, there are various special cases to check. These are either completely obvious, since the charts on X are compatible, or they follow from the observation that if

$$h : D^n_+(a_1,\varepsilon_1) \longrightarrow D^n_+(a_2,\varepsilon_2)$$

is a homeomorphism into, i.e. injective and open, then we can reflect h to obtain a diagram

$$\begin{array}{ccc} D^n(a_1,\varepsilon_1) & \xrightarrow{\;f_1\;} & D^n_+(a_1,\varepsilon_1) \\ {\scriptstyle h'}\downarrow & & \downarrow{\scriptstyle h} \\ D^n(a_2,\varepsilon_2) & \xrightarrow{\;f_2\;} & D^n_+(a_2,\varepsilon_2) \end{array}$$

when the f_i are folding maps, $i = 1,2$, and h' is a homeomorphism into. Thus, the φ' 's generate an atlas for X .

Remarks

(1) By mapping $\underline{E}_\Gamma$ into the category of topological spaces and local homeomorphisms, instead of into Top , one can arrange matters so that the $\underline{E}_\Gamma$-objects are exactly the Γ-manifolds. In addition to being somewhat artificial, this has several other drawbacks. For one thing, since topological spaces and local homeomorphisms do not have arbitrary small colimits, the realization fuctor does not exist in general, although the model induced cotriple does. Also, by doing this one excludes from consideration many interesting examples of non-regular $\underline{E}_\Gamma$-objects such as the previous two.

(2) Since we have presented Γ-manifolds as coalgebras over the model induced cotriple, the morphisms that we obtain are morphisms of coalgebras - i.e. morphisms that preserve the structure. It is easy to see that these are maps which are locally like elements of Γ . These are useful for some purposes, e.g. for the existence of certain adjoint functors that we will discuss in a separate paper. However, it is clear that this is not a wide enough class for the general study of manifolds. One can obtain the proper notion of morphism by considering subdivisions of atlases, which we will do elsewhere.

(4) G-bundles

For the moment, let $\underline{A}$ be an arbitrary category and $B \in \underline{A}$. Consider the comma category $(\underline{A},B)$, i.e. the category of objects over B . As the terminology indicates, an object of $(\underline{A},B)$ is an $\underline{A}$-morphism

$$p : A \longrightarrow B$$

and a morphism $f : p_1 \longrightarrow p_2$ is a commutative triangle

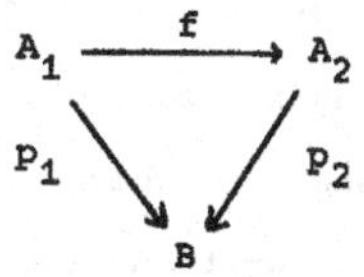

in $\underline{A}$. There is the obvious (faithful) functor

$$\partial_o : (\underline{A},B) \longrightarrow \underline{A}$$

given by $\partial_o(p : A \longrightarrow B) = A$, $\partial_o f = f$. This functor has the important property that it <u>creates</u> <u>colimits</u> . That is, let

$$D : \underline{J} \longrightarrow (\underline{A},B)$$

be a functor, and suppose

$$\gamma : \partial_o D \longrightarrow A$$

is a colimit of $\partial_o D$ in $\underline{A}$. There is a natural transformation

$$d : \partial_o D \longrightarrow B$$

given by $dj = Dj : \partial_o Dj \longrightarrow B$ for $j \in \underline{J}$. Hence there exists a unique $\underline{A}$-morphism $p : A \longrightarrow B$ such that

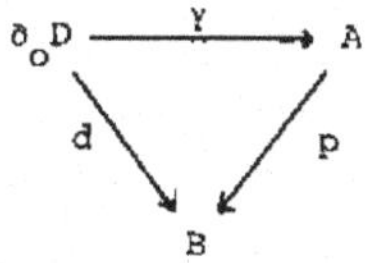

Thus, $\gamma : D \longrightarrow p$ in $(\underline{A},B)$, and it is trivial to verify that (p,γ) is a colimit of D in $(\underline{A},B)$. In particular, if $\underline{A}$ has small colimits so does $(\underline{A},B)$ for any $B \in \underline{A}$, and ∂_o preserves them. If $\underline{A}$ has a colimit preserving underlying set functor $U : \underline{A} \longrightarrow \mathcal{S}$, so does $(\underline{A},B)$, namely $U\partial_o$. Note also, for what follows, that ∂_o reflects equalizers.

Now let B be a fixed space in <u>Top</u>. By the above discussion, (<u>Top</u>,B) has small colimits and a faithful colimit preserving underlying set functor. Let G be a fixed topological group, and let Y be a fixed left G-space on which G operates effectively. That is, there is an action $\xi : G\times Y \longrightarrow Y$, written $\xi(g,y) = g\cdot y$, for which $e\cdot y = y$ (e the identity of G), $(g_1g_2)\cdot y = g_1\cdot(g_2\cdot y)$, and $g\cdot y = y$ for all $y \in Y$ implies $g = e$.

Define a model category $\underline{M}$ as follows: an object of $\underline{M}$ is an open set U of B. If $V \subset U$, then a morphism $V \longrightarrow U$ in $\underline{M}$ is a triple (V,α,U) where $\alpha : V \longrightarrow G$ is a continuous map. There are no morphisms $V \longrightarrow U$ if $V \not\subset U$. If $(V,\alpha,U) : V \longrightarrow U$ and $(U,\beta,W) : U \longrightarrow W$, then let

$$(U,\beta,W)\cdot(v,\alpha,u) \quad = \quad (V,\beta\alpha,W)$$

where $\beta\alpha : V \longrightarrow G$ is the map $(\beta\alpha)(b) = \beta(b)\alpha(b)$ for $b \in V$. Identities $U \longrightarrow U$ for this composition are given by (U,e,U) where $e : U \longrightarrow G$ is the constant map $e(b) = e$ for all $b \in U$. $\underline{M}$ is clearly a category (and small). In the notation (V,α,U), V and U serve to fix domain and codomain. When these are evident, we will denote the morphism (V,α,U) by α alone.

We define a functor

$$I : \underline{M} \longrightarrow (\underline{\text{Top}},B)$$

by setting

$$IU \quad = \quad U\times Y \longrightarrow B \quad ,$$

(projection onto U followed by inclusion into B). If $(V,\alpha,U) : V \longrightarrow U$, then in

$$\begin{array}{ccc} V\times Y & \xrightarrow{\ I\alpha\ } & U\times Y \\ & \searrow \quad \swarrow & \\ & B & \end{array}$$

we let $I\alpha$ be the map $I\alpha(b,y) = (b,\alpha(b)\cdot y)$ for $(b,y) \in V\times Y$. I is clearly a functor, and each $I\alpha$ is open (being the composite of a homeomorphism $V\times Y \longrightarrow V\times Y$

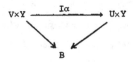

and the inclusion $V \times Y \longrightarrow U \times Y$). Note that since G acts effectively on Y, I is faithful.

Let us verify condition (a) for $I : \underline{M} \longrightarrow (\underline{Top}, B)$. So, suppose we have a diagram

$$\begin{array}{ccc} & & IU_1 \\ & & \downarrow{\scriptstyle I\alpha_1} \\ IU_2 & \xrightarrow[I\alpha_2]{} & IU \end{array}$$

in $(\underline{Top}, B)$ together with points $(b_i, y_i) \in U_i \times Y$ for $i = 1,2$ such that $I\alpha_1(b_1,y_1) = I\alpha_2(b_2,y_2)$, i.e. $(b_1, \alpha_1(b_1)\cdot y_1) = (b_2, \alpha_2(b_2)\cdot y_2)$. Then $b_1 = b_2$ (let $b \in U_1 \cap U_2$ denote the common value) and $\alpha_1(b)\cdot y_1 = \alpha_2(b)\cdot y_2$. Thus we have, say, $y_1 = (\alpha_1(b)^{-1}\alpha_2(b))\cdot y_2$. Now $\alpha_1^{-1}\alpha_2 : U_1 \cap U_2 \longrightarrow G$ is continuous, so $(U_1 \cap U_2, \alpha_1^{-1}\alpha_2, U_1) : U_1 \cap U_2 \longrightarrow U_1$ is a morphism of $\underline{M}$. Clearly,

$$\begin{array}{ccc} U_1 \cap U_2 & \xrightarrow{\alpha_1^{-1}\alpha_2} & U_1 \\ {\scriptstyle e}\downarrow & & \downarrow{\scriptstyle \alpha_1} \\ U_2 & \xrightarrow[\alpha_2]{} & U \end{array}$$

commutes in $\underline{M}$, so we have

$$\begin{array}{ccc} I(U_1 \cap U_2) & \xrightarrow{I(\alpha_1^{-1}\alpha_2)} & IU_1 \\ {\scriptstyle Ie}\downarrow & & \downarrow{\scriptstyle I\alpha_1} \\ IU_2 & \xrightarrow[I\alpha_2]{} & IU \end{array}$$

in $(\underline{Top}, B)$. By the above, $I(\alpha_1^{-1}\alpha_2)(b,y_2) = (b,y_1)$ and $Ie(b,y_2) = (b,y_2)$.

Since $(\underline{Top}, B)$ has small colimits, we have the realization

$$r : (\underline{M}^*, \mathfrak{S}) \longrightarrow (\underline{Top}, B) .$$

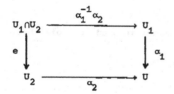

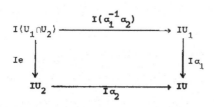

We write, for $F : \underline{M}^* \longrightarrow \mathcal{S}$,

$$rF = (E_F \xrightarrow{\pi_F} B) .$$

Since colimits in $(\underline{Top},B)$ are computed in $\underline{Top}$, E_F consists of equivalence classes $|U,x,(b,y)|$ where $U \subset B$ is open, $x \in FU$, and $(b,y) \in U\times Y$. The equivalence relation is the obvious one since I satisfies (a). π_F is given by $\pi_F(|U,x,(b,y)|) = b$. For $x \in FU$, we have

$$\begin{array}{ccc} U\times Y & \xrightarrow{i_x} & E_F \\ & \searrow \quad \swarrow {\scriptstyle \pi_F} & \\ & B & \end{array}$$

given by $i_x(b,y) = |U,x,(b,y)|$. Furthermore, by 5.8, each i_x is open and a basis for the topology of E_F is given by the collection of all images of open sets under these maps.

Let $\mathbb{G} = (rs,\varepsilon,r\eta s)$ be, as usual, the model induced cotriple. Let (p,θ) be a $\mathbb{G}$-coalgebra, and consider

$$rj(p,\theta) : r\bar{s}(p,\theta) \longrightarrow rsp .$$

For $U \in \underline{M}$ and

$$\begin{array}{ccc} U\times Y & \xrightarrow{\varphi} & X \\ & \searrow \quad \swarrow {\scriptstyle p} & \\ & B & \end{array}$$

a morphism of coalgebras, we have

$$\begin{array}{ccc} U\times Y & & \\ {\scriptstyle i_\varphi}\downarrow & \searrow^{i_\varphi} & \\ E_{\bar{s}(p,\theta)} & \xrightarrow{rj(p,\theta)} & E_{sp} \end{array} ,$$

and hence $rj(p,\theta)$ is open. Condition (a) shows that the underlying set of

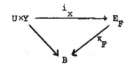

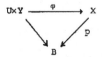

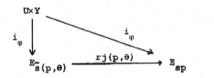

$$E_{\bar{s}(p,\theta)} \xrightarrow{rj(p,\theta)} E_{sp} \underset{\delta p}{\overset{rs\theta}{\rightrightarrows}} E_{s\pi_{sp}}$$

is an equalizer, and this together with $rj(p,\theta)$ open makes it an equalizer in <u>Top</u> . ∂_o reflects equalizers, so

$$r\bar{s}(p,\theta) \xrightarrow{rj(p,\theta)} rsp \underset{\delta p}{\overset{rs\theta}{\rightrightarrows}} (rs)^2 p$$

is an equalizer in (<u>Top</u>,B) for all G-coalgebras (p,θ) . By 3.3 then,

$$\bar{\varepsilon} : \overline{r}\,\overline{s} \xrightarrow[\sim]{} 1 ,$$

and the <u>M</u>-objects of (<u>Top</u>,B) are exactly those $X \xrightarrow{p} B$ admitting a G-coalgebra structure

$$\begin{array}{ccc} X & \xrightarrow{\theta} & E_{sp} \\ & \searrow \quad \swarrow{\scriptstyle \pi_{sp}} & \\ & B & \end{array}$$

In terms of atlases, the <u>M</u>-objects of (<u>Top</u>,B) are characterized as follows. $X \xrightarrow{p} B$ is an <u>M</u>-object iff there is a family $\mathcal{F}$ of morphisms φ in (<u>Top</u>,B) where

$$\begin{array}{ccc} U\times Y & \xrightarrow{\varphi} & X \\ & \searrow \quad \swarrow{\scriptstyle p} & \\ & B & \end{array}$$

such that

(i) $\mathcal{F}$ covers X . That is, if $x \in X$ then there is a $\varphi \in \mathcal{F}$ and a point $(b,y) \in U\times Y$ such that $\varphi(b,y) = x$.

(ii) $\mathcal{F}$ is a compatible family. That is if

$$\begin{array}{ccc} & & U\times Y \\ & & \downarrow{\scriptstyle\varphi} \\ V\times Y & \xrightarrow{\psi} & X \end{array}$$

is a pair of morphisms in $\mathcal{F}$ (where we have omitted the components over B for simplicity of notation), and $\varphi(b,y) = \psi(b',y')$ for some $(b,y) \in U\times Y$ and $(b',y') \in V\times Y$, then there is a diagram

$$\begin{array}{ccc} W\times Y & \xrightarrow{I\alpha} & U\times Y \\ {\scriptstyle I\beta}\downarrow & & \downarrow{\scriptstyle\varphi} \\ V\times Y & \xrightarrow[\psi]{} & X \end{array}$$

in $(\underline{\text{Top}},B)$, and a point $(b_o,y_o) \in W\times Y$ such that $I\beta(b_o,y_o) = (b',y')$, $I\alpha(b_o,y_o) = (b,y)$.

(iii) $\mathcal{F}$ is open, i.e. each $\varphi \in \mathcal{F}$ is an open map.

The proof that these are indeed the $\underline{M}$-objects is essentially the same as that for simplicial spaces and manifolds, and hence details will be left to the reader. We remark only that the correspondence between $\underline{M}$-objects and $X \xrightarrow{p} B$ possessing such a family $\mathcal{F}$ is given as follows. If $\mathcal{G} \longrightarrow sp$ is an $\underline{M}$-atlas for $X \xrightarrow{p} B$, then any generating family for $\mathcal{G}$ satisfies (i)-(iii). On the other hand, if there is such a family $\mathcal{F}$ for $X \xrightarrow{p} B$, and if $\mathcal{G}$ is the preatlas generated by $\mathcal{F}$, then it is easy to see that (i)-(iii) for $\mathcal{F}$ imply

$$\begin{array}{ccc} E_{\mathcal{G}} & \xrightarrow[\sim]{e} & X \\ & {\scriptstyle\pi}\searrow \quad \swarrow{\scriptstyle p} & \\ & B & \end{array}$$

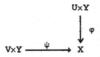

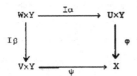

Given this description of $\underline{M}$-objects, it follows that regular $\underline{M}$-objects are $X \xrightarrow{p} B$ in (Top,B) possessing a family $\mathcal{F}$ satisfying (i), (iii), and (ii') For any pair

$$\begin{array}{ccc} & & U\times Y \\ & & \downarrow{\scriptstyle\varphi} \\ V\times Y & \xrightarrow[\psi]{} & X \end{array}$$

in $\mathcal{F}$ there is a pullback diagram in (Top,B) of the form

$$\begin{array}{ccc} W\times Y & \xrightarrow{I\alpha} & U\times Y \\ {\scriptstyle I\beta}\big\downarrow & & \big\downarrow{\scriptstyle\varphi} \\ V\times Y & \xrightarrow[\psi]{} & X \end{array} .$$

By 2.2 each $\varphi \in \mathcal{F}$ is injective, and hence a homeomorphism into.

We show that fibre bundles are regular M-objects. Recall that a fibre bundle with base B, fibre Y, and structural group G, is an object $X \xrightarrow{p} B$ in (Top,B) for which there exists an open covering $\{U_i\}$ of B with the following properties. For each U_i there is a homeomorphism

$$\begin{array}{ccc} U_i\times Y & \xrightarrow[\sim]{\Phi_i} & p^{-1}(U_i) \\ & \searrow \quad \swarrow{\scriptstyle p} & \\ & U_i & \end{array}$$

where the unnamed map is projection onto U_i. Furthermore, the Φ_i are required to be compatible in the following sense. Namely, if U_j is another element of the covering, then we have a diagram

$$\begin{array}{ccccc} (U_i\cap U_j)\times Y & \xrightarrow[\sim]{\Phi_i} & p^{-1}(U_i\cap U_j) & \xrightarrow[\sim]{\Phi_j^{-1}} & (U_i\cap U_j)\times Y \\ & \searrow & \big\downarrow{\scriptstyle p} & \swarrow & \\ & & U_i\cap U_j & & \end{array} .$$

And we require that there exist an $\alpha_{ij} : U_i\cap U_j \longrightarrow G$ (necessarily unique) such that

$$\Phi_j^{-1}\Phi_i = I\alpha_{ij} .$$

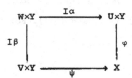

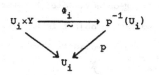

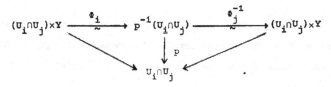

Now if $X \xrightarrow{p} B$ is such a fibre bundle, choose a covering $\{U_i\}$ as above, and let $\mathcal{F}$ consist of all

$$\begin{array}{ccc} U_i \times Y & \xrightarrow{\Phi_i} & X \\ & \searrow \quad \swarrow{\scriptstyle p} & \\ & B & \end{array}$$

in the chosen system. (We use the same letter Φ_i to denote also the composite $U_i \times Y \xrightarrow{\Phi_i} p^{-1}(U_i) \longrightarrow X$.) Clearly, $\mathcal{F}$ satisfies (i) and (iii). For any pair Φ_i , Φ_j in $\mathcal{F}$, we have the diagram

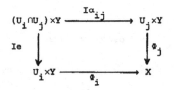

$$\begin{array}{ccc} (U_i \cap U_j) \times Y & \xrightarrow{I\alpha_{ij}} & U_j \times Y \\ {\scriptstyle Ie}\downarrow & & \downarrow{\scriptstyle \Phi_j} \\ U_i \times Y & \xrightarrow[\Phi_i]{} & X \end{array}$$

in (Top,B) . (α_{ij} as above), and it is trivial to verify that this is a pullback. (If $U_i \cap U_j = \emptyset$, the obvious modifications are to be made in all the preceding.) Thus, $\mathcal{F}$ satisfies (ii'), making $X \xrightarrow{p} B$ a regular $\underline{M}$-object. Note, however, that the converse is <u>not</u> true here. That is, not every regular $\underline{M}$-object is a fibre bundle. In fact, the fibre bundles can be characterized as those $\underline{M}$-objects having an atlas with regular generating set $\mathcal{F}$ such that for each $\varphi \in \mathcal{F}$,

$$\begin{array}{ccc} U \times Y & \xrightarrow{\varphi} & X \\ \downarrow & & \downarrow{\scriptstyle p} \\ U & \hookrightarrow & B \end{array}$$

is a pullback diagram in <u>Top</u>. The regular $\underline{M}$-objects are a common generalization of sheaves and fibre bundles. Sheaves are obtained by choosing Y = point, $G = (e)$. We shall give a separate treatment of these in § 6, since the model induced cotriple is idempotent in this case.

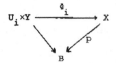

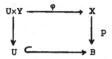

(5) G-spaces

Let G be a topological group, and denote by $\underline{G}$ the category with one object G and morphisms the elements $g \in G$. Define

$$I : \underline{G} \longrightarrow \underline{\mathrm{Top}}$$

by $IG = G$, and $Ig : G \longrightarrow G$ is left translation by $g \in G$. Consider a pair of maps

$$\begin{array}{ccc} & & G \\ & & \downarrow{\scriptstyle Ig_1} \\ G & \xrightarrow[Ig_2]{} & G \end{array}$$

and elements $g_1', g_2' \in G$ such that $Ig_1(g_1') = Ig_2(g_2')$ i.e. $g_1g_1' = g_2g_2'$. Then

$$\begin{array}{ccc} G & \xrightarrow{Ig_1^{-1}} & G \\ {\scriptstyle Ig_2^{-1}}\downarrow & & \downarrow{\scriptstyle Ig_1} \\ G & \xrightarrow[Ig_2]{} & G \end{array}$$

commutes trivially, and $g = g_1g_1' = g_2g_2'$ provides the element necessary for condition (a).

In the adjointness

$$\underline{\mathrm{Top}} \underset{r}{\overset{s}{\rightleftarrows}} (\underline{G}^*, \mathcal{S})$$

we can identify $(\underline{G}^*, \mathcal{S})$ as the category of right G-sets and equivariant functions. For a functor $F : \underline{G}^* \longrightarrow \mathcal{S}$ is determined by the set $FG = X$ and the operations $Fg : X \longrightarrow X$ for $g \in G$. Writing these as $Fg(x) = x\cdot g$, functoriality is simply $x\cdot 1 = x$ and $(x\cdot g_1)\cdot g_2 = x\cdot(g_1g_2)$. A natural transformation $F_1 \longrightarrow F_2$ is simply a G-equivariant function $X_1 \longrightarrow X_2$. If $F : \underline{G}^* \longrightarrow \mathcal{S}$ is a functor (or G-set), then rF consists of equivalence classes $|x,g|$ where $x \in FG = X$, and $g \in G$. The equivalence relation is determined by $|x\cdot g_1, g_2| = |x, g_1\cdot g_2|$. If $g_2 = 1$ we have $|x\cdot g, 1| = |x,g|$, so the functions $rF \longrightarrow X$ by $|x,g| \rightsquigarrow x\cdot g$ and

$X \longrightarrow rF$ by $x \longrightarrow |x,1|$ provide a bijection of sets. Under this identification, the canonical maps

$$i_x : G \longrightarrow X$$

for $x \in X$ become simply $i_x(g) = x\cdot g$. By 5.8, these maps are open, and a basis for the topology on X (or rF) is given by all images of open sets under these maps. The image of i_x is just $x\cdot G$ - the orbit of x . For $x \in X$, let $G_x = \{g \in G \; x\cdot g = x\}$ be the isotropy subgroup of x . i_x induces a map $\bar{i}_x : G/G_x \longrightarrow x\cdot G$ so that

$$\begin{array}{ccc} G & & \\ {\scriptstyle p_x}\downarrow & \searrow{\scriptstyle i_x} & \\ G/G_x & \xrightarrow[\bar{i}_x]{} & x\cdot G \end{array}$$

commutes, with p_x the natural projection. Since i_x is open and p_x is onto, $\bar{i}_x$ is open, and hence a homeomorphism. Thus, the topology of rF (or X) is completely determined. Furthermore, if $\gamma : F_1 \longrightarrow F_2$ is a natural transformation (equivariant function) then $r\gamma : rF_1 \longrightarrow rF_2$ is open. This follows, since for $x_1 \in F_1G = X_1$, the diagram

$$\begin{array}{ccc} G & & \\ {\scriptstyle i_{x_1}}\downarrow & \searrow{\scriptstyle i_{\gamma(x_1)}} & \\ X_1 & \xrightarrow{r\gamma} & X_2 \end{array}$$

commutes. This remark together with condition (a) gives $\bar{\varepsilon} : \bar{r}\bar{s} \xrightarrow[\sim]{} 1$ in

$$\underline{\mathrm{Top}}_{\mathbb{G}} \underset{\bar{r}}{\overset{\bar{s}}{\rightleftarrows}} (\underline{G}^*, \mathfrak{S}) \quad .$$

$\mathbb{G}$, of course, is the model induced cotriple $\mathbb{G} = (G,\varepsilon,\delta)$. Here, if $X \in \underline{\mathrm{Top}}$ then $sX : \underline{G}^* \longrightarrow \mathfrak{S}$ is the G-set $sX(G) = (G,X)$ with G-action $\varphi\cdot g = \varphi\cdot Ig$. Thus, $GX = (G,X)$ as a set, with the above described topology.

Finally, it is obvious that r reflects isomorphisms. Namely, suppose

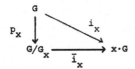

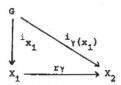

$\gamma : F_1 \longrightarrow F_2$ is a natural transformation. Then $r\gamma : rF_1 \longrightarrow rF_2$ is just the G-equivariant function $\gamma G : F_1 G = X_1 \longrightarrow X_2 = F_2 G$ which is continuous in the above topology. It is a homeomorphism iff γG is injective and surjective, iff γ is an equivalence. Thus, by 4.4, $(\bar{\varepsilon},\bar{\eta}) : \bar{r} \longrightarrow\!\!| \;\bar{s}$ is an equivalence of categories.

In a subsequent paper, we will show that one obtains all G-spaces as coalgebras if one considers the singular functor as taking values in $(\underline{G^*},\underline{Top})$.

6. Idempotent cotriples

Let $\mathbb{G} = (G,\varepsilon,\delta)$ be a cotriple in a category $\underline{A}$. We say $\mathbb{G}$ is *idempotent*, if $\delta : G \xrightarrow{\approx} G^2$. Later in this section, we consider categories with models for which the model induced cotriple is idempotent. As will be seen from the remarks below, much of the analysis of § 3 and § 4 becomes trivial in this case. For now, however, let $\mathbb{G}$ denote an arbitrary cotriple in $\underline{A}$.

Proposition 6.1

$\mathbb{G}$ is idempotent iff

$$G\varepsilon = \varepsilon G : G^2 \longrightarrow G \ .$$

Proof: Suppose $\mathbb{G}$ is idempotent. Since $\delta : G \longrightarrow G^2$ is an equivalence, and $G\varepsilon\cdot\delta = \varepsilon G\cdot\delta = 1_G$, we get $G\varepsilon = \delta^{-1} = \varepsilon G$. On the other hand, assume $G\varepsilon = \varepsilon G$. By naturality of ε , we have a diagram

$$\begin{array}{ccc} G^2 & \xrightarrow{G\delta} & G^3 \\ {\scriptstyle \varepsilon G}\downarrow & & \downarrow{\scriptstyle \varepsilon G^2} \\ G & \xrightarrow{\delta} & G^2 \end{array} \ .$$

Now $G\varepsilon = \varepsilon G$ given $G\varepsilon G = \varepsilon G^2$, so that

$$\delta\cdot\varepsilon G \;=\; \varepsilon G^2\cdot G\delta \;=\; G\varepsilon G\cdot G\delta \;=\; G(\varepsilon G\cdot\delta) \;=\; 1_{G^2}$$

and δ is an equivalence, since $\varepsilon G\cdot\delta = 1_G$ always.

<u>Proposition 6.2</u>

$\mathbb{G}$ is idempotent iff for all $(A,\theta) \in \underline{A}_{\mathbb{G}}$, $\varepsilon A : GA \longrightarrow A$ is a monomorphism.

<u>Proof:</u> Suppose εA is a monomorphism for all coalgebras (A,θ) . Since $\varepsilon A \cdot \theta = 1_A$, εA is also a split epimorphism. But then εA is an isomorphism with inverse θ . In particular, $(GA,\delta A)$ is always a coalgebra, so εGA is an isomorphism with inverse δA , and $\mathbb{G}$ is idempotent. Suppose $\mathbb{G}$ is idempotent. Naturality of ε gives for each $(A,\theta) \in \underline{A}_{\mathbb{G}}$,

$$\begin{array}{ccc} GA & \xrightarrow{G\theta} & G^2A \\ {\scriptstyle \varepsilon A}\downarrow & & \downarrow{\scriptstyle \varepsilon GA} \\ A & \xrightarrow{\theta} & GA \end{array} .$$

By 6.1, $\varepsilon GA = G\varepsilon A$. Thus,

$$\theta \cdot \varepsilon A = \varepsilon GA \cdot G\theta = G\varepsilon A \cdot G\theta = G(\varepsilon A \cdot \theta) = 1_{GA} ,$$

and εA is an isomorphism.

<u>Remark</u>.

Equivalent to 6.2 is: $\mathbb{G}$ is idempotent iff $\theta : A \longrightarrow GA$ is epic for all $(A,\theta) \in \underline{A}_{\mathbb{G}}$.

<u>Proposition 6.3</u>

$\mathbb{G}$ is idempotent iff

$$L : \underline{A}_{\mathbb{G}} \longrightarrow \underline{A}$$

is full.

<u>Proof:</u> Suppose L is full, then for each $(A,\theta) \in \underline{A}_{\mathbb{G}}$ $\varepsilon A : GA \longrightarrow A$ is a morphism of coalgebras i.e.

$$\begin{array}{ccc} GA & \xrightarrow{\varepsilon A} & A \\ {\scriptstyle \delta A}\downarrow & & \downarrow{\scriptstyle \theta} \\ G^2A & \xrightarrow{G\varepsilon A} & GA \end{array}$$

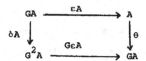

commutes. But then εA is an isomorphism, and $\mathbb{G}$ is idempotent by 6.2. For the other direction, suppose $\mathbb{G}$ is idempotent, (A,θ) and (A',θ') are coalgebras, and $f : A \longrightarrow A'$ is an arbitrary $\underline{A}$-morphism. Consider the diagram

$$\begin{array}{ccc} A & \xrightarrow{f} & A' \\ {\scriptstyle \theta}\downarrow & & \downarrow{\scriptstyle \theta'} \\ GA & \xrightarrow{Gf} & GA' \\ {\scriptstyle \varepsilon A}\downarrow & & \downarrow{\scriptstyle \varepsilon A'} \\ A & \xrightarrow{f} & A' \end{array} .$$

The whole diagram (without the arrow Gf) clearly commutes, as does the bottom by naturality of ε. But by 6.2, $\varepsilon A'$ is monic, so the top commutes also, and f is a morphism of coalgebras. Since f was arbitrary, L is full.

Putting 6.2 and 6.3 together it follows that $\mathbb{G}$ is idempotent iff

$$L : \underline{A}_{\mathbb{G}} \longrightarrow \underline{A}$$

provides an equivalence between $\underline{A}_{\mathbb{G}}$ and the full subcategory of $\underline{A}$ consisting of objects $A \in \underline{A}$ such that $\varepsilon A : GA \xrightarrow{\sim} A$.

<u>Proposition 6.4</u>

$\mathbb{G}$ is idempotent iff for all $(A,\theta) \in \underline{A}_{\mathbb{G}}$,

$$G\theta = \delta A : GA \longrightarrow G^2A .$$

<u>Proof:</u> We know that for any cotriple $\mathbb{G}$, and for any coalgebra (A,θ) ,

$$A \xrightarrow{\theta} GA \underset{\delta A}{\overset{G\theta}{\rightrightarrows}} G^2A$$

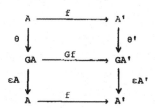

is an equalizer diagram. Now if $\mathbb{G}$ is idempotent, then θ is an isomorphism and it follows that $G\theta = \delta A$. On the other hand, if $G\theta = \delta A$ then the equalizer condition provides a morphism $f : GA \longrightarrow A$ such that $\theta \cdot f = 1_{GA}$. Obviously, $f = \varepsilon A$, so εA is monic and $\mathbb{G}$ is idempotent by 6.2.

Suppose now that $\underline{A}$ is a category with models $I : \underline{M} \longrightarrow \underline{A}$. If $\underline{A}$ has enough colimits, let $\mathbb{G} = (G,\varepsilon,\delta)$ be the model induced cotriple.

<u>Proposition 6.5</u>

$\mathbb{G}$ is idempotent iff for all $(A,\theta) \in \underline{A}_{\mathbb{G}}$,

$$rj(A,\theta) : r\bar{s}(A,\theta) \xrightarrow{\sim} GA .$$

<u>Proof</u>: Consider in $(\underline{A}^*,\mathfrak{S})$ the monomorphism

$$j(A,\theta) : \bar{s}(A,\theta) \longrightarrow sA$$

for $(A,\theta) \in \underline{A}_{\mathbb{G}}$. If $\mathbb{G}$ is idempotent, then by 6.3 $j(A,\theta)$ is also epic, and hence an equivalence, making $rj(A,\theta)$ an isomorphism. On the other hand, $rj(A,\theta)$ equalizes the pair

$$GA \underset{\delta A}{\overset{G\theta}{\rightrightarrows}} G^2A .$$

Thus, if $rj(A,\theta)$ is an isomorphism, we have $G\theta = \delta A$ and $\mathbb{G}$ idempotent by 6.4.

Summarizing, if the model induced cotriple $\mathbb{G}$ is idempotent, then for all $(A,\theta) \in \underline{A}_{\mathbb{G}}$, we have $G\theta = \delta A$, and $rj(A,\theta)$ an isomorphism. But then, it is trivial to verify that

$$r\bar{s}(A,\theta) \xrightarrow{rj(A,\theta)} GA \underset{\delta A}{\overset{G\theta}{\rightrightarrows}} G^2A$$

is an equalizer diagram. Thus, by 3.3, $\bar{\varepsilon} : \bar{r}\bar{s} \xrightarrow{\sim} 1$ in the adjoint pair

$$\underline{A}_{\mathbb{G}} \underset{\bar{r}}{\overset{\bar{s}}{\rightleftarrows}} (\underline{M}^*,\mathfrak{S}) .$$

This in turn shows, as we have seen in § 3, that the class of $\underline{M}$-objects of $\underline{A}$ is exactly the class of objects $A \in \underline{A}$ admitting a $\mathbb{G}$-coalgebra structure. This, by the above, is the class of those $A \in \underline{A}$ such that $\varepsilon A : GA \xrightarrow{\approx} A$. Morphisms of $\underline{M}$-objects are arbitrary $\underline{A}$-morphisms by 6.3 .

Examples.

(1.) Let $\underline{C}$ be the category of compact Hausdorff spaces and continuous maps. Let $I : \underline{C} \longrightarrow \underline{Top}$ be the inclusion. $\underline{C}$ is not small, so $(\underline{C}^*, \mathfrak{S})$ is an illegitimate category. Therefore, since $\underline{Top}$ has only set indexed colimits, we must be careful about constructing

$$r : (\underline{C}^*, \mathfrak{S}) \longrightarrow \underline{Top} \quad ,$$

i.e. we cannot simply write down the usual colimit expression for rF in $\underline{Top}$. What we shall do is to construct r legally by another method, and then show $r \longrightarrow\!\!\!\mid s$. Thus, in fact, the requisite colimits will exist in $\underline{Top}$.

To construct r , let $*$ be a fixed choice of a one point space. Clearly $* \in \underline{C}$. (We drop I from the notation.) Let

$$e : (*,X) \xrightarrow{\approx} X$$

be the evaluation map onto the underlying set of X . If $x \in X$, let $\tilde{x} : * \longrightarrow X$ denote the unique map such that $e(\tilde{x}) = x$. Now suppose $F : \underline{C}^* \longrightarrow \mathfrak{S}$ is an arbitrary functor. For each $C \in \underline{C}$ and $y \in F(C)$ we define a set theoretical function

$$i(C,y) : C \longrightarrow F(*)$$

by $i(C,y)(c) = F\tilde{c}(y)$ for $c \in C$, i.e. $\tilde{c} : * \longrightarrow C$, and $i(C,y)(c)$ is the image of $y \in F(C)$ under the function $F\tilde{c} : F(C) \longrightarrow F(*)$. Let rF be the set $F(*)$ with the weak topology determined by the $i(C,y)$. Thus, $U \subset F(*)$ is open iff $i(C,y)^{-1}U$ is open in C for all $C \in \underline{C}$ and $y \in FC$. Equivalently, if $X \in \underline{Top}$ a function $f : F(*) \longrightarrow X$ is continuous iff each composite $f \cdot i(C,y)$ is. If

$\gamma : F' \longrightarrow F$ is a natural transformation, put

$$r\gamma \;=\; \gamma(*) : F'(*) \longrightarrow F(*) \; .$$

$r\gamma$ is continuous, since for each $C \in \underline{C}$, $y \in F'C$, and $c \in C$, we have the diagram

$$\begin{array}{ccc} F'(C) & \xrightarrow{\gamma(C)} & F(C) \\ {\scriptstyle F'\tilde{c}}\downarrow & & \downarrow{\scriptstyle F\tilde{c}} \\ F'(*) & \xrightarrow[\gamma(*)]{} & F(*) \end{array} \quad ,$$

and hence the diagram

$$\begin{array}{ccc} C & & \\ {\scriptstyle i(C,y)}\downarrow & \searrow^{i(C,\gamma(C)(y))} & \\ F'(*) & \xrightarrow[\gamma(*)]{} & F(*) \end{array} \quad .$$

With this definition, it is clear that r is a functor.

As always, we have

$$s : \underline{Top} \longrightarrow (\underline{C}^*, \mathfrak{S})$$

by $sX.C = (C,X)$ for $X \in \underline{Top}$ and $C \in \underline{C}$. We want to show that $r \longrightarrow\!\!\!\!| \; s$. For this, define natural transformations

$$\varepsilon : rs \longrightarrow 1$$
$$\eta : 1 \longrightarrow sr$$

as follows. If $X \in \underline{Top}$, let

$$\varepsilon X \;=\; e : rsX \;=\; (*,X) \longrightarrow X \; .$$

εX is continuous, since for each $C \in \underline{C}$, $\varphi : C \longrightarrow X$ in $sX.C$, and $c \in C$, we have

$$i(C,\varphi)(c) \;=\; sX(\tilde{c})(\varphi) \;=\; \varphi\cdot\tilde{c} \; ,$$

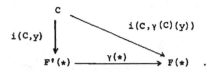

and hence

$$\begin{array}{ccc} C & & \\ {\scriptstyle i(C,\varphi)}\downarrow & \searrow^{\varphi} & \\ (*,X) & \xrightarrow{\ e\ } & X \end{array} \quad .$$

εX is clearly natural in X. If $F : \underline{C}^* \longrightarrow \mathcal{S}$, and $C \in \underline{C}$, let

$$\eta F : F \longrightarrow srF$$

be defined by:

$$\eta F(C)(y) = i(C,y) : C \longrightarrow rF = F(*)$$

for $y \in F(C)$. It is immediate that $\eta F(C)$ is natural in both C and F. Consider the composites

$$s \xrightarrow{\ \eta s\ } srs \xrightarrow{\ s\varepsilon\ } s$$

and $$r \xrightarrow{\ r\eta\ } rsr \xrightarrow{\ \varepsilon r\ } r \quad .$$

That the first is 1_s follows from the computation used to prove εX continuous. For the second, let $F : \underline{C}^* \longrightarrow \mathcal{S}$ be an arbitrary functor. Then

$$rF \xrightarrow{\ r\eta F\ } rsrF \xrightarrow{\ \varepsilon rF\ } rF$$

is the composite

$$F(*) \xrightarrow{\ \eta F(*)\ } (*,F(*)) \xrightarrow{\ e\ } F(*) \quad .$$

If $x \in F(*)$, then

$$\eta F(*)(x) = i(*,x) : * \longrightarrow F(*)$$

is simply $\tilde{x}$, for

$$i(*,x)(*) = F\tilde{*}(x) = x \,,$$

because $\tilde{*} : * \longrightarrow *$ is 1_*.

But then $(e\cdot\eta F(*))(x) = x$ and we are done. Thus, we have

$$(\varepsilon,\eta) : r \dashv s ,$$

and in the usual way, we obtain a 1-1 correspondence

$$(rF,X) \sim (F,sX)$$

for $F : \underline{C}^* \longrightarrow \mathcal{S}$ and $X \in \underline{\text{Top}}$. In particular, the latter is a set.

Let $\mathbb{G} = (rs,\varepsilon,r\eta s) = (G,\varepsilon,\delta)$ be the model induced cotriple in $\underline{\text{Top}}$. Then if $X \in \underline{\text{Top}}$, $GX = (*,X)$ with the above topology, and

$$\varepsilon X = e : (*,X) \longrightarrow X .$$

εX is clearly a monomorphism, so by 6.2 $\mathbb{G}$ is idempotent. Therefore, the category $\underline{\text{Top}}_{\mathbb{G}}$ of $\mathbb{G}$-coalgebras is the full subcategory of $\underline{\text{Top}}$ consisting of all X for which $\varepsilon X : GX \longrightarrow X$ is a homeomorphism. That is, spaces X having the weak topology with respect to continuous maps $\varphi : C \longrightarrow X$ where $C \in \underline{C}$. Such spaces are called <u>compactly generated weakly Hausdorff</u>.

Since $\mathbb{G}$ is idempotent, it follows that we have $\bar{\varepsilon} : \bar{r}\bar{s} \xrightarrow{\sim} 1$ in the adjoint pair

$$\underline{\text{Top}}_{\mathbb{G}} \underset{\bar{r}}{\overset{\bar{s}}{\rightleftarrows}} (\underline{C}^*,\mathcal{S}) .$$

By considering the category $\underline{Q}$ of quasi-spaces and quasi-continuous maps, we will show that $\bar{\eta} : 1 \longrightarrow \bar{s}\bar{r}$ is <u>not</u> an equivalence. Recall from [7] that a <u>quasi-space</u> is a set X together with a family $\mathcal{A}(C,X)$ of <u>admissible</u> functions $C \longrightarrow X$ for each $C \in \underline{C}$. These families satisfy the following axioms:

(i) Any constant map $C \longrightarrow X$ is in $\mathcal{A}(C,X)$.

(ii) If $\alpha : C' \longrightarrow C$ is in $\underline{C}$ and $\varphi \in \mathcal{A}(C,X)$, then $\varphi\cdot\alpha \in \mathcal{A}(C',X)$.

(iii) If C is the disjoint union of C_1 and C_2 in $\underline{C}$, then $\varphi \in \mathcal{A}(C,X)$ iff $\varphi|C_i \in \mathcal{A}(C_i,X)$ for $i = 1,2$.

(iv) If $\alpha : C_1 \longrightarrow C_2$ is surjective for $C_1, C_2 \in \underline{C}$, then $\varphi \in \mathcal{A}(C_2,X)$ iff $\varphi \cdot \alpha \in \mathcal{A}(C_1,X)$.

A function $f : X \longrightarrow Y$ is quasi-continuous iff for $C \in \underline{C}$ and $\varphi \in \mathcal{A}(C,X)$, $f \cdot \varphi \in \mathcal{A}(C,Y)$.

The admissible maps provide an embedding

$$\mathcal{A} : \underline{Q} \longrightarrow (\underline{C}^*, \mathfrak{S})$$

defined by $\mathcal{A}X.C = \mathcal{A}(C,X)$ for $X \in \underline{Q}$ and $C \in \underline{C}$. The effect on morphisms is composition in both variables, which makes sense by axiom (ii) and the above definition of quasi-continuous. (Note that axioms (iii) and (iv) can then be combined to: $\mathcal{A}X : \underline{C}^* \longrightarrow \mathfrak{S}$ preserves finite limits.) $\mathcal{A}$ is faithful, since if $f,g : X \longrightarrow Y$ are quasi-continuous and $\mathcal{A}f = \mathcal{A}g : \mathcal{A}X \longrightarrow \mathcal{A}Y$, then, in particular, $\mathcal{A}f.* = \mathcal{A}g.* : \mathcal{A}(*,X) \longrightarrow \mathcal{A}(*,Y)$. However, by (i), if $x \in X$ then $\tilde{x} \in \mathcal{A}(*,X)$. Thus, $f \cdot \tilde{x} = g \cdot \tilde{x}$ so $f(x) = g(x)$ and $f = g$. Clearly, if $\mathcal{A}X = \mathcal{A}Y$ then $X = Y$ as quasi spaces, so we may regard $\underline{Q}$ as a (non-full) subcategory of $(\underline{C}^*, \mathfrak{S})$ by means of $\mathcal{A}$.

Now if $\bar{\eta} : F \xrightarrow[\sim]{} \bar{s}\bar{r}F$ for all $F : \underline{C}^* \longrightarrow \mathfrak{S}$, then for each quasi-space X we must have

$$\bar{\eta}\mathcal{A}X : \mathcal{A}X \xrightarrow[\sim]{} \bar{s}\bar{r}\mathcal{A}X .$$

But $\bar{r}\mathcal{A}X$ is just

$$r\mathcal{A}X = \mathcal{A}(*,X)$$

with the weak topology determined by the maps

$$i(C,\varphi) : C \longrightarrow \mathcal{A}(*,X)$$

for $C \in \underline{C}$ and $\varphi \in \mathcal{A}(C,X)$. For $c \in \underline{C}$ we have

$$i(C,\varphi)(c) = \mathcal{A}X(\tilde{c})(\varphi) = \varphi \cdot \tilde{c} ,$$

so that

$$\begin{array}{ccc} C & & \\ {\scriptstyle i(C,\varphi)}\downarrow & \searrow^{\varphi} & \\ (*,X) & \xrightarrow{\ e\ } & X \end{array} \quad .$$

By axiom (i) for quasi-spaces, e is a bijection of sets. Making e a homeomorphism, we provide X with the topology : $U \subset X$ is open iff $\varphi^{-1}U$ is open in C for all $C \in \underline{C}$ and $\varphi \in \mathcal{A}(C,X)$. With this topology, X is clearly compactly generated weakly Hausdorff - i.e. a $\mathcal{C}$-coalgebra. Under the identification e , the natural transformation

$$\bar{\eta}\mathcal{A}X : \mathcal{A}X \longrightarrow \bar{s}\bar{r}\mathcal{A}X$$

becomes simply the inclusion

$$\mathcal{A}X \longrightarrow sX ,$$

which expresses the fact that every admissable map is continuous. (Note that we can replace $\bar{s}$ by s since every continuous map of coalgebras is a coalgebra morphism.) Thus, if $\bar{\eta}\mathcal{A}X$ is an equivalence we must have

$$\mathcal{A}X = sX$$

for all $X \in \underline{Q}$. However, Spanier in [7] provides an example of a quasi-space X such that for <u>no</u> topology on X is

$$\mathcal{A}X = sX ,$$

in particular, not for the above topology. Hence, $\bar{\eta} : 1 \longrightarrow \bar{s}\bar{r}$ is not an equivalence.

(2.) Sheaves

Let X be a fixed topological space, and denote by $\underline{X}$ the category of open sets of X . That is, an object of $\underline{X}$ is an open set U of X , and a morphism $U \longrightarrow V$ is an inclusion. Let

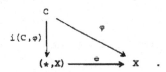

$$I : \underline{X} \longrightarrow (\underline{Top}, X)$$

be the functor which assigns to each open set $U \subset X$ the inclusion $i_U : U \longrightarrow X$, and to each inclusion $U \longrightarrow V$ the triangle

$$\begin{array}{ccc} U & \hookrightarrow & V \\ & {\scriptstyle i_U}\searrow \quad \swarrow{\scriptstyle i_V} & \\ & X & \end{array} \quad .$$

(For properties of $(\underline{Top}, X)$ see § 5 example (4.).) I trivially satisfies condition (a).

Since $(\underline{Top}, X)$ has small colimits, we have the usual adjoint pair

$$(\underline{Top}, X) \underset{r}{\overset{s}{\rightleftarrows}} (\underline{X}^*, \mathfrak{S}) \ .$$

Here $(\underline{X}^*, \mathfrak{S})$ is the category of pre-sheaves of sets over X [4]. The singular functor s is just the section functor, i.e. if $p : E \longrightarrow X$ is in $(\underline{Top}, X)$, then $sp : \underline{X}^* \longrightarrow \mathfrak{S}$ is the functor

$$sp(U) \;=\; \{\varphi : U \longrightarrow E \mid p \cdot \varphi = i_U\}$$

and $sp(j)(\varphi) = \varphi \cdot j$ for $j : V \longrightarrow U$ an inclusion. The realization r is the étalé space functor described in [4, p. 110]. To see this, let $F : \underline{X}^* \longrightarrow \mathfrak{S}$ be a pre-sheaf of sets. Let

$$rF \;=\; (E_F \xrightarrow{\ \pi_F\ } X) \ .$$

Since colimits in $(\underline{Top}, X)$ are computed in $\underline{Top}$, E_F can be described as follows. Consider all triples (U, s, x) for $U \in \underline{X}$, $s \in FU$, $x \in U$. Let $\equiv$ be the equivalence relation (4.1) $(U_1, s_1, x_1) \equiv (U_2, s_2, x_2)$ iff there are $j_1 : V \longrightarrow U_1$ and $j_2 : V \longrightarrow U_2$ in $\underline{X}$, and $x \in V$ such that $j_1 x = x_1$, $j_2 x = x_2$, and $Fj_1(s_1) = Fj_2(s_2)$, i.e. iff $x_1 = x_2 \in V \subset U_1 \cap U_2$, and $Fj_1(s_1) = Fj_2(s_2)$. Then E_F is the set of equivalence classes $|U, s, x|$ with the weak topology determined by the functions

$$i(U,s) : U \longrightarrow E_F$$

given by $i(U,s)(x) = |U,s,x|$. π_F is given by $\pi_F|U,s,x| = x$. By 5.8, each $i(U,s)$ is open, and their images form a basis for the topology of E_F . Thus, π_F is a local homeomorphism. Comparing this description with that of [4, p. 110], one sees immediately that rF is the étalé space over X associated to the pre-sheaf F .

Let $\mathbb{G} = (rs,\varepsilon,r\eta s)$ be the model induced cotriple in (Top,X) . For $p : E \longrightarrow X$ in (Top,X) , the points of E_{sp} are equivalence classes $|U,\varphi,x|$ where $\varphi : U \longrightarrow E$ is a section of p over U . The counit $\varepsilon p : rsp \longrightarrow p$ is given by $\varepsilon p|U,\varphi,x| = \varphi x$. Let (p,θ) be a $\mathbb{G}$-coalgebra. Then

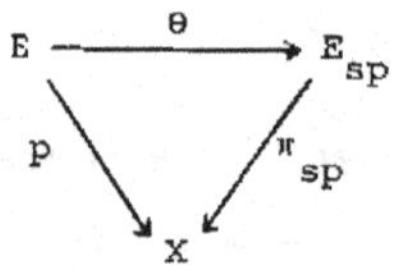

with the usual properties on θ . We show εp is monic. Namely, suppose $\varepsilon p|U_1,\varphi_1,x_1| = \varepsilon p|U_2,\varphi_2,x_2|$, i.e. $\varphi_1 x_1 = \varphi_2 x_2$. Since φ_1 and φ_2 are sections of p , it follows that $x_1 = x_2$. Furthermore, $\theta\cdot\varphi_1$ and $\theta\cdot\varphi_2$ are clearly sections of π_{sp} , and they agree at $x_1 = x_2$. Since π_{sp} is a local homeomorphism, there is an open neighborhood $W \subset U_1 \cap U_2$ of $x_1 = x_2$ such that

$$\theta\cdot\varphi_1|W = \theta\cdot\varphi_2|W .$$

But then $\varphi_1|W = \varphi_2|W$, so $|U_1,\varphi_1,x_1| = |U_2,\varphi_2,x_2|$. By 6.2, $\mathbb{G}$ is idempotent, and $\bar{\varepsilon} : \overline{rs} \xrightarrow{\approx} 1$. Thus, we may identify $(\text{Top},X)_{\mathbb{G}}$ with a full subcategory of the category of pre-sheaves $(\underline{X}^*,\mathcal{S})$. This is the usual identification of an étalé space with its sheaf of sections.

(3.) Schemes

One of the most interesting examples of a model induced cotriple is obtained by choosing $\underline{M} = \underline{R}^*$ - the dual of the category of commutative rings with unit, $\underline{A} = \underline{LRS}$ - the category of local ringed spaces, and $I = \text{Spec} : \underline{R}^* \longrightarrow \underline{LRS}$. This example has been independently considered by Gabriel [3]. If $\mathbb{G} = (G,\varepsilon,\delta)$ is the model induced cotriple in this situation, then Gabriel considers the full subcategory of $\underline{LRS}$ consisting of local ringed spaces X , for which $\varepsilon X : GX \xrightarrow{\approx} X$. A scheme is shown to be such an object. By the remark following 6.2 and 6.3, in order to bring this treatment in line with ours, it suffices to show that $\mathbb{G}$ is idempotent. This can be done, but due to limitations of space and time we will save the details of this example for a separate paper.

Bibliography

[1] S. Eilenberg - J. Moore, Adjoint functors and triples, Ill. Jour. Math. 9 (1965) 381 - 389.

[2] S. Eilenberg - J.A. Zilber, Semi-simplicial complexes and singular homology, Ann. of Math. 51 (1950) 499 - 513.

[3] P. Gabriel, Séminaire Heidelberg - Strasbourg Exposé 1, Mimeographed Notes Strasbourg University.

[4] R. Godement, Théorie des faisceaux, Hermann, Paris 1958.

[5] D. Kan, Functors involving c.s.s. complexes, Trans. AMS 87 (1958) 330 - 346.

[6] J. Milnor, The geometric realization of a semi-simplicial complex, Ann. of Math. 65 (1957) 357 - 362.

[7] E. Spanier, Quasi-topologies, Duke Math. J. 30 (1963) 1 - 14.

HOMOLOGY AND STANDARD CONSTRUCTIONS

by Michael Barr [1] and Jon Beck [2]

Introduction

In ordinary homological algebra, if M is an R-module, the usual way of starting to construct a projective resolution of M is to let F be the free R-module generated by the elements of M and F → M the epimorphism determined by (m) → m . One then takes the kernel of F → M and continues the process. But notice that in the construction of F → M a lot of structure is customarily overlooked. F is actually a functor MG of M , F → M is an instance of a natural transformation G → (identity functor) ; there is also a "comultiplication" G → GG which is a little less evident. The functor G , equipped with these structures, is an example of what is called a standard construction or "cotriple".

In this paper we start with a category $\underline{C}$, a cotriple **G** in $\underline{C}$, and show how resolutions and derived functors or homology can be constructed by means of this tool alone. The category $\underline{C}$ will be non-abelian in general (note that even for modules the cotriple employed fails to respect the additive structure of the category), and the coefficients will consist of an arbitrary functor $E : \underline{C} \to \mathcal{A}$ where $\mathcal{A}$ is an abelian category. For ordinary homology and cohomology theories, E will be tensoring, homming or deriving with or into a module of some kind.

To summarize the contents of the paper: In § 1 we define the derived functors and give several examples of categories with cotriples. In § 2 we study the derived

1) Partially supported by NSF Grant GP 5478

2) Supported by an NAS-NRC Postdoctoral Fellowship

functors $H_n(\ ,E)_{\mathbb{G}}$ as functors on $\underline{C}$ and give several of their properties. In § 3 we fix a first variable $X \in \underline{C}$ and study $H_n(X,\ )_{\mathbb{G}}$ as a functor of the abelian variable E. As such it admits a simple axiomatic characterization. § 4 considers the case in which $\underline{C}$ is additive and shows that the general theory can always, in effect, be reduced to that case. In § 5 we study the relation between cotriples and projective classes (defined - essentially - by Eilenberg-Moore [16]) and show that the homology only depends on the projective class defined by the cotriple. §§ 6-9 are concerned largely with various special properties that these derived functors possess in well known algebraic categories (groups, modules, algebras, ...). In § 10 we consider the problem of defining a cotriple to produce a given projective class (in a sense, the converse problem to that studied in § 5) by means of "models". We also compare the results with other theories of derived functors based on models. § 11 is concerned with some technical items on acyclic models.

Before beginning the actual homology theory, we give some basic definitions concerning the simplicial objects which will be used. Let $\mathbb{G} = (G,\varepsilon,\delta)$ be a cotriple in $\underline{C}$, that is,

$$\underline{C} \xrightarrow{\ G\ } \underline{C}$$

$$G \xrightarrow{\ \varepsilon\ } \underline{C} \quad , \quad G \xrightarrow{\ \delta\ } GG$$

and the unitary and associative laws hold, as given in the Introduction to this volume. (Note that here and throughout we identify identity maps with the corresponding objects; thus $\underline{C}$ denotes the identity functor $\underline{C} \to \underline{C}$.) If X is an object in $\underline{C}$, the following is an augmented simplicial object in $\underline{C}$:

$$X \xleftarrow{\ \varepsilon_0\ } XG \underset{\delta_0}{\overset{\varepsilon_0,\varepsilon_1}{\rightleftarrows}} XG^2 \rightleftarrows \cdots \quad XG^{n+1} \quad \cdots$$

XG^{n+1} is the n-dimensional component, $\varepsilon_i = G^i\varepsilon G^{n-i} : G^{n+1} \to G^n$, $\delta_i = G^i\delta G^{n-i} : G^{n+1} \to G^{n+2}$ for $0 \leq i \leq n$, and the usual simplicial identities hold:

$$\varepsilon_i\varepsilon_j = \varepsilon_{j+1}\varepsilon_i, \quad i < j,$$

$$\delta_i\delta_j = \delta_{j-1}\delta_i, \quad i < j,$$

$$\delta_i\varepsilon_j = \begin{cases} \varepsilon_{j-1}\delta_i, & i < j-1 \\ \text{identity}, & i = j-1, j \\ e_j\delta_{i-1}, & i > j \end{cases}$$

(composition is from left to right).

If X admits a map $s : X \to XG$ such that $s.X\varepsilon = X$ (such X are called $\mathfrak{G}$-projective, see (2.1)), then the above simplicial object develops a <u>contraction</u>

$$X \xrightarrow{h_{-1}} XG \xrightarrow{h_o} XG^2 \longrightarrow \ldots \longrightarrow XG^{n+1} \xrightarrow{h_n} \ldots ,$$

namely $h_n = sG^{n+1}$. These operators satisfy the equations

$h_n\varepsilon_o = XG^{n+1}$, $h_n\varepsilon_i = \varepsilon_{i-1}h_{n-1}$ for $0<i\leq n+1$, $n\geq -1$. They express the fact that the simplicial object $(XG^{n+1})_{n\geq o}$ is homotopically equivalent to the constant simplicial object which has X in all dimensions.

If $(X_n)_{n\geq -1}$ is a simplicial set with such a contraction, we conclude $\Pi_n(X) = 0$, $n>o$, $\Pi_o(X) = X_{-1}$.

On the other hand, if $E : \underline{C} \to \mathcal{A}$ is a functor into any other category and E possesses a natural transformation $\theta : E \to GE$ such that $\theta.\varepsilon E = E$, then $(XG^{n+1}E)_{n\geq -1}$ also has a contraction

$$XE \xrightarrow{h_{-1}} XGE \xrightarrow{h_o} XG^2E \longrightarrow \ldots \longrightarrow XG^{n+1}E \xrightarrow{h_n} \ldots$$

Here $h_n = XG^{n+1}\theta$ and the identities satisfied are a little different (This is a "right" homotopy [30]): $h_n\varepsilon_i = \varepsilon_i h_{n-1}$, $0\leq i\leq n$, $h_n\varepsilon_{n+1} = XG^{n+1}E$, $n\geq -1$. Both here and above some equations involving degeneracies also hold, but our concern is usually with homology so we omit them.

If the functor E takes values in an abelian category, then as follows from a well known theorem of J.C. Moore [21] the homotopy in any sense of $(XG^{n+1}E)_{n\geq 0}$ is the same as the homology of the associated chain complex

$$0 \leftarrow XGE \xleftarrow{\partial_1} XG^2E \xleftarrow{\partial_2} \cdots \xleftarrow{\partial_n} XG^{n+1}E \leftarrow \cdots$$

where $\partial_n = \Sigma(-1)^i \varepsilon_i E$. If there is a contraction, $H_n = 0$ for $n>0$, $H_0 = XE$.

1. Definition of the homology theory $H_n(X,E)_{\mathbb{G}}$.

Let $X \in \underline{C}$, let $\mathbb{G} = (G,\varepsilon,\delta)$ be a cotriple in $\underline{C}$, and let $E : \underline{C} \to \mathcal{A}$ be a functor into an abelian category. Applying E to $(XG^{n+1})_{n\geq -1}$ we get an augmented simplicial object in $\mathcal{A}$:

$$XE \leftarrow XGE \leftleftarrows XG^2E \leftleftarrows \cdots$$

The homotopy of this simplicial object, or what is the same thing by Moore's theorem, the homology of the associated chain complex

$$0 \leftarrow XGE \xleftarrow{\partial_1} XG^2E \xleftarrow{\partial_2} \cdots,$$

is denoted by $H_n(X,E)_{\mathbb{G}}$, $n\geq 0$. These are the homology groups (objects) of X with coefficients in E relative to the cotriple $\mathbb{G}$. Often $\mathbb{G}$ is omitted from the notation if it is clear from the context.

The homology is functorial with respect to maps $X \to X_1$ in $\underline{C}$ and natural transformations of the coefficient functors $E \to E_1$.

A natural transformation (augmentation)

$$H_0(\ ,E)_{\mathbb{G}} \xrightarrow{\lambda=\lambda E} E$$

is defined by the fact that H_o is a cokernel:

$$\begin{array}{ccccccc} XG^2E & \xrightarrow{\partial_1} & XGE & \longrightarrow & H_o(X,E) & \longrightarrow & o \\ & & & \searrow^{\partial_o} & \downarrow{\scriptstyle\lambda} & & \\ & & & & XE & & \end{array}$$

$\lambda(H_o(\ ,E))$ and $H_o(\ ,\lambda E)$ coincide since they both fit in the diagram

$$\begin{array}{ccccccc} H_o(XG^2,E) & \longrightarrow & H_o(XG,E) & \longrightarrow & H_o(X,H_o(\ ,E)) & \longrightarrow & o \\ \downarrow{\scriptstyle\lambda} & & \downarrow{\scriptstyle\lambda} & & \downarrow & & \\ XG^2E & \longrightarrow & XGE & \longrightarrow & H_o(X,E) & \longrightarrow & o \end{array}$$

Thus λ can be viewed as a reflection into the subcategory of all functors $E : \underline{C} \to \mathcal{A}$ with $\lambda : H_o(\ ,E) \cong E$. These are the functors which transform $XG^2 \rightrightarrows XG \to X$ into a coequalizer diagram in $\mathcal{A}$, for all $X \in \underline{C}$, a sort of right exactness property.

The following variations occur. If, dually, $\mathbb{T} = (T,\eta,\mu)$ is a triple in $\underline{C}$ and $E : \underline{C} \to \mathcal{A}$ is a coefficient functor, <u>cohomology groups</u> $H^n(X,E)_{\mathbb{T}}$, $n \geq o$, are defined by means of the cochain complex

$$o \longrightarrow XTE \xrightarrow{d^1} XT^2E \xrightarrow{d^2} \cdots \xrightarrow{d^n} XT^{n+1}E \longrightarrow \cdots$$

where $d^n = \Sigma(-1)^i X\eta_i E$, $o \leq i \leq n$, $\eta_i = T^i\eta T^{n-i}$.

If $\mathbb{G} = (G,\varepsilon,\delta)$ is a cotriple and $E : \underline{C}^* \to \mathcal{A}$ is a functor (or $E : \underline{C} \to \mathcal{A}$ is contravariant), the complex would take the form

$$o \longrightarrow XGE \longrightarrow XG^2E \longrightarrow \cdots \longrightarrow XG^{n+1}E \longrightarrow \cdots$$

In effect this is cohomology with respect to the triple $\mathbb{G}^*$ in the dual category. However, we write the theory as $H^n(X,E)_{\mathbb{G}}$.

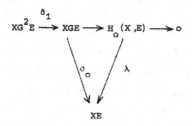

For the most part we will only state theorems about the cotriple-covariant functor situation and leave duals to the reader. Usually cotriples arise from adjoint functors, although another method of construction will be essayed in §10. If $F : \underline{A} \to \underline{C}$ is left adjoint to $U : \underline{C} \to \underline{A}$, there are well known natural transformations $\eta : \underline{A} \to FU$, $\varepsilon : UF \to \underline{B}$. If we set $G = UF$, we have $\varepsilon : G \to \underline{B}$, and if we set $\delta = U\eta F$, then $\delta : G \to G^2$. The relations obeyed by η, ε

$$\begin{array}{ccc} F & \xrightarrow{\eta F} & FUF \\ & \searrow\!\!\!\!= & \downarrow F\varepsilon \\ & & F \end{array} \qquad \begin{array}{ccc} U & \xrightarrow{U\eta} & UFU \\ & \searrow\!\!\!\!= & \downarrow \varepsilon U \\ & & U \end{array}$$

imply that $\mathbb{G} = (G,\varepsilon,\delta)$ is a cotriple in $\underline{C}$. This fact was first recognized by Huber [14].

(1.1) <u>Additive example: Homology of modules .</u>

Let R-<u>Mod</u> be the category of (left) R-modules. Let $\mathbb{G} = (G,\varepsilon,\delta)$ be the cotriple generated by the adjoint pair

$$\begin{array}{ccc} R\text{-}\underline{Mod} & \xrightarrow{\quad G \quad} & R\text{-}\underline{Mod} \\ & {\scriptstyle U}\searrow \qquad \nearrow{\scriptstyle F} & \\ & \mathcal{S} & \end{array}$$

U is the usual underlying set function, $F \dashv U$ is the free R-module functor. Thus we have $MG = R(M)$, the free R-module with the elements of M as basis, and the counit $M\varepsilon : MG \to M$ is the map which takes each basis element into the same element in M (just the usual way of starting to construct an R-free resolution of M). The comultiplication $M\delta : MG \to MG^2$ we leave to the reader.

Later we shall show that the complex

$$0 \leftarrow M \xleftarrow{\partial_0} MG \xleftarrow{\partial_1} MG^2 \leftarrow \cdots \xleftarrow{\partial_n} MG^{n+1} \leftarrow \cdots$$

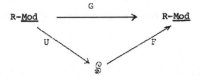

where $\partial_n = \Sigma(-1)^i M\varepsilon_i$, $0 \leq i \leq n$, is an R-free resolution of M (the only issue is exactness). Taking as coefficient functors $ME = A \otimes_R N$ or $ME = \mathrm{Hom}_R(M,A)$, we obtain $H_n(M, A \otimes_R\ )$ and $H^n(M, \mathrm{Hom}_R(\ ,A))$ as n-th homology or cohomology of

$$0 \longleftarrow A \otimes_R MG \longleftarrow A \otimes_R MG^2 \longleftarrow \cdots$$

$$0 \longrightarrow \mathrm{Hom}_R(MG,A) \longrightarrow \mathrm{Hom}_R(MG^2,A) \longrightarrow \cdots$$

That is, $H_n = \mathrm{Tor}_n^R(A,M)$, $H^n = \mathrm{Ext}_R^n(M,A)$.

Since R-modules are an additive category and the coefficient functors considered were additive, we could form the alternating sum of the face operators to obtain a chain complex in R-Mod before applying the coefficient functor.

As another example of this we mention the Eckmann-Hilton homotopy groups $\Pi_n(M,N)$ (as re-indexed in accordance with [14]). These are the homology groups of the complex

$$0 \longleftarrow \mathrm{Hom}_R(M,NG) \longleftarrow \mathrm{Hom}_R(M,NG^2) \longleftarrow \cdots$$

Of course, in these examples the homology should have a subscript $\mathbb{G}$ to indicate that the cotriple relative to the underlying category of sets was used to construct the resolution. Other underlying categories and cotriples are possible. For example, if

$$K \xrightarrow{\ \varphi\ } R$$

is a ring map, we get an adjoint pair

$$\begin{array}{ccc} R\text{-}\underline{\mathrm{Mod}} & \xrightarrow{\ G_\varphi\ } & R\text{-}\underline{\mathrm{Mod}} \\ & \searrow \qquad \nearrow {\scriptstyle \otimes_K R} & \\ & K\text{-}\underline{\mathrm{Mod}} & \end{array}$$

where the underlying is restriction of operators to K by means of φ . We have $MG_\varphi = M \otimes_K R$. The standard resolution is

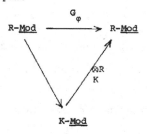

$$M \longleftarrow M \underset{K}{\otimes} R \longleftarrow M \underset{K}{\otimes} R \underset{K}{\otimes} R \Lleftarrow \cdots$$

Using the above coefficient functors we will find that the homology and cohomology are Hochschild's K-relative Tor and Ext [46]:

$$H_n(M, A \otimes_R \ ) = \mathrm{Tor}_n^{\varphi}(A,M)$$

$$H^n(M, \mathrm{Hom}_R(\ ,A)) = \mathrm{Ext}_{\varphi}^n(M,A)$$

Hochschild actually considered a subring $K \to R$ and wrote $\mathrm{Tor}^{(R,K)}$, etc.

We now turn to homology of groups and algebras. A useful device in the non-additive generalizations of homology theory is the "comma category" $(\underline{C},X)$ of all objects (of a given category $\underline{C}$) over a fixed object X . That is, an object of $(\underline{C},X)$ is a map $C \to X$, and a map of $(\underline{C},X)$ is a commutative triangle

$$\begin{array}{ccc} C_0 & \longrightarrow & C_1 \\ & \searrow \quad \swarrow & \\ & X & \end{array}$$

A cotriple $\mathbb{G} = (G,\varepsilon,\delta)$ in $\underline{C}$ naturally operates in $(\underline{C},X)$ as well. The resulting cotriple $(\mathbb{G},X)$ has

$$(C \xrightarrow{p} X)(G,X) = CG \xrightarrow{C\varepsilon} C \xrightarrow{p} X$$

$$(C \longrightarrow X)(\varepsilon,X) = \text{the map} \quad \begin{array}{ccc} CG & \xrightarrow{C\varepsilon} & C \\ & \searrow \quad \swarrow & \\ & X & \end{array}$$

$$(C \longrightarrow X)(\delta,X) = \text{the map} \quad \begin{array}{ccc} CG & \xrightarrow{C\delta} & CG^2 \\ & \searrow \quad \swarrow & \\ & X & \end{array}$$

The standard $(\mathbb{G},X)$-resolution of an object over $X, C \to X$, comes out in the form

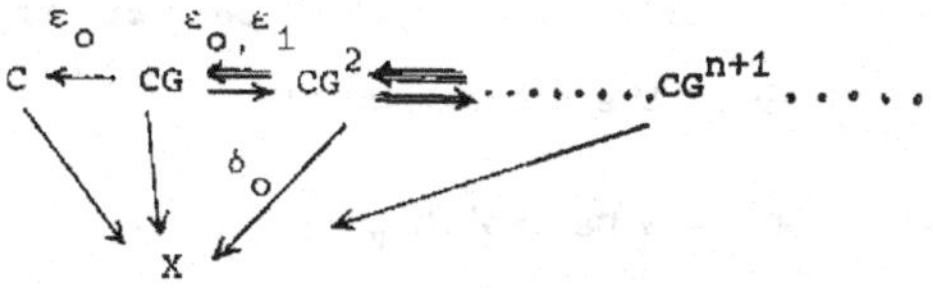

in other words, the usual faces and degeneracies turn out to be maps over X .

Homology groups $H_n(C,E)_{(\mathbb{G},X)}$ are then defined, when $E : (\underline{C},X) \to \mathcal{A}$ is a coefficient functor. We could write, with greater precision, $H_n(p,E)_{(\mathbb{G},X)}$, or with less, $H_n(C,E)_X$ or $H_n(C,E)$, leaving X understood.

Usually the coefficient functors involve a module over the terminal object X . This can be treated as a module over all the objects of $(\underline{C},X)$ simultaneously, by pullback via the structural maps to X . For example, derivations or differentials with values in an X-module become functors on the category of all algebras over X . This is the way in which homology and cohomology of algebras arise.

(1.2) <u>Homology of groups</u>.

Let $\mathcal{G}$ be the category of groups and $\mathbb{G}$ the cotriple arising from

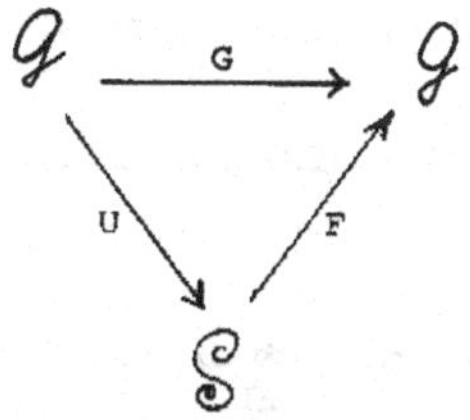

Thus ΠG is the free group on the underlying set of Π , and the counit $\Pi G \to \Pi$ is the natural surjection of the free group onto Π .

If $W \to \Pi$ is a group over Π and M is a left Π-module, a <u>derivation</u> $f : W \to M$ (<u>over</u> Π) is a function such that $(ww')f = w.w'f + wf$ ($W \to \Pi$ allows W to act on M). The abelian group of such derivations, $\mathrm{Der}(W,M)_\Pi$, gives a functor

$(\mathcal{G},\Pi)^* \to Ab$. We define the cohomology of $W \to \Pi$ with coefficients in M, $H^n(W,M)_\Pi$ (relative to $\mathbb{G}$) as the cohomology of the cochain complex

$$o \longrightarrow Der(WG,M)_\Pi \longrightarrow Der(WG^2,M)_\Pi \longrightarrow \cdots$$
$$\longrightarrow Der(WG^{n+1},M)_\Pi \longrightarrow \cdots$$

It is known that this theory coincides with Eilenberg-Mac Lane cohomology except for a shift in dimension [4]

$$H^n(W,M)_\Pi \xrightarrow{\sim} \begin{cases} Der(W,M)_\Pi, & n = 0, \\ H^{n+1}_{E-M}(W,M), & n > 0 . \end{cases}$$

Derivations $W \to M$ are represented by a Π-module of <u>differentials of</u> W (<u>over</u> Π) which we write as $Diff_\Pi(W)$:

$$Der(W,M)_\Pi \cong Hom_W(IW,M)$$
$$= Hom_\Pi(\mathbb{Z}\Pi \otimes_W IW,M),$$

hence $Diff_\Pi(W) = \mathbb{Z}\Pi \otimes_W IW$. (It is well known that the augmentation ideal $IW = ker(ZW \to Z)$ represents derivations of W into W-modules [10,11]. This is fudged by $Z\Pi\otimes_W -$ to represent derivations into Π-modules.)

The <u>homology of</u> $W \to \Pi$ <u>with coefficients in a right</u> Π-<u>module</u> M is defined as the homology of

$$o \longleftarrow M\otimes_{Z\Pi} Diff_\Pi(WG) \longleftarrow M\otimes_{Z\Pi} Diff_\Pi(WG^2) \longleftarrow \cdots$$
$$\cdots \longleftarrow M\otimes_{Z\Pi} Diff_\Pi(WG^{n+1}) \longleftarrow \cdots$$

Then

$$H_n(W,M)_\Pi \xrightarrow{\sim} \begin{cases} M\otimes_{Z\Pi} Diff_\Pi(W) = M\otimes_{Z\Pi} IW , & n=o , \\ H^{E-M}_{n+1}(W,M) , & n>o ; \end{cases}$$

this is because $(\mathrm{Diff}_\Pi(WG^{n+1}))_{n \geq -1}$ is a Π-free resolution of $\mathrm{Diff}_\Pi(W)$, and as [10],[11] show, the Eilenberg-Mac Lane homology can be identified with $\mathrm{Tor}^{ZW}_{n+1}(Z,N) = \mathrm{Tor}^{ZW}_n(IW,N)$. Π-Freeness is because $\mathrm{Diff}_\Pi(WG) = Z\Pi \otimes_W I(WG)$, and $I(WG)$ is well known to be WG-free. As for acyclicity, the cohomology of

$$0 \longrightarrow \mathrm{Hom}_\Pi(\mathrm{Diff}_\Pi(W),Q) \longrightarrow \mathrm{Hom}_\Pi(\mathrm{Diff}_\Pi(WG),Q) \longrightarrow \cdots\cdots$$

is zero in all dimensions ≥ -1 , if Q is an injective Π-module; this is true because the cohomology agrees with the Eilenberg-Mac Lane theory, which vanishes on injective coefficient modules. A direct acyclic - models proof of the coincidence of the homology theories can also be given.

As special cases note: if Π is regarded as a group over Π by means of the identity map $\Pi \to \Pi$, the $H^n(\Pi,M)_\Pi$, $H_n(\Pi,M)_\Pi$ are the ordinary (co) homology groups of Π with coefficients in a Π-module. On the other hand, if $\Pi = 1$, any W can be considered as a group over Π . Since a 1-module is just an abelian group, $\mathrm{Diff}_1(W) = W/[W,W]$, W abelianized, i.e., with its commutator subgroup divided out. The (co-)homology is that of W with coefficients in a trivial module.

<u>Remark</u>. ([20],[4]). Via interpretation as split extensions, Π-modules can be identified with the abelian group objects in the category $(\mathcal{G},\Pi)$. $\mathrm{Der}(W,M)_\Pi$ is then the abelian group of maps in $(\mathcal{G},\Pi)$:

$$\begin{array}{ccc} W & \longrightarrow & \Pi \times M \\ & \searrow \quad \swarrow & \\ & \Pi & \end{array}$$

Diff_Π is just the free abelian group functor, that is, the left adjoint of the forgetful functor

$$(\mathcal{G},\Pi) \longleftarrow \text{abelian groups in } (\mathcal{G},\Pi) = \Pi\text{-Modules} .$$

For general triple cohomology this interpretation is essential. In particular, the analogue of Diff exists for any category tripleable over sets provided the triple has a rank in the sense of [12] .

For the next example we need the comma category $(X,\underline{C})$ of objects and maps in $\underline{C}$ under X. An object of this category is a map $X \to Y$, a map is a commutative triangle

$$\begin{array}{ccc} & X & \\ \swarrow & & \searrow \\ Y_0 & \longrightarrow & Y_1 \end{array}$$

Assuming $\underline{C}$ has coproducts $X*Y$, a cotriple $\mathbb{G} = (G,\varepsilon,\delta)$ in $\underline{C}$ naturally induces a cotriple $(X,\mathbb{G}) = ((X,G),\ldots)$ in $(X,\underline{C})$:

$$(X \xrightarrow{f} Y)(X,G) = X \longrightarrow X * YG \quad \text{(coproduct injection)}$$

$$(X \longrightarrow Y)(X,\varepsilon) = \text{the map} \quad \begin{array}{ccc} & X & \\ \swarrow & & \searrow \\ X*YG & \xrightarrow[(f,Y\varepsilon)]{} & Y \end{array}$$

$$(X \longrightarrow Y)(X,\delta) = \text{the map} \quad \begin{array}{ccc} & X & \\ \swarrow & & \searrow \\ X*YG & \xrightarrow[X*(Y\delta . jG)]{} & X*(X*YG)G \end{array}$$

Where $j : YG \to X*YG$ is a coproduct injection.

Actually, the coproduct $X*(\ )$ defines an adjoint pair of functors $(X,\underline{C}) \to \underline{C} \to (X,\underline{C})$; the right adjoint is $X \to C \mapsto C$, the left adjoint is $C \mapsto X \to X*C$. By a general argument [14], the composition

$$(X,\underline{C}) \longrightarrow \underline{C} \xrightarrow{G} \underline{C} \xrightarrow{X*(\ )} (X,\underline{C})$$

is then a cotriple in $(X,\underline{C})$, namely $(X,\mathbb{G})$.

Replacing $(X,\underline{C}) \longrightarrow \underline{C} \longrightarrow (X,\underline{C})$ by an arbitrary adjoint pair and specializing $\mathbb{G}$ to the identity cotriple proves the remark preceding (1.1).

Homology and cohomology relative to the cotriple $(X,\mathbb{G})$ will be studied in more detail in §8. This cotriple enters in a rather mild way into:

(1.3) Homology of commutative rings and algebras.

Let Comm be the category of commutative rings. For $A \in$ Comm let (A,Comm) be the category of commutative rings under A, that is, maps $A \to B \in$ Comm. Thus (A,Comm) is our notation for the category of commutative A-algebras. We review the notions of differentials and derivations in this category.

For the same reason as in the category of groups we place ourselves in a category of algebras over a fixed commutative ring D, that is, in a double comma category (A,Comm,D); here an object is an A-algebra $A \to B$ equipped with a map $B \to D$, and a map is a commutative diagram

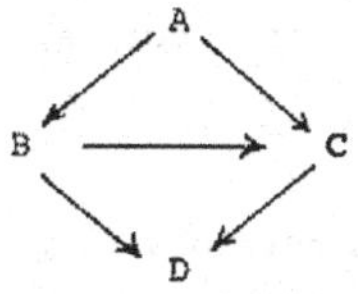

If M is a D-module, an A-derivation $B \to M$ is an A-linear function satisfying $(bb')f = b'.bf + b.fb'$, where $B \in$ (A,Comm,D) and A and B act on M via the given maps $A \to B \to D$. Such modules of derivations define a functor

$$(A,\underline{\text{Comm}},D)^* \xrightarrow{\text{A-Der}(\ ,M)_D} D\text{-}\underline{\text{Mod}}$$

This eventually gives rise to cohomology.

As is well known, any A-derivation $B \to$ a B-module M factors uniquely through a B-module map

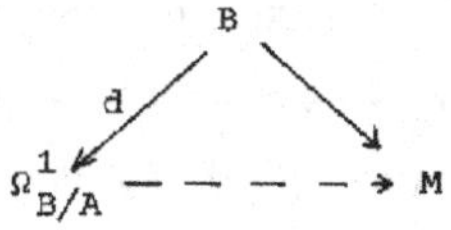

where $\Omega^1_{B/A}$ is the B-module of A-differentials of B, and d is the universal such derivation. $\Omega^1_{B/A}$ can be viewed as I/I^2 where $I = \ker(B \otimes_A B \to B)$ and $db = b \otimes 1 - 1 \otimes b$, or as the free B-module on symbols db modulo $d(b+b') = db + db'$, $d(ab) = a.db$, $d(bb') = b'.db + b.db'$ [1],[2]. Universal for A-derivations of $B \to$ D-modules M is then

$$\mathrm{Diff}_D(A \to B) = \Omega^1_{B/A} \otimes_B D$$

The functor which is usually used as coefficients for homology is

$$(A,\underline{\mathrm{Comm}},D) \xrightarrow{\mathrm{Diff}_D(A \to (\ )) \otimes_D M = \Omega^1_{(\ )/A} \otimes_{()} M} D\text{-}\underline{\mathrm{Mod}}.$$

There are two natural ways of defining homology in the category of A-algebras (over D), depending on the choice of cotriple, or equivalently, choice of the underlying category.

First let $\mathbb{G} = (G,\varepsilon,\delta)$ be the cotriple in the category of commutative rings arising from the adjoint pair

$$\begin{array}{ccc} \underline{\mathrm{Comm}} & \xrightarrow{\ G\ } & \underline{\mathrm{Comm}} \\ & {}_U\searrow \quad \nearrow_F & \\ & \mathcal{S} & \end{array}$$

Then $CG = Z[C]$, the polynomial ring with the elements of C as variables; the counit $CG \to C$ is the map defined by sending the variable $c \to c \in C$. This cotriple operates in $(\underline{\mathrm{Comm}},D)$ in the natural fashion described before (1.2).

Now consider the category $(A,\underline{\mathrm{Comm}})$ of commutative A-algebras. According to the remarks preceding this section, $\mathbb{G}$ gives rise to a cotriple $(A,\mathbb{G})$ in this category. Since the coproduct in the category $\underline{\mathrm{Comm}}$ is $A \otimes_Z B$, we have

$$\begin{aligned} (A \to C)(A,G) &= A \to A \otimes_Z CG \\ &= A \to A \otimes_Z Z[C] \\ &= A \to A[C] , \end{aligned}$$

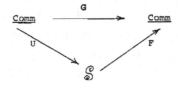

the polynomial A-algebra with the elements of C as variables. This cotriple is just that which is induced by the underlying set and free A-algebra functors

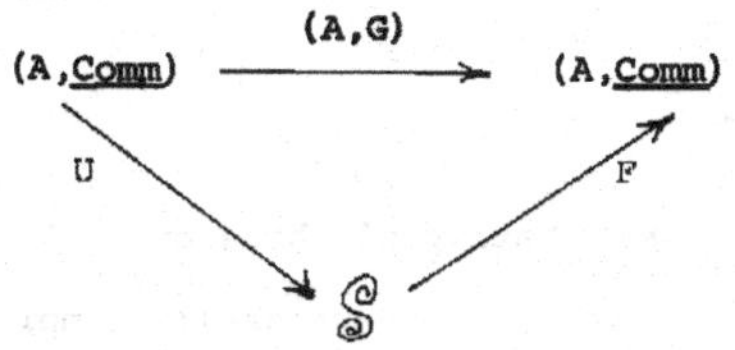

Furthermore, $(A,\mathbb{G})$ operates in the category of A-algebras over D , (A,Comm,D) , the values of $(A,\mathbb{G},D)$ being given by:

$$\begin{pmatrix} A \\ \downarrow \\ C \\ \downarrow \\ D \end{pmatrix} \longmapsto \begin{pmatrix} A \\ \downarrow \\ A[C] \\ \downarrow \\ D \end{pmatrix}$$

The counit is:

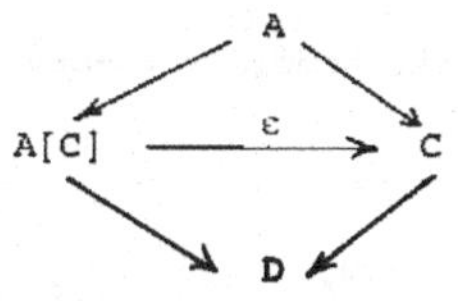

If M is a D-module we thus have homology and cohomology D-modules $H_n(C,M)$, $H^n(C,M)$, $n \geq o$, writing simply C for an A-algebra over D . These are defined by

$$H_n(C,M) = H_n\,[(\mathrm{Diff}_D(C.(A,G)^{p+1}) \otimes_D M)_{p \geq o}]$$

$$= H_n\,[\Omega^1_{A[\ldots[C]\ldots]/A} \otimes_A M]$$

where there are p + 1 applications of the A-polynomial operation to C in dimension p , and by

$$H^n(C,M) = H^n[A\text{-}Der_D(C(A,G)^{p+1},M)_{p>o}]$$

$$= H^n[A\text{-}Der_D(A[\ldots[C]\ldots],M)_{p>o}]$$

again with $p + 1$ A[ ]'s .

This homology theory of commutative algebras over D coincides with those considered in [19] , [28] ; of course, one generally simplifies the setting slightly by taking $C = D$ above. Both of these papers contain proofs that the cotriple theory coincides with theirs. The homology theory of [1] also agrees.

This theory, however it is described, is called the "absolute" homology theory of commutative algebras. The term arises as a reference to the underlying category which is involved, namely that of sets; no underlying object functor could forget more structure. But it also seems germane to consider so-called relative homology theories of algebras for which the underlying category is something else, usually a category of modules.

As an example of this, consider the homology theory in (A,Comm) comming from the adjoint functors

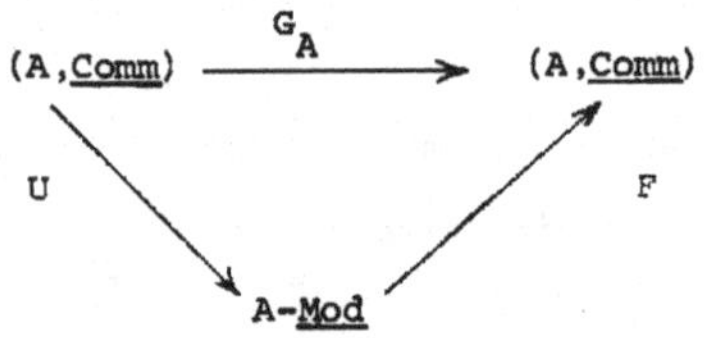

That is,

$$(A \to C)G_A = A + C + \frac{C \otimes_A C}{S(2)} + \frac{C \otimes_A C \otimes_A C}{S(3)} + \ldots,$$

the symmetric A-algebra on C (the S's are the symmetric groups). Note that this cotriple is not of the form $(A,\mathbb{G})$ for any cotriple $\mathbb{G}$ on the category of commutative rings. Exactly as above we now have homology and cohomology groups

$$H_n(C,M) = H_n\ [\mathrm{Diff}_D(CG_A^{p+1}) \otimes_D M)_{p>o}]$$

$$= H_n\ [\Omega^1_{CG_A^{p+1}/A} \otimes_D M)_{p>o}]\ ,$$

$$H^n(C,M) = H^n[(A\text{-}\mathrm{Der}(CG_A^{p+1},M)_D)_{p>o}]\ ,$$

where M is a D-module, and we are writing C instead of $A \to C$ for an A-algebra.

These two cohomology theories should really be distinguished by indicating the cotriples used to define them:

$H^n(C,M)_{(A,\mathbb{G})}$ = absolute theory, relative to sets,

$H^n(C,M)_{\mathbb{G}_A}$ = theory relative to A-modules .

The following is an indication of the difference between them: if $C = D$ and M is a C-module, $H^1(C,M)_{(A,\mathbb{G})}$ classifies commutative A-algebra extension $E \to C$ such that $I = \ker(E \to C)$ is an ideal of E with $I^2 = 0$, and such that there exists a lifting of the counit

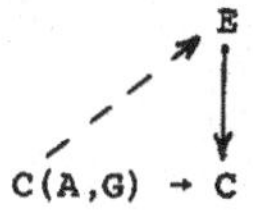

$H^1(C,M)_{\mathbb{G}_A}$ classifies those extensions with kernel of square zero that have liftings

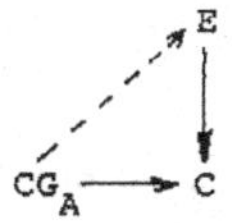

The absolute lifting condition is equivalent to the existence of a set section of $E \to C$, i.e. to surjectivity, the A-relative condition to the existence of an A-linear splitting of $E \to C$, as one can easily check. The relative theory is thus insensitive to purely A-linear phenomena, while the absolute theory takes all the structure into account. (We refer to [20] for details on classification of extensions).

The A-relative cohomology theory has been studied but little. Harrison has given an A-relative theory in [24] (A was a ground field but his formulas are meaningful for any commutative ring). Barr [33] has proved that

$$H^n(C,M) \simeq \begin{cases} \mathrm{Der}(C,M), & n = 0 \\ \mathrm{Harr}^{n+1}(C,M), & n > 0 \end{cases}$$

if A is a field of characteristic zero.

(1.4) <u>Homology of associative K-algebras</u>.

Let $\mathbb{G}_K$ be the cotriple relative to the underlying category of K-modules:

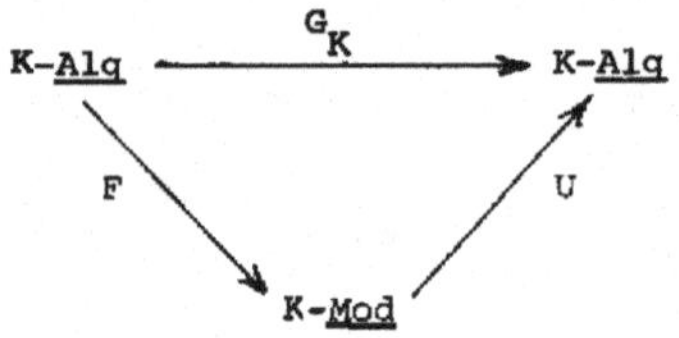

Thus if Λ is an associative algebra with unit over the commutative ring K, then

$$\Lambda G_K = K + \Lambda + \Lambda \otimes \Lambda + \ldots ,$$

the K-tensor algebra.

If $\Gamma \to \Lambda$ and M is a $\Lambda - \Lambda$-bimodule, we define $H^n(\Gamma,M)_\Lambda$ as the cohomology of the cosimplicial object

$$0 \longrightarrow \mathrm{Der}(\Gamma G_K, M)_\Lambda \Longrightarrow \mathrm{Der}(\Gamma G_K^2, M)_\Lambda \Rrightarrow$$

$$\ldots \longrightarrow \mathrm{Der}(\Gamma G_K^{n+1}, M)_\Lambda \longrightarrow \ldots$$

It is known that this coincides with Hochschild cohomology [3],[4] :

$$H^n(\Gamma,M)_\Lambda \xrightarrow{\sim} \begin{cases} \mathrm{Der}(\Gamma,M) , & n = o , \\ \mathrm{Hoch}^{n+1}(\Gamma,M) , & n > o . \end{cases}$$

The universal object for K-linear derivations $\Gamma \to M$, where M is a two-sided Λ-module, is $\mathrm{Diff}_\Lambda(\Gamma) = \mathrm{Diff}_\Gamma(\Gamma)\otimes_\Gamma e\Lambda^e = \Lambda\otimes_\Gamma J\Gamma\otimes_\Gamma\Lambda$; $J\Gamma$ is the kernel of the multiplication $\Gamma^e = \Gamma\otimes_K\Gamma^{op} \to \Gamma$ and represents derivations of Γ into Γ-modules [10], [11] . The <u>homology</u> of $\Gamma \to \Lambda$ with coefficients in M is defined as the homology of the complex

$$o \longleftarrow \mathrm{Diff}_\Lambda(\Gamma G_K) \otimes_{\Lambda^e} M \xleftarrow{\partial_1} \mathrm{Diff}_\Lambda(\Gamma G_K^2) \otimes_{\Lambda^e} M \xleftarrow{\partial_2} \cdots$$

[3] proves that $(\mathrm{Diff}_\Lambda(\Gamma G_K^{n+1}))_{n\geq -1}$ is a K-contractible complex of Λ^e-modules which are free relative to the underlying category of K-modules. Thus

$$H_n(\Gamma,M) \xrightarrow{\sim} \begin{cases} \mathrm{Diff}_\Lambda(\Gamma) \otimes_{\Lambda^e} M , & n = o , \\ \mathrm{Hoch}_{n+1}(\Gamma,M) , & n > o , \end{cases}$$

the last being Hochschild homology as defined in [MacH, Chapter X] .

The foregoing is a K-relative homology theory for associate K-algebras, in the sense of (1.3) . There is also an absolute theory, due to Shukla [22] , which Barr has proved coincides with the cotriple theory relative to the category of sets (with the usual dimension shift) [23] . We shall not deal with this absolute theory in his paper.

This concludes the present selection of examples. A further flock of examples will appear in §10.

2. <u>Properties of the</u> $H_n(X,E)_{\mathbb{G}}$ <u>as functors of</u> X , <u>including exact sequences</u>.

Objects of the form XG , that is, values of the cotriple $\mathbb{G}$, can be thought of as <u>free</u> relative to the cotriple. Free objects are acyclic:

(2.1) <u>Proposition</u>.

$$H_o(XG,E)_{\mathbb{G}} \xrightarrow{\widetilde{\lambda}} XGE ,$$

$$H_n(XG,E)_{\mathbb{G}} = o , \quad n > o .$$

An object P is called $\mathbb{G}$-<u>projective</u> if P is a retract of some value of G , or equivalently, if there is a map $s : P \to PG$ such that $s \cdot P\varepsilon = P$. $\mathbb{G}$-projectives obviously have the same acyclicity property.

To prove (2.1) we just recall from the Introduction that the simplicial object $(XGG^{n+1}E)_{n \geq -1}$ has a contraction.

If $f : X \to Y$ in <u>C</u> , we define "relative groups" or homology groups of the map, $H_n(f,E)_{\mathbb{G}}$, $n \geq o$, such that the following holds:

(2.2) <u>Proposition</u>.

If $X \to Y$ in <u>C</u> , there is an exact sequence

$$\cdots \longrightarrow H_n(X,E)_{\mathbb{G}} \longrightarrow H_n(Y,E)_{\mathbb{G}} \longrightarrow H_n(X \to Y,E)_{\mathbb{G}} \xrightarrow{\partial} H_{n-1}(X,E)_{\mathbb{G}} \longrightarrow \cdots \longrightarrow H_o(X \to Y,E)_{\mathbb{G}} \to o$$

(2.3) <u>Proposition</u>

If $X \to Y \to Z$ in <u>C</u> , there is an exact sequence

$$\cdots \quad H_n(X \to Y,E)_{\mathfrak{G}} \to H_n(X \to Z,E)_{\mathfrak{G}} \to H_n(Y \to Z,E)_{\mathfrak{G}} \xrightarrow{\partial} H_{n-1}(X \to Y,E)_{\mathfrak{G}} \to \cdots\cdots \to H_o(Y \to Z,E)_{\mathfrak{G}} \to o$$

If o is an initial object in $\underline{C}$, that is, if there is a unique map $o \to X$ for every X, then o is $\mathfrak{G}$-projective and

$$H_n(X,E) \xrightarrow{\simeq} H_n(o \to X,E) \ ,$$

$$H_n(X,E) \xrightarrow{\simeq} H_n(XG \xrightarrow{\varepsilon} X,E) \ , \quad n > o \ .$$

Examples of these sequences will be deferred to §8. There we will show that under certain conditions the homology group $H_n(X \to Y,E)$ can be interpreted as a cotriple homology group relative to the natural cotriple in the category $(X,\underline{C})$. For one thing, it will turn out that the homology of a map of commutative rings, $H_n(A \to B)$, is just the homology of B as an A-algebra.

Imitative though these sequences may be of theorems in algebraic topology, we don't know how to state a uniqueness theorem for $\mathfrak{G}$-homology in our present context.

As to the definition of the relative groups, we just let

$$H_n(X \xrightarrow{f} Y,E)_G = H_n(Cf)$$

where Cf is the mapping cone of the chain transformation

$$fG^{n+1}E : XG^{n+1}E \longrightarrow YG^{n+1}E \ , \quad n > o \ . \quad \text{That is,}$$

$$(Cf)_n = \begin{cases} YG^{n+1}E \oplus XG^nE \ , & n > o \\ YGE \ , & n = o \ , \end{cases}$$

$$\partial_n = \begin{pmatrix} \partial_Y & o \\ fG^{n+1}E & -\partial_X \end{pmatrix} : (Cf)_n \to (Cf)_{n-1} \, , \quad n \geq 2 \, ,$$

$$\partial_1 = \begin{pmatrix} \partial_Y \\ fGE \end{pmatrix} : (Cf)_1 \to (Cf)_o \, .$$

(These matrices act on row vectors from the right, ∂_X , ∂_Y indicate boundary operators in the standard complexes of X,Y .)

(2.2) follows from the exact sequence of chain complexes

$$o \to (YG^{n+1}E)_{n \geq o} \longrightarrow Cf \xrightarrow{\Pi} (XG^{n+1}E)_{n \geq o} \to o$$

where the projection Π is a chain transformation of degree -1 .

(2.3) follows from (2.2) by routine algebraic manipulation ([36],[37]) .

3. <u>Axioms for the</u> $H_n(X,E)_{\mathbb{G}}$ <u>as functors of the abelian variable</u> E .

In this section we show that the functors $H_n(\ ,E)_{\mathbb{G}} : \underline{C} \to \mathcal{A}$ are characterized by the following two properties. (In §4 it will appear that they are characterized by a little bit less.)

(3.1) $\mathbb{G}$-<u>acyclicity.</u>

$$H_o(\ ,GE)_{\mathbb{G}} \xrightarrow[\sim]{\lambda} GE \, ,$$

$$H_n(\ ,GE)_{\mathbb{G}} = 0\ ,\quad n > 0\ .$$

(3.2) $\mathbb{G}$-connectedness.

If $0 \to E' \to E \to E'' \to 0$ is a $\mathbb{G}$-short exact sequence of functors $\underline{C} \to \mathcal{A}$, then there is a long exact sequence in homology:

$$\cdots\cdots \longrightarrow H_n(\ ,E')_{\mathbb{G}} \to H_n(\ ,E)_{\mathbb{G}} \to H_n(\ ,E'')_{\mathbb{G}} \xrightarrow{\partial} H_{n-1}(\ ,E')_{\mathbb{G}} \to \cdots\cdots \to H_0(\ ,E'')_{\mathbb{G}} \to 0$$

The acyclicity is trivial; as mentioned in the Introduction, the simplicial object XG^*GE always has a contraction by virtue of

$$GE \xrightarrow{\delta E} G(GE)\ .$$

For the homology sequence, we define a sequence of functors $0 \to E' \to E \to E'' \to 0$ to be $\mathbb{G}$-exact if it is exact in the (abelian) functor category $(\underline{C},\mathcal{A})$ after being composed with $G : \underline{C} \to \underline{C}$, i.e., if and only if $0 \to XGE' \to XGE \to XGE'' \to 0$ is an exact sequence in $\mathcal{A}$ for every object $X \in \underline{C}$. In this event we get a short exact sequence of chain complexes in $\mathcal{A}$,

$$0 \to (XG^{n+1}E') \to (XG^{n+1}E) \to (XG^{n+1}E'') \to 0\ ,\quad n > -1\ ,$$

from which the homology sequence is standard.

Next we show that properties (3.1), (3.2) are characteristic of the homology theory $H_{\mathbb{G}}$. Define $\mathbb{L} = (L_n,\lambda,\partial)$ to be a theory of $\mathbb{G}$-left derived functors if:

(1) $\mathbb{L}$ assigns to every functor $E : \underline{C} \to \mathcal{A}$ a sequence of functors $L_nE : \underline{C} \to \mathcal{A}$, and to every natural transformation $\theta: E \to E_1$ a sequence of natural transformations $L_n\theta : L_nE \to L_nE_1$, $n > 0$, $L_n(\theta\theta_1) = L_n(\theta).L_n(\theta_1)$;

(2) λ is a natural transformation $L_0E \to E$ which has property (3.1) for every

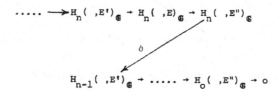

functor which is of the form GE;

(3) whenever $o \to E' \to E \to E'' \to o$ is a $\mathfrak{G}$-exact sequence of functors $\underset{\mathfrak{D}}{\underline{C}} \to \mathcal{A}$, then there is a long exact homology sequence

$$\begin{array}{l} \cdots\cdots \longrightarrow L_nE' \longrightarrow L_nE \longrightarrow L_nE'' \\ \qquad \swarrow \partial \\ L_{n-1}E' \to \cdots\cdots \to L_oE'' \to o , \end{array}$$

where ∂ actually depends on the given sequence, of course, and

$$\begin{array}{ccc} L_nE'' & \xrightarrow{\partial} & L_{n-1}E' \\ \downarrow & & \downarrow \\ L_nF'' & \xrightarrow{\partial} & L_{n-1}F' \end{array}$$

commutes for every map of $\mathfrak{G}$-short exact sequences

$$\begin{array}{ccccccccc} o & \to & E' & \to & E & \to & E'' & \to & o \\ & & \downarrow & & \downarrow & & \downarrow & & \\ o & \to & F' & \to & F & \to & F'' & \to & o . \end{array}$$

We now prove a uniquenesss theorem for $\mathfrak{G}$-left derived functors. A proof in purel abelian-category language exists also, in fact, has existed for a long time (cf.[42] and F. Ulmer's paper in this volume.)

(3.3) <u>Theorem</u>.

If $\mathbb{L}$ is a theory of $\mathfrak{G}$-left derived functors, then there exists a unique famil of natural isomorphisms

$$L_nE \xrightarrow{\sigma_n} H_n(\ ,E)_{\mathfrak{G}} , \quad n > o ,$$

which are natural in E, and are compatible with the augmentations

$$\begin{array}{ccc} L_oE & \xrightarrow{\sigma_o} & H_o(\ ,E)_{\mathfrak{G}} \\ & \searrow \quad \swarrow & \\ & E & \end{array}$$

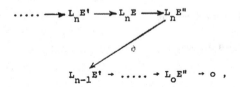

and connecting homomorphisms

$$\begin{array}{ccc} L_nE'' & \xrightarrow{\ \partial\ } & L_{n-1}E' \\ \sigma_n \downarrow & & \downarrow \sigma_{n-1} \\ H_n(\ ,E)_{\mathbb{G}} & \xrightarrow{\ \partial\ } & H_{n-1}(\ ,E')_{\mathbb{G}} \end{array}$$

corresponding to $\mathbb{G}$-short exact sequences.

Proof. In this proof we write $H_n(\ ,E)$ for $H_n(\ ,E)_{\mathbb{G}}$.

As we shall prove in a moment, the following is a consequence of $\mathbb{G}$-connectedness:

(3.4) Lemma: $L_o(G^2E) \xrightarrow{L_o\partial_1} L_o(GE) \xrightarrow{L_o\partial_o} L_oE \to o$ is an exact sequence in the functor category $(\underline{C},\mathcal{A})$.

Supposing that λ is a natural transformation $L_oE \to E$ which is natural in E as well, we get an unique map of the cokernels

(3.5)
$$\begin{array}{ccccccc} L_o(G^2E) & \to & L_o(GE) & \to & L_oE & \to & 0 \\ \downarrow \lambda & & \downarrow \lambda & & \downarrow \sigma_o & & \\ G^2E & \longrightarrow & GE & \longrightarrow & H_o(\ ,E) & \longrightarrow & o \end{array}$$

which is compatible with the augmentations:

$$\begin{array}{ccc} L_oE & \xrightarrow{\ \sigma_o\ } & H_o(\ ,E) \\ & \lambda \searrow \quad \swarrow \lambda & \\ & E & \end{array}$$

Now extend σ_o inductively to a map of $\mathbb{G}$-connected theories, $\sigma_n : L_nE \to H_n(\ ,E)$ for all $n > o$, as follows. Let $N = \ker\ (GE \to E)$ so that

$$o \to N \xrightarrow{\ i\ } GE \to E \to o\ .$$

is an exact sequence of functors, a fortiori $\mathbb{G}$-exact as well.

As σ_o is obviously natural in the E variable, we get a diagram

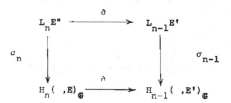

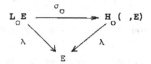

$$\begin{array}{ccccccc} & & L_1E & \xrightarrow{\partial} & L_oN & \longrightarrow & L_o(GE) \\ & & \downarrow{\scriptstyle\sigma_1} & & \downarrow{\scriptstyle\sigma_o} & & \downarrow{\scriptstyle\sigma_o} \\ o & \to & H_1(\ ,E) & \xrightarrow[\partial]{} & H_o(\ ,N) & \longrightarrow & H_o(\ ,GE) \end{array}$$

the bottom row being exact by virtue of $H_1(\ ,GE) = o$. This defines σ_1 . For σ_n , $n > 2$, use the diagram

(3.6)
$$\begin{array}{ccccccc} & & L_nE & \xrightarrow{\partial} & L_{n-1}N & & \\ & & \downarrow{\scriptstyle\sigma_n} & & \downarrow{\scriptstyle\sigma_{n-1}} & & \\ o & \to & H_n(\ ,E) & \xrightarrow{\partial} & H_{n-1}(\ ,N) & \to & o \end{array}$$

This defines all of the maps σ_n . But to have a map of $\mathbb{G}$-connected homology theories, we must verify that each square

$$\begin{array}{ccc} L_nE'' & \xrightarrow{\partial} & L_{n-1}E' \\ \downarrow{\scriptstyle\sigma_n} & & \downarrow{\scriptstyle\sigma_{n-1}} \\ H_n(\ ,E'') & \xrightarrow[\partial]{} & H_{n-1}(\ ,E') \end{array}$$

corresponding to a $\mathbb{G}$-exact sequence $o \to E' \to E \to E'' \to 0$ commutes.

We prove this first for σ_1 , σ_o , using what is basically the classical abelian-categories method. We are indebted to F. Ulmer for pointing it out.
Form the diagram

$$\begin{array}{ccccccccc} o & \to & M & \to & GE & \to & E'' & \to & o \\ & & \downarrow & & \downarrow & & \| & & \\ o & \to & N'' & \to & GE'' & \to & E'' & \to & o \end{array}$$

where M is ker $(GE \to E'')$. The left vertical arrow exists by virtue of N'' being ker $(GE'' \to E'')$. This induces

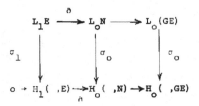

(3.7)

$$\begin{array}{ccccc} L_1E'' & \longrightarrow & L_oM & \longrightarrow & L_oN'' \\ \downarrow{\scriptstyle \sigma_1} & & \downarrow{\scriptstyle \sigma_o} & & \downarrow{\scriptstyle \sigma_o} \\ o \to H_1(\ ,E'') & \xrightarrow[\partial]{} & H_o(\ ,M) & \longrightarrow & H_o(\ ,N'') \end{array}$$

Since the map labeled ∂ is the kernel of $H_o(\ ,M) \to H_o(\ ,GE)$, there exists a map $L_1E'' \to H_1(\ ,E'')$ such that the left square commutes. As the right square commutes by naturality of σ_o , the outer rectangle commutes when this unknown map $L_1E'' \to H_1(\ ,E'')$ is inserted. But this is exactly the property which determines σ_1 uniquely. Thus the left square commutes with σ_1 put in. As there is obviously a map

$$\begin{array}{ccccccccc} o & \to & M & \to & GE & \to & E'' & \to & o \\ & & \downarrow & & \downarrow & & \| & & \\ o & \to & E' & \to & E & \to & E'' & \to & o \end{array}$$

a prism is induced:

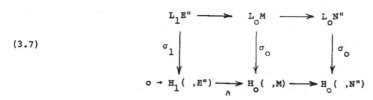

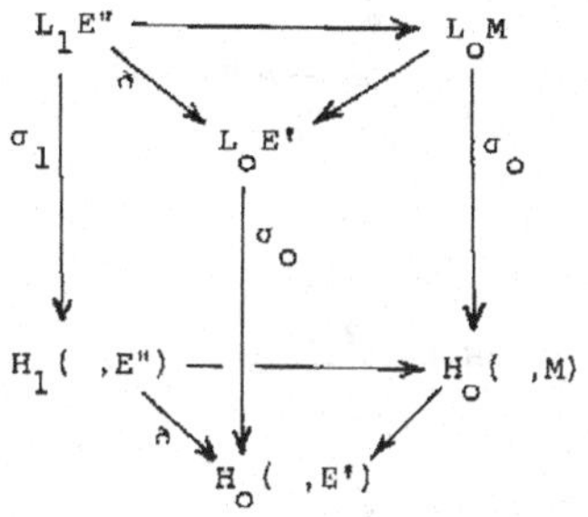

The top and bottom commute by naturality of the connecting homomorphisms in the L - and $H(\ ,\ )$ - theories, the right front face commutes by naturality of σ_o , the back commutes as it is the left square of (3.7) , so the left front face also commutes, q.e.d.

The proof that the σ_n are compatible with connecting homomorphisms in dimensions > 1 is similar.

Finally, assuming that the theory $\mathbb{L}$ also satisfies the acyclicity condition (2) (or (3.1)):

$$L_o(GE) \xrightarrow[\sim]{\lambda} GE$$

$$L_n(GE) = o\ , \qquad n > o\ ,$$

then from (3.5), $\sigma_o : L_oE \xrightarrow{\sim} H_o(\ ,E)$, and inductively from (3.7), $\sigma_n : L_nE \xrightarrow{\sim} H_n(\ ,E)$ for $n > o$. This completes the uniqueness proof, except for lemma (3.4).

Let $o \to N \to GE \to E \to o$ be exact, as above, and let $\nu : G^2E \to N$ be defined by the kernel property:

(3.8)
$$\begin{array}{ccccccccc} & & & & G^2E & & & & \\ & & & \swarrow^{\nu} & \downarrow{\scriptstyle \partial_1} & & & & \\ o & \longrightarrow & N & \xrightarrow[i]{} & GE & \xrightarrow[\partial_o]{} & E & \longrightarrow & o\ . \end{array}$$

Then

$$\begin{array}{ccccccc} & & H_o(\ ,G^2E) & & & & \\ & \swarrow^{H_o(\ ,\nu)} & \downarrow{\scriptstyle H_o(\ ,\partial_1)} & & & & \\ H_o(\ ,N) & \longrightarrow & H_o(\ ,GE) & \xrightarrow[H_o(\ ,\partial_o)]{} & H_o(\ ,E) & \to & o \end{array}$$

has an exact bottom row as $o \to N \to \ldots.$ is $\mathbb{G}$-exact as well. To prove $H_o(\ ,\partial_1)$, $H_o(\ ,\partial_o)$ exact it suffices to prove that $H_o(\ ,\nu)$ is onto. Let $K = \ker \nu$, so that $o \to K \to G^2E \to N$ is exact. Composing this with G, it is enough to show that

$$o \to GK \to GG^2E \xrightarrow{G\nu} GN \to o$$

is exact, which just means $G\nu$ onto (in fact it turns out to be split).

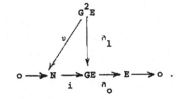

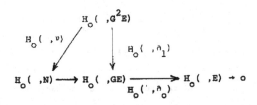

If we apply G to (3.8), we get

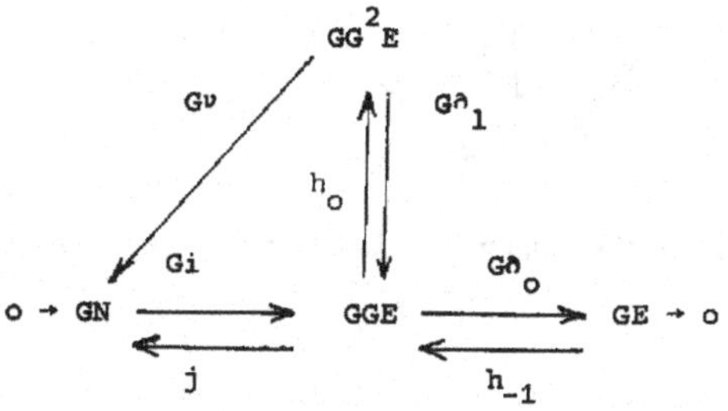

where the contracting maps h_{-1} , h_o obey $G\partial_o.h_{-1} + h_o.G\partial_1 = GGE$, among other things, and the bottom row is split $(h_{-1} = \delta E,\ h_o = \delta GE)$. Now $G\nu$ splits, for $Gi.h_o.G\nu = GN$. Since Gi is a monomorphism it suffices to prove $Gi.h_o.G\nu.Gi = Gi$. But

$$\begin{aligned} Gi \,.\, h_o \,.\, G(\nu i) &= Gi \,.\, h_o\, \partial_1 \\ &= Gi\,(GGE - G\partial_o .\, h_{-1}) \\ &= Gi - G(i\partial_o) \,.\, h_{-1} \\ &= Gi \end{aligned}$$

since $i\partial_o$ is zero.

4. <u>Homology in additive categories</u>.

Now we assume that $\underline{C}$ is an additive category and $\mathbb{G} = (G,\varepsilon,\delta)$ is a cotriple in $\underline{C}$. It is not necessary to suppose that $G : \underline{C} \to \underline{C}$ is additive or even that $OG = 0$.

If $E : \underline{C} \to \mathcal{A}$ is a coefficient functor, and <u>this</u> is assumed to be additive, the homology functors $H_n(\ ,E)_{\mathbb{G}} : \underline{C} \to \mathcal{A}$ are defined as before, and are additive. They admit of an axiomatic characterization like that in homological algebra (cf. (4.5)).

$\mathbb{G}$-projectives play a big role in the additive case. We recall $P \in \underline{C}$ is $\mathbb{G}$-<u>projective</u> if there is a map $s : P \to PG$ such that $s.P\varepsilon = P$. A useful fact, holding in any category, is that the <u>coproduct</u> $P * Q$ <u>of</u> $\mathbb{G}$-<u>projectives is again</u> $\mathbb{G}$-<u>projective.</u>

$$\begin{array}{ccccc} P * Q & \xrightarrow{s * t} & PG * QG & \longrightarrow & (P * Q)G \\ & \searrow\!\!\searrow & \downarrow{\scriptstyle \varepsilon * \varepsilon} & \swarrow{\scriptstyle \varepsilon} & \\ & & P * Q & & \end{array}$$

In an additive category the coproduct is $P \oplus Q$. We assume from now on that $\underline{C}$ is additive.

(4.1) <u>Definition</u>.

$$X' \xrightarrow{i} X \xrightarrow{j} X''$$

is $\mathbb{G}$-<u>exact</u> ($\mathbb{G}$-<u>acyclic</u>) if $ij = 0$ and $(AG,X') \to (AG,X) \to (AG,X'')$ is an exact sequence of abelian groups for all $A \in \underline{C}$, or equivalently, if $ij = o$ and $(P,X') \to (P,X) \to (P,X'')$ is an exact sequence of abelian groups for every $\mathbb{G}$-projective P . A $\mathbb{G}$-<u>resolution</u> of X is a sequence $o \longleftarrow X \longleftarrow X_o \longleftarrow X_1 \longleftarrow \ldots.$ which is $\mathbb{G}$-acyclic and in which $X_o, X_1, \ldots.$ are $\mathbb{G}$-projective.

The usual facts about $\mathbb{G}$-resolutions can be proved:

(4.2) <u>Existence and comparison theorem.</u>

$\mathbb{G}$-resolutions always exist. If $o \longleftarrow X \longleftarrow X_o \longleftarrow X_1 \longleftarrow \ldots..$ is a $\mathbb{G}$-projective complex and $o \longleftarrow Y \longleftarrow Y_o \longleftarrow Y_1 \longleftarrow \ldots..$ is a $\mathbb{G}$-acyclic complex then any $f : X \to Y$ can be extended to a map of complexes

$$\begin{array}{ccccccccc} o & \longleftarrow & X & \longleftarrow & X_o & \longleftarrow & X_1 & \longleftarrow & \ldots. \\ & & \downarrow{\scriptstyle f} & & \downarrow{\scriptstyle f_o} & & \downarrow{\scriptstyle f_1} & & \\ o & \longleftarrow & Y & \longleftarrow & Y_o & \longleftarrow & Y_1 & \longleftarrow & \ldots. \end{array}$$

Any two such extensions are chain homotopic.

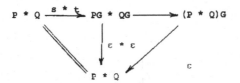

In fact,

$$0 \longleftarrow X \overset{\partial_0}{\longleftarrow} XG \overset{\partial_1}{\longleftarrow} XG^2 \longleftarrow \cdots$$

is a $\mathbb{G}$-resolution of X if we let $\partial_n = \Sigma(-1)^i X\varepsilon_i$. It is a $\mathbb{G}$-projective complex, and if AG is hommed into the underlying augmented simplicial object $X\mathbb{G}^*$, the resulting simplicial set has a contraction $(AG,X) \xrightarrow{h_{-1}} (AG,XG) \xrightarrow{h_0} (AG,XG^2) \xrightarrow{h_1} \cdots$ defined by $x \cdot h_n = A\delta \cdot xG$ for $x : AG \to XG^{n+1}$. Thus the simplicial group $(AG,X\mathbb{G}^*)$ has no homotopy, or homology, with respect to the boundary operators (AG,∂_n) .

The rest of the comparison theorem is proved just as in homological algebra.

Now we characterize the homology theory $H_n(\ ,E)_{\mathbb{G}} : \underline{C} \to \mathcal{A}$ by axioms on the $\underline{C}$ variable. In doing this we use finite projective limits in $\underline{C}$, although we still refrain from assuming $\mathbb{G}$ additive. We do assume that the coefficient functor $E : \underline{C} \to \mathcal{A}$ is additive, which forces additivity of the homology functors. The axioms we get are:

(4.3) $\mathbb{G}$-<u>acyclicity</u>.

If P is $\mathbb{G}$-projective, then

$$H_0(P,E)_{\mathbb{G}} \xrightarrow[\approx]{\lambda} PE ,$$

$$H_n(P,E)_{\mathbb{G}} = 0 , \qquad n > 0 .$$

(4.4) $\mathbb{G}$-<u>connectedness</u>.

If $0 \to X' \to X \to X'' \to 0$ is a $\mathbb{G}$-exact sequence in $\underline{C}$, then there is a long exact sequence in homology:

$$\cdots \longrightarrow H_n(X',E)_{\mathbb{G}} \longrightarrow H_n(X,E)_{\mathbb{G}} \longrightarrow H_n(X'',E)_{\mathbb{G}} \xrightarrow{\partial} H_{n-1}(X',E)_{\mathbb{G}} \longrightarrow \cdots \longrightarrow H_0(X'',E)_{\mathbb{G}} \to 0 .$$

The connecting maps are natural with respect to maps of $\mathbb{G}$-exact sequences

$$\begin{array}{ccccccccc} o & \to & X' & \to & X & \to & X'' & \to & o \\ & & \downarrow & & \downarrow & & \downarrow & & \\ o & \to & Y' & \to & Y & \to & Y'' & \to & o \end{array}.$$

It follows from (4.4) that if $X = X' \oplus X''$, then the canonical map $H_n(X',E)_{\mathbb{G}} \oplus H_n(X'',E)_{\mathbb{G}} \to H_n(X' \oplus X'',E)_{\mathbb{G}}$ is an isomorphism, $n \geq o$. Thus the $H_n(\ ,E)_{\mathbb{G}}$ are additive functors.

We are able to prove the following characterization:

(4.5) Uniqueness.

If $E_o \xrightarrow{\lambda} E$ is a natural transformation, and $E_1, E_2, \ldots, \partial$ is a sequence of functors together with a family of connecting homomorphisms satisfying (4.3) and (4.4), then there is a unique isomorphism of connected sequences $\sigma_n : E_n \xrightarrow{\simeq} H_n(\ ,E)_{\mathbb{G}}$, $n \geq o$, which commutes with the augmentations $E_o \to E, H_o(\ ,E)_{\mathbb{G}} \to E$.

For the proofs of the above, (4.3) = (2.1) . For (4.4), we assume that $\underline{C}$ has splitting idempotents. This causes no difficulty as $\mathbb{G}$ can clearly be extended to the idempotent completion of $\underline{C}$ and any abelian category valued functor can be likewise extended [5] . Moreover, it is clear that this process does not affect the derived functors. (Or assume that $\underline{C}$ has kernels).

Now if $o \to X' \to X \to X'' \to o$ is $\mathbb{G}$-exact it follows from exactness of $(X''G,X) \to (X''G,X'') \to o$ that there is a map $X''G \to X$ whose composite with $X \to X''$ is $X''\varepsilon$. Applying G to it we have $X''G \xrightarrow{X''\delta} X''G^2 \longrightarrow XG$ which splits $XG \to X''G$. By our assumption we can find X_o so that

$$o \longrightarrow X_o \longrightarrow XG \longrightarrow X''G \longrightarrow o$$

is split exact. X_o being presented as a retract of a free is $\mathbb{G}$-projective. Also, the composite

$$\begin{array}{ccc} X_o & \longrightarrow & XG \\ & & \downarrow \\ & & X \longrightarrow X'' \end{array}$$

is zero and we can find $X_o \to X'$ so that

$$\begin{array}{ccc} X_o & \longrightarrow & XG \\ \downarrow & & \downarrow \\ X' & \longrightarrow & X \end{array}$$

commutes. Continuing in this fashion we have a weakly split exact sequence of complexes

$$\begin{array}{ccccccccc} & & \downarrow & & \downarrow & & \downarrow & & \\ o & \to & X_n & \to & XG^{n+1} & \to & X''G^{n+1} & \to & o \\ & & \downarrow & & \downarrow & & \downarrow & & \\ o & \to & X_o & \to & XG & \to & X''G & \to & o \\ & & \downarrow & & \downarrow & & \downarrow & & \\ o & \to & X' & \to & X & \to & X'' & \to & o \\ & & \downarrow & & \downarrow & & \downarrow & & \\ & & o & & o & & o & & \end{array}$$

Homming a YG into it produces a weakly split exact sequence of abelian group complexes, two of which are exact, and so, by the exactness of the homology triangle, is the third. But then the first column is a $\mathcal{G}$-projective resolution of X' and the result easily follows.

For uniqueness, (4.5), $\underline{C}$ must have kernels, and the argument follows the classical prescription (§3). This is reasonable, for otherwise there wouldn't be enough exact sequences for (4.4) to be much of a restriction. First, $XG^2E_o \to XGE_o \to XE_o \to o$ is exact in $\mathcal{A}$. Using $\lambda : E_o \to E$ one gets a unique $\sigma_o : E_o \to H_o(\ ,E)$ which is compatible with the augmentations:

$$\begin{array}{ccc} E_o & \longrightarrow & H_o(\ ,E) \\ & {\scriptstyle\lambda}\searrow \quad \swarrow{\scriptstyle\lambda} & \\ & E & \end{array}$$

Letting $N = \ker (XG \to X)$, the sequence $o \to N \to NG \to X \to o$ is $\mathcal{G}$-exact.

$\sigma_1 : E_1 \to H_1(\ ,E)$ is uniquely determined by

$$\begin{array}{ccccc} XE_1 & \to & NE_o & \to & XGE_o \\ \downarrow{\scriptstyle \sigma_1} & & \downarrow{\scriptstyle \sigma_o} & & \downarrow{\scriptstyle \sigma_o} \\ o \to H_1(X,E) & \to & H_o(N,E) & \to & H_o(XG,E) \end{array} ,$$

σ_n similarly. Now the argument of the uniqueness part of (3.3) goes through and shows that the σ's commute with all connecting maps. Finally, if the E_n are $\mathbb{G}$-acyclic (4.3), the σ_n are isomorphisms.

As examples we cite $\mathrm{Tor}_n^R(A,M)$, $\mathrm{Ext}_R^n(M,A)$ obtained as $\mathbb{G}$-derived functors, or $\mathbb{G}$-homology, of the coefficient functors

$$\underline{R\text{-}Mod} \xrightarrow{A\otimes_R} \underline{Ab}$$

$$\underline{R\text{-}Mod}^* \xrightarrow{\mathrm{Hom}_R(\ ,A)} \underline{Ab}$$

relative to the free R-module cotriple (1.1). Proved in this section are additivity of these functors and their usual axiomatic characterizations.

Similarly one gets axioms for the K-relative Tor and Ext (1.1), and for the pure Tor and Ext defined in §10.

(4.6) Application to §3.

Let $\underline{C}$ be arbitrary, $\mathbb{G}$ a cotriple in $\underline{C}$. Let $\mathbb{G}$ operate in the functor category $(\underline{C},\mathcal{Q})$ by composition. The resulting cotriple is called $(\mathbb{G},\mathcal{Q})$:

$$(E)(G,\mathcal{Q}) = GE \xrightarrow{(E)(\varepsilon,\mathcal{Q}) = \varepsilon E} E ,$$

$$(E)(G,\mathcal{Q}) = GE \xrightarrow{(E)(\delta,\mathcal{Q}) = \delta E} G^2E = (E)(G,\mathcal{Q})^2 .$$

Iterating this cotriple in the usual way, we build up a simplicial functor

$$E \xleftarrow{(\varepsilon,\mathcal{A})_0} (E)(G,\mathcal{A}) \underset{(\delta,\mathcal{A})_0}{\overset{(\varepsilon,\mathcal{A})_{0,1}}{\leftleftarrows\!\rightarrow}} (E)(G,\mathcal{A})^2 \quad\cdots$$

from $\underline{C} \to \mathcal{A}$. Rewritten, this is

$$E \leftarrow GE \leftleftarrows\!\rightarrow G^2E \quad\cdots$$

Note that the i-th operator $(\varepsilon,\mathcal{A})_i : (E)(G,\mathcal{A})^{n+1} \longrightarrow (E)(G,\mathcal{A})^n$ is actually $\varepsilon_{n-i}E$ using the notation of the Introduction (dual spaces cause transposition). But reversing the numbering of face and degeneracy operators in a simplicial object does not change homotopy or homology. Therefore

$$H_n(E,\mathrm{id.})_{(\mathbb{G},\mathcal{A})} = H_n(\ ,E)_{\mathbb{G}}\ , \quad n > 0\ ;$$

on the left coefficients are in the identity functor $(\underline{C},\mathcal{A}) \to (\underline{C},\mathcal{A})$.

Thus the homology theory $H(\ ,E)_{\mathbb{G}}$ can always be obtained from a cotriple on an additive (even abelian) category, and the cotriple can be assumed additive. How can the axioms of this section be translated into axioms for the $H_n(\ ,E)_{\mathbb{G}}$ in general?

The $(\mathbb{G},\mathcal{A})$-projective functors are just the retracts of functors of the form GE . Thus the acyclicity axiom (4.3) becomes:

$$H_0(\ ,E)_{\mathbb{G}} \xrightarrow[\sim]{\lambda} E\ ,$$

$$H_n(\ ,E)_{\mathbb{G}} = 0\ , \qquad n > 0\ ,$$

if E is $(\mathbb{G},\mathcal{A})$-projective; this is equivalent to (3.1).

For the homology sequence, $0 \to E' \to E \to E'' \to 0$ is $(\mathbb{G},\mathcal{A})$-exact $\Longleftrightarrow$ $0 \to GE' \to GE \to GE'' \to 0$ is <u>split exact</u> in the functor category. (Prove this considering the picture

$$0 \longrightarrow E' \longrightarrow E \overset{s}{\dashleftarrow\!\!\rightarrow} E'' \longrightarrow 0\ . \qquad (GE'' \to E'')$$

$GE" \xrightarrow{\delta E"} G^2E" \xrightarrow{Gs} GE$ splits the sequence).
$(\mathbb{G},\mathcal{A})$-exactness $\Rightarrow$ $\mathbb{G}$-exactness as defined in (3.2). The homology sequence axiom of this § is weaker than that of §3: it requires the exact homology sequence to be produced for a smaller class of short exact sequences.

Concepts equivalent to $(\mathbb{G},\mathcal{A})$-projectivity and -exactness have recently been employed by Mac Lane to give a projective complex → acyclic complex form to the cotriple acyclic-models comparison theorem (11.1) (unpublished). In particular, $\mathbb{G}$-representability in the acyclic-models sense (existence of $\theta : E \to GE$ splitting the counit $\varepsilon E : GE \to E$) is the same thing as $(\mathbb{G},\mathcal{A})$-projectivity, $\mathbb{G}$-contractibility is the same as $(\mathbb{G},\mathcal{A})$-acyclicity.

(4.7) <u>Application to extensions</u>.

Let an n-<u>dimensional</u> $\mathbb{G}$-<u>extension</u> of X by Y be a $\mathbb{G}$-exact sequence

$$o \to Y \to X_{n-1} \to \dots \to X_o \to X \to o \, , \quad n > o \, .$$

Under the usual Yoneda equivalence these form a set $E^n(X,Y)_{\mathbb{G}}$. $E^o(X,Y)_{\mathbb{G}} = (X,Y)$, the hom set in $\underline{C}$ (which is independent of $\mathbb{G}$). Using the comparison theorem (4.2), an extension gives rise to a map of complexes

$$\begin{array}{ccccccccccc} \dots \longrightarrow & XG^{n+1} & \longrightarrow & XG^n & \longrightarrow & \dots & \longrightarrow & XG & \longrightarrow & X & \longrightarrow o \\ & a \downarrow & & \downarrow & & & & \downarrow & & \| & \\ o \longrightarrow & Y & \longrightarrow & X_{n-1} & \longrightarrow & \dots & \longrightarrow & X_o & \longrightarrow & X & \longrightarrow o \end{array}$$

The map a is an n-cocycle of X with values in the representable functor $(\ ,Y) : \underline{C}^* \longrightarrow Ab$. We get in this way a map

$$E^n(X,Y)_{\mathbb{G}} \longrightarrow H^n(X,Y)_{\mathbb{G}} \, , \quad n > o \, ,$$

(in dimension o, any $X \to Y$ determines a o-cocycle $XG \to X \to Y$).

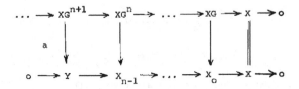

In practice cotriples often have the property that $XG^2 \rightrightarrows XG \rightarrow X$ is always a coequalizer diagram. In this case,

$$E^n(X,Y)_{\mathbb{G}} \rightarrow H^n(X,Y)_{\mathbb{G}}$$

is an isomorphism for $n = o$, and a monomorphism for $n > o$. If $\mathbb{G}$ is the free cotriple in a tripleable adjoint pair $\underline{C} \rightarrow \underline{A} \rightarrow \underline{C}$ this coequalizer condition holds; in fact, in that case

$$E^n(X,Y)_{\mathbb{G}} \xrightarrow{\sim} H^n(X,Y)_{\mathbb{G}} \ , \quad n > o \ ,$$

as is proved in [20] .

In categories of modules or abelian categories with projective generators [15] , this gives the usual cohomological classification of extensions.

5. <u>General notion of a $\mathbb{G}$-resolution and the fact that the homology depends on the $\mathbb{G}$-projectives alone</u>

There is no shortage of resolutions from which the $\mathbb{G}$-homology can in principle be computed, as the standard one always exists. But it would be nice to be able to choose more convenient resolutions in particular problems, and have available something like the additive comparison theorem (§4) in order to relate them to the standard resolutions. In fact a simplicial comparison theorem does exist ([30]) , but we can get by with something much easier. Any category can be made freely to generate an additive category by a well known construction and we find the solution to our problem by transferring it to this additive context. This is the same technique as is used by André [19 ,§4] .

The free additive category on $\underline{C}$, $Z\underline{C}$, has formal sums and differences of maps in $\underline{C}$ as its maps. Exact definitions and properties connected with $\underline{ZC}$ are given after the following definitions.

(5.1) **Definitions.**

A $\mathbb{G}$-resolution of X is a complex

$$X \xleftarrow{\partial_o} X_o \xleftarrow{\partial_1} X_1 \leftarrow \dots \xleftarrow{\partial_n} X_n \leftarrow \dots\dots \text{ in } Z\underline{C}$$

in which all X_n, $n \geq o$, are $\mathbb{G}$-projective and which is $\mathbb{G}$-acyclic in the sense that

$$o \leftarrow (AG,X)_{Z\underline{C}} \leftarrow (AG,X_o)_{Z\underline{C}} \leftarrow \dots\dots \leftarrow (AG,X_n)_{Z\underline{C}} \leftarrow \dots$$

has zero homology in all dimensions for all values AG of the cotriple $\mathbb{G}$.

A simplicial $\mathbb{G}$-resolution of X is an augmented simplicial object in $\underline{C}$

$$X \xleftarrow{\varepsilon_o} X_o \underset{\delta_o}{\overset{\varepsilon_o,\varepsilon_1}{\rightleftarrows}} X_1 \cdots\cdots$$

such that the associated chain complex in $Z\underline{C}$

$$X \xleftarrow{\partial_o = (\varepsilon_o)} X_o \xleftarrow{\partial_1 = (\varepsilon_o) - (\varepsilon 1)} X_1 \leftarrow \cdots \xleftarrow{\partial_n} X_n \leftarrow \cdots$$

is a $\mathbb{G}$-resolution as defined above (as usual, $\partial_n = \Sigma(-1)^i (X\varepsilon_i))$.
In particular, the standard complex

$$X \xleftarrow{X\varepsilon_o} XG \underset{X\delta_o}{\overset{X\varepsilon_o\, X\varepsilon_1}{\rightleftarrows}} XG^2 \cdots$$

is a $\mathbb{G}$-resolution of X, since the simplicial set $(AG,XG^{n+1})_{n\geq -1}$ has the contraction given in the proof of (4.2).

To be precise about $Z\underline{C}$, its objects are the same as those of $\underline{C}$, while a map $X \to Y$ in $Z\underline{C}$ is a formal Z-linear combination of such maps in $\underline{C}$, i.e., if $n_i \in Z$, $f_i : X \to Y \in C$, we get a map

$$X \xrightarrow{\Sigma\, n_i(f_i)} Y \qquad \text{in } ZC .$$

(We enclose the free generators in parentheses for clarity in case $\underline{C}$ is already additive.) Composition is defined like multiplication in a group ring, $(\Sigma\, m_i(f_i))(\Sigma\, n_j(g_j)) = \Sigma\Sigma\, m_i n_j (f_i g_j)$.

The natural inclusion of the basis $\underline{C} \to Z\underline{C}$ can be used to express the following universal mapping property. If $E : \underline{C} \to \mathcal{A}$ is a functor into an additive category, there is a unique additive functor $\overline{E} : Z\underline{C} \to \mathcal{A}$ such that

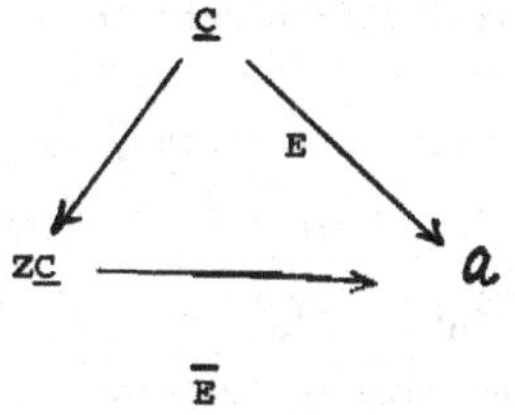

commutes. Explicitly, $X\overline{E} = XE$ and $(\Sigma\, n_i(f_i))\overline{E} = \Sigma\, n_i f_i E$.

Let $\mathbf{G} = (G,\varepsilon,\delta)$ be a cotriple in $\underline{C}$. Thinking of G as taking values in $Z\underline{C}$ we get an additive extension

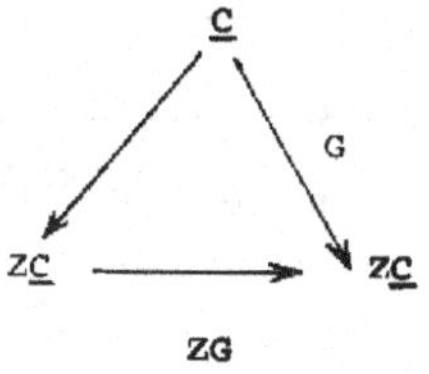

which is a cotriple $Z\mathbf{G} = (ZG, Z\varepsilon, Z\delta)$ in $Z\underline{C}$. Explicitly, $X \,.\, ZG = XG$, and the counit and comultiplication are

$$XG \xrightarrow{(X\varepsilon)} X ,$$
$$XG \xrightarrow{(X\delta)} XG^2 .$$

Although there are more maps in $Z\underline{C}$, the notion of object does not change, and neither does the notion of projective object. For $P \in \underline{C}$ is $\mathbf{G}$-projective $\Longleftrightarrow$ P regarded as an object in $Z\underline{C}$ is ZG-projective. The forward implication is evident, and if

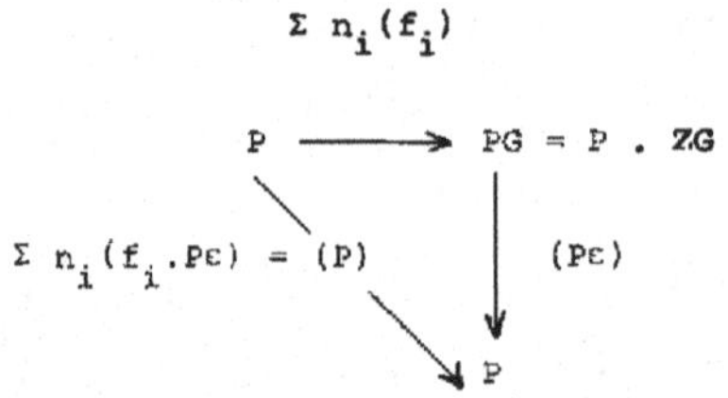

then $f_i \cdot P\varepsilon = P$ for some i, as $(P,P)'_{Z\underline{C}}$ is a free abelian group on a basis of which both $f_i \cdot P\varepsilon$ and P are members; this proves the other implication.

Thus the $\mathbb{G}$-resolutions of (5.1) are exactly the $Z\mathbb{G}$-resolutions relative to the co-triple in the additive category $Z\underline{C}$, in the sense of (4.1). Invoking the comparison theorem (4.2), we see that if (X_n), (Y_n) are $\mathbb{G}$- or equivalently $Z\mathbb{G}$-resolutions of $X_{-1} = Y_{-1} = X$, then there is a chain equivalence

$$(X_n) \xrightarrow{\ \sim\ } (Y_n) \qquad \text{in } Z\underline{C} .$$

Finally, let $E : \underline{C} \to \mathcal{A}$ be a coefficient functor and $\overline{E} : Z\underline{C} \to \mathcal{A}$ its additive extension constructed above. As the following complexes are identical ---

$$\begin{array}{l}
X(ZG)\overline{E} \xleftarrow{\partial_1} X(ZG)^2\overline{E} \longleftarrow \cdots\cdots \xleftarrow{\partial_n = (\sum_0^n (-1)^i (X\varepsilon_i))\overline{E}} X(ZG)^{n+1}\overline{E} \ \cdots\cdots \\
XGE \xleftarrow[\partial_1]{} XG^2E \longleftarrow \cdots\cdots \xleftarrow[\partial_n = \sum_0^n (-1)^i X\varepsilon_i E]{} XG^{n+1}E \ \cdots\cdots
\end{array}$$

we conclude that

$$H_n(X,E)_{\mathbb{G}} = H_n(X,\overline{E})_{Z\mathbb{G}} , \qquad n > o ,$$

another reduction of the general homology theory to the additive theory of §4. The last equation states that the diagram

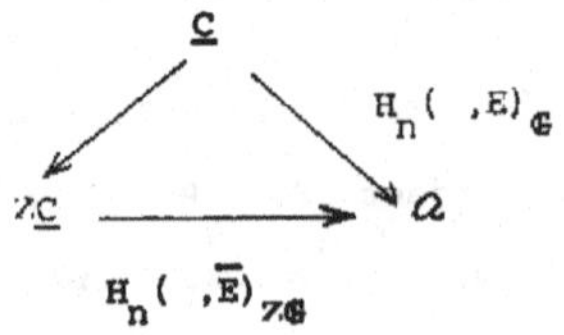

commutes, that is, the $H_n(\ ,\bar{E})_{Z\mathbb{G}}$ are the additive extensions of the $H_n(\ ,E)_{\mathbb{G}}$.

Parenthetically, an additive structure on $\underline{C}$ is equivalent to a unitary, associative functor $\theta : Z\underline{C} \to \underline{C}$, that is, $Z(\ )$ is a triple in the universe, and its algebras are the additive categories; if $\underline{C}$ is additive, θ is $(\Sigma\, n_i(f_i))\theta = \Sigma\, n_i f_i$.

$$\begin{array}{ccc} Z\underline{C} & \xrightarrow{H_n(\ ,\bar{E})_{Z\mathbb{G}}} & \mathcal{A} \\ & {\scriptstyle\theta}\searrow \quad \nearrow{\scriptstyle H_n(\ ,E)_{\mathbb{G}}} & \\ & \underline{C} & \end{array}$$

also commutes. In fact, this commutativity relation is equivalent to additivity of the homology functors, which in turn is equivalent to the homology functors' being $Z(\ )$ - algebra maps.

In the general case ——— $\underline{C}$ arbitrary ——— the above gives the result that $\mathbb{G}$-homology depends only on the $\mathbb{G}$-projectives:

(5.2) <u>Theorem</u>.

Let $\mathbb{G},\mathbb{K}$ be cotriples in $\underline{C}$ such that the classes of $\mathbb{G}$-projectives and $\mathbb{K}$-projectives coincide. Then $\mathbb{G}$ and $\mathbb{K}$ determine the same homology theory, that is, there is an isomorphism

$$H_n(X,E)_{\mathbb{G}} \xrightarrow{\ \sim\ } H_n(X,E)_{\mathbb{K}} , \qquad n \geq 0 ,$$

which is natural in both variables $X \in \underline{C}$, $E \in (\underline{C},\mathcal{A})$.

The same isomorphism holds for homology groups of a map $X \to Y$ (see §2) .

If $(\mathbb{G},X)$, $(\mathbb{K},X)$ are $\mathbb{G}$ and $\mathbb{K}$ lifted to the category of objects over X , $(\underline{C},X)$, then the $(\mathbb{G},X)$-and $(\mathbb{K},X)$-projectives also coincide. Thus if $E : (\underline{C},X) \to \mathcal{A}$ is a coefficient functor, there is an isomorphism

$$H_n(W,E)_{(\mathbb{G},X)} \xrightarrow{\ \sim\ } H_n(W,E)_{(\mathbb{K},X)} , \qquad n \geq 0 ,$$

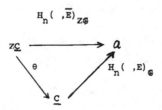

natural with respect to the variables $W \to X \in (\underline{C},X)$ and $E : (\underline{C},X) \to \mathcal{A}$.

Proof. $P \in \underline{C}$ is $\mathbb{Z}\mathbb{G}$-projective $\Longleftrightarrow$ P is $\mathbb{Z}K$-projective. The augmented complexes

$$X \leftarrow XG \leftarrow XG^2 \leftarrow \cdots$$

$$X \leftarrow XK \leftarrow XK^2 \leftarrow \cdots \qquad \text{in } \mathbb{Z}\underline{C}$$

are thus projective and acyclic with respect to the same projective class in $\mathbb{Z}\underline{C}$. The comparison theorem yields a chain equivalence

$$(XG^{n+1}) \xrightarrow{\simeq} (XK^{n+1})_{n \geq -1}$$

As to naturality in X, if $X \to X_1$ in $\underline{C}$, the comparison theorem also says that

$$\begin{array}{ccc} (XG^{n+1}) & \xrightarrow{\simeq} & (XK^{n+1}) \\ \downarrow & & \downarrow \\ (X_1G^{n+1}) & \xrightarrow{\simeq} & (X_1K^{n+1}) \end{array}$$

commutes up to chain homotopy. The comment about homology of a map follows from homotopy-invariance of mapping cones. $W \to X$ being $(\mathbb{G},X)$-projective $\Longleftrightarrow$ W is $\mathbb{G}$-projective is a trivial calculation.

(5.2) can also be proved through the intermediary of homology in categories with models ([18],[19,§12], and §10 below), as well as by a derived-functors argument (Ulmer).

To conclude this section we state the criteria for $\mathbb{G}$-resolutions and $(\mathbb{G},X)$-resolutions which will be used in §§6-9.

(5.3) Proposition.

$$X \leftarrow X_o \Leftarrow X_1 \Lleftarrow \cdots$$

is a simplicial $\mathbb{G}$-resolution of $X = X_{-1}$ if the X_n are $\mathbb{G}$-projective for $n \geq o$ and the following condition, which implies $\mathbb{G}$-acyclicity, holds: the cotriple $\mathbb{G}$ factors through an adjoint pair

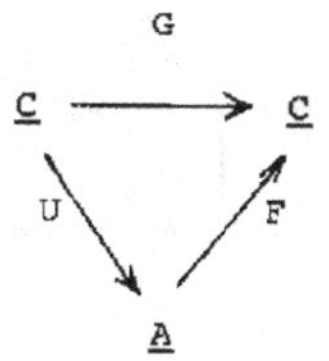

and the simplicial object $(X_nU)_{n \geq -1}$ in the underlying category $\underline{A}$ has a contraction

$$XU \xrightarrow{h_{-1}} X_oU \xrightarrow{h_o} X_1U \xrightarrow{h_1} \cdots \longrightarrow X_nU \xrightarrow{h_n} \cdots$$

(satisfying $h_n.\varepsilon_iU = \varepsilon_iU.h_{n-1}$ for $o \leq i \leq n$, $h_n.\varepsilon_{n+1}U = X_nU$). In particular, the standard $\mathbb{G}$-resolution $(XG^{n+1})_{n \geq -1}$ then has such a contraction:

$$XU \xrightarrow{h_1} XGU \xrightarrow{h_o} \cdots \longrightarrow XG^{n+1}U \xrightarrow{h_n} \cdots$$

to wit, $h_n = XG^nU\eta$ where η is the adjointness unit $\eta : \underline{A} \to FU$.

<u>Complement</u>. Let

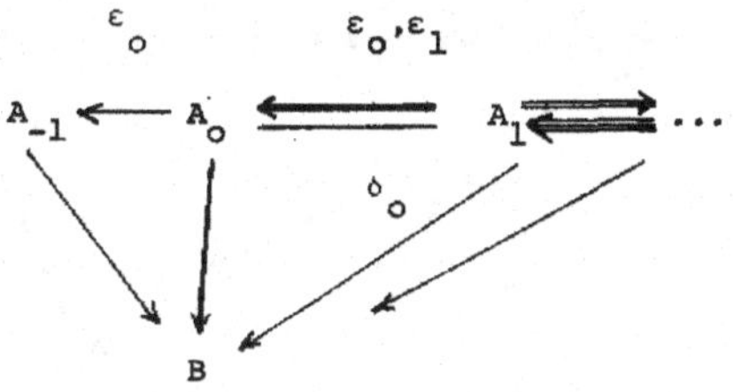

be a simplicial object in a category of objects over B , $(\underline{A},B)$. If

$$A_{-1} \xrightarrow{h_{-1}} A_o \xrightarrow{h_o} A_1 \xrightarrow{h_1} \cdots$$

is a contraction of the simplicial object sans B , then it is also a contraction of the simplicial object in $(\underline{A},B)$, that is, the h_n commute with the structural maps into B :

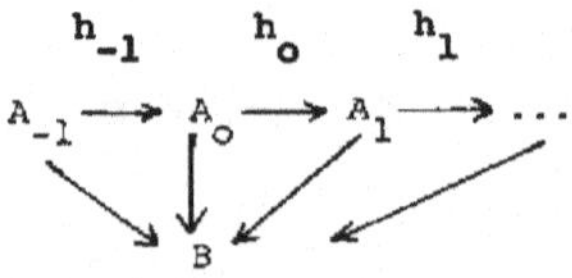

Thus when searching for contractions in categories of objects over a fixed object, the base object can be ignored.

<u>Proof</u>. If the stated condition holds, the simplicial set

$$(AG, X_n)_{n \geq -1}$$

has a contraction, so the free abelian group complex $(AG, X_n)_{Z\underline{C}}$ has homology zero in all dimensions ≥ -1. Indeed,

$$(AG, X_n) \xrightarrow{k_n} (AG, X_{n+1}) , \qquad n \geq 1 ,$$

is defined by

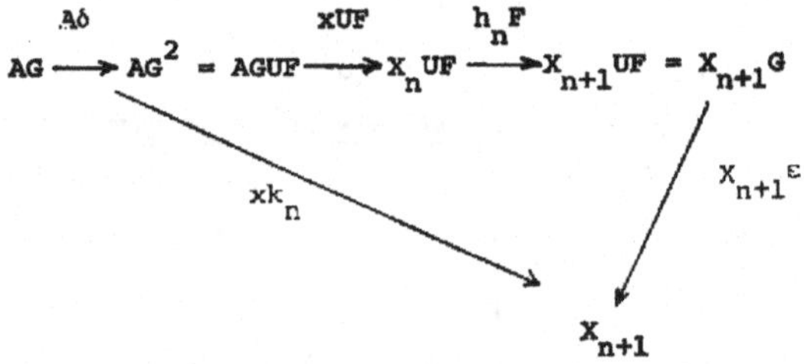

if $x : AG \to X_n$.

As to the Complement, if the maps into B are $p_n : A_n \to B$ then

$$h_n \cdot p_{n+1} = h_n \varepsilon_{n+1} p_n = A_n p_n = p_n$$

so in view of h_n's satisfying the identity $h_n \cdot \varepsilon_{n+1} = A_n$, it is a map over B .

6. Acyclicity and coproducts.

Given a $\mathbb{G}$-resolution

$$X = X_{-1} \leftarrow X_o \leftarrow X_1 \Leftarrow \ldots ,$$

is its term-by-term coproduct with a fixed object Y,

$$X * Y \leftarrow X_o * Y \leftarrow X_1 * Y \Leftarrow \ldots ,$$

still a $\mathbb{G}$-resolution? (The new face operators are of the form $\varepsilon_i * Y$.) If Y is $\mathbb{G}$-projective, so are all the $X_n * Y$, $n \geq o$. The problem is, is $\mathbb{G}$-acyclicity preserved? In this section we consider the examples of groups, commutative algebras and (associative) algebras, and prove that acyclicity is preserved, sometimes using supplementary hypotheses. The cotriples involved come from adjoint functors

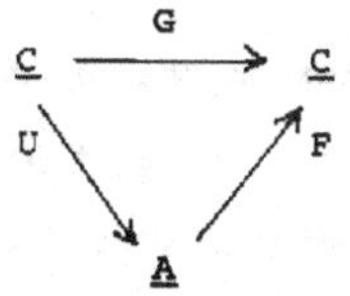

The general idea is to assume that (X_n) has a contraction in $\underline{A}$ and then show that this contraction somehow induces one in $(X_n * Y)$, even though the coproduct $*$ is not usually a functor on the underlying category level.

(6.1) Groups.

Let $(\Pi_n)_{n \geq -1}$ be an augmented simplicial group and $U : \mathcal{G} \to \mathcal{S}$ the usual underlying set functor where $\mathcal{G}$ is the category of groups.

From simplicial topology we know that the underlying simplicial set $(\Pi_n U)$ has a contraction $\Leftrightarrow$ the natural map into the constant simplicial set

$$(\Pi_n U)_{n \geq o} \longrightarrow (\Pi_{-1} U)$$

is a homotopy equivalence $\Leftrightarrow$ the set of components of $\Pi_* U$ is $\Pi_{-1} U$, and $H_n(\Pi_* U) = o$ for $n > o$ $(\Pi_* = (\Pi_n)_{n > o})$. (This is because simplicial groups satisfy the Kan extension condition, hence Whitehead's Theorem; $\pi_1 = H_1$ by the group property, so the fundamental group is zero, above).

Now suppose that $(\Pi_n U)_{n > -1}$ is acyclic, or has a contraction, and Π is another group. We shall prove that $((\Pi_n * \Pi)U)_{n > -1}$ also has a contraction;

We do this by considering the group ring functor $Z(\ ) : g \to$ <u>Rings</u> . The simplicial ring $Z\Pi_*$ obtained by applying the group ring functor in dimensionwise fashion has a contraction, namely the additive extension of the given set contraction in Π_* . In (6.3) below we shall show that the coproduct of this simplicial ring with $Z\Pi$ in the category of rings, $(Z\Pi_*) * Z\Pi$, where the n-dimensional component is $Z\Pi_n * Z\Pi$, also has a contraction. But as the group ring functor is a left adjoint,

$$(Z\Pi_*) * Z\Pi \xrightarrow{\sim} Z(\Pi_* * \Pi) .$$

Thus the set of components of the complex on the right is just $Z(\Pi_{-1} * \Pi)$ and its n-th homotopy is zero for $n > o$. This implies that $\Pi_* * \Pi$ has $\Pi_{-1} * \Pi$ as its set of components and has no higher homotopy. (This is equivalent to the curiosity that $Z(\ )$ as an endofunctor on sets satisfies the hypotheses of the "precise" tripleableness theorem ([26] or [27]).)

(6.2) <u>Commutative algebras</u>.

First let $\mathbb{G}_A$ be the cotriple relative to A-modules:

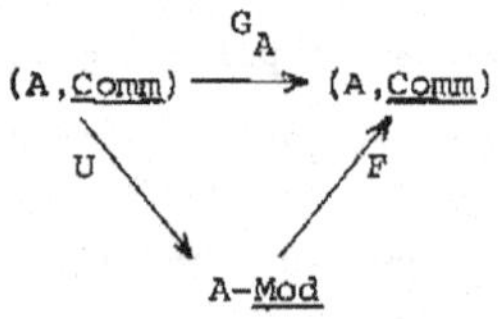

$\mathbb{G}_A$-resolutions behave very well with respect to coproducts of commutative A-algebras, $B \otimes_A C$. Indeed, as the standard resolution

$$B \longleftarrow BG_A \Leftarrow BG_A^2 \Lleftarrow \cdots\cdots$$

has an <u>A-linear</u> contraction (5.3) , so has

$$B \otimes_A C \longleftarrow BG_A \otimes_A C \Leftarrow BG_A^2 \otimes_A C \Lleftarrow \ldots$$

On the other hand, let $\mathbb{G}$ be the absolute cotriple

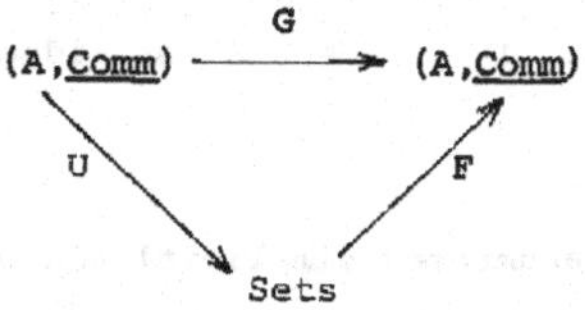

The standard $\mathbb{G}$-resolution has a contraction on the underlying set level (5.3). Thus the chain complex of A-modules associated to

$$B \longleftarrow BG \Leftarrow BG^2 \Lleftarrow \ldots$$

is an A-free resolution of B as an A-module in the usual homological sense. Thus the nonnegative-dimensional part of

$$B \otimes_A C \longleftarrow BG \otimes_A C \Leftarrow BG^2 \otimes_A C \Lleftarrow \ldots$$

has $H_n = \mathrm{Tor}_n^A(B,C)$, $n > o$. Since it is also a group complex, this simplicial object will have a contraction as a simplicial set $\longleftrightarrow \mathrm{Tor}_n^A(B,C) = o$, $n > o$.

(6.3) <u>Resolutions and coproducts of associative algebras</u>.

Let K-<u>Alg</u> be the category of associative K-algebras with identity. We are interested in resolutions relative to the adjoint pair

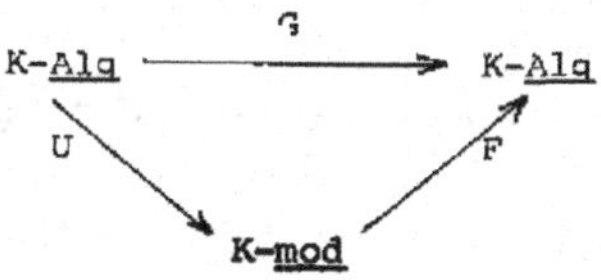

These will give rise to Hochschild homology. Here F is the K-tensor algebra $MF = K + M + M\otimes M + M\otimes M\otimes M + \ldots\ldots$ If Λ,Γ are K-algebras, their coproduct

$$\Lambda * \Gamma = (\Lambda + \Gamma)F/I$$

where I is the 2-sided ideal generated by the elements $\lambda_1 \otimes \lambda_2 - \lambda_1\lambda_2$, $\gamma_1 \otimes \gamma_2 - \gamma_1\gamma_2$, $1_K - 1_\Lambda$, $1_K - 1_\Gamma$ (1_K is in the summand of degree o , 1_Λ and 1_Γ are in the summand of degree 1). The K-linear maps $\Lambda,\Gamma \to (\Lambda \oplus \Gamma)F$ become algebra maps when I is divided out and these two maps are the coproduct injections $\Lambda,\Gamma \to \Lambda * \Gamma$. (In fact, I is the smallest ideal which makes these maps of unitary K-algebras.)

Let $(\Lambda_n)_{n \geq -1}$ be an augmented simplicial algebra which is U-contractible, i.e., there exists a K-linear contraction

$$\Lambda_{-1}U \xrightarrow{h_{-1}} \Lambda_o U \xrightarrow{h_o} \Lambda_1 U \to \cdots$$

We want to know that such a contraction continues to exist in the simplicial algebra $(\Lambda_n * \Gamma)_{n \geq -1}$. But we can only prove this in a special case.

An algebra Λ is called K-<u>supplemented</u> if there is a K-linear map $\Lambda \to K$ such that $K \to \Lambda \to K$ is the identity of K. (The first map sends $1_K \to 1_\Lambda$) . An algebra map $\Lambda \to \Lambda_1$ is called K-<u>supplemented</u> if Λ,Λ_1 are K-supplemented and

$$\begin{array}{ccc} \Lambda & \longrightarrow & \Lambda_1 \\ & \searrow \quad \swarrow & \\ & K & \end{array}$$

commutes.

We will show that if Λ,Γ are K-supplemented, then the coproduct of the canonical resolution of Λ with Γ ,

$$\Lambda * \Gamma \longleftarrow \Lambda G * \Gamma \Longleftarrow \Lambda G^2 * \Gamma \Lleftarrow \cdots\cdots,$$

possesses a K-contraction. We refer to (5.3) for the fact that $(\Lambda G^{n+1})_{n \geq -1}$ always has a K-linear contraction, and we prove that this contraction survives into the coproduct of the resolution with Γ .

The cotriple $\mathbf{G}$ operates in a natural way in the category of K-supplemented algebras. For if Λ is K-supplemented, the composition $\Lambda G \to \Lambda \to K$ defines a K-supplementation of ΛG . If $\Lambda \to \Lambda_1$ is K-supplemented, so is the induced $\Lambda G \to \Lambda_1 G$,

$$\begin{array}{ccc} \Lambda G & \longrightarrow & \Lambda_1 G \\ \downarrow & & \downarrow \\ \Lambda & \longrightarrow & \Lambda_1 \\ & \searrow \quad \swarrow & \\ & K & \end{array}$$

and if Λ is K-supplemented, the counit and comultiplication maps $\Lambda G \to \Lambda$, $\Lambda G \to \Lambda G^2$ are also K-supplemented.

When Λ is K-supplemented let $\overline{\Lambda} = \ker(\Lambda \to K)$. If $f : \Lambda \to \Lambda_1$ is K-supplemented, then $f = K \oplus \overline{f}$ where $\overline{f} : \overline{\Lambda} \to \overline{\Lambda}_1$ is induced in the obvious way. This means that if we write $f : K \oplus \overline{\Lambda} \to K \oplus \overline{\Gamma}$ in the form of 2×2 matrix, the matrix is diagonal:

$$\begin{pmatrix} K & o \\ o & \overline{f} \end{pmatrix}$$

Using the above supplementation and writing $\Lambda G^{n+1} = K \oplus \overline{\Lambda G^{n+1}}$, all of the face operators in the standard resolution $(\Lambda G^{n+1})_{n \geq -1}$ will be diagonal:

$$K \oplus \overline{\Lambda G^{n}} \xleftarrow{\begin{pmatrix} K & o \\ o & \overline{\varepsilon}_i \end{pmatrix}} K \oplus \overline{\Lambda G^{n+1}} \;, \quad o \leq i \leq n .$$

The K-linear contraction $h_n : \Lambda G^{n+1} \longrightarrow \Lambda G^{n+2}$ is given by a 2×2 matrix

$$K \oplus \overline{\Lambda G^{n+1}} \xrightarrow{\begin{pmatrix} h_{11} & h_{12} \\ h_{21} & h_{22} \end{pmatrix}} K \oplus \overline{\Lambda G^{n+2}}$$

The relation $h_n \varepsilon_{n+1} = \Lambda G^{n+1}$ is equivalent to $h_{11} = K$, $h_{21} = o$, $h_{12}\bar{\varepsilon}_{n+1} = o$, $h_{22}\bar{\varepsilon}_{n+1} = \overline{\Lambda G^{n+1}}$ (the matrix acts on row vectors from the right). The relation $h_n \varepsilon_i = \varepsilon_i h_n$, $o \leq i \leq n$, is equivalent to $h_{22}\bar{\varepsilon}_i = \bar{\varepsilon}_i h_{22}$ as another matrix calulation shows. Thus the contraction matrix has the form

$$\begin{pmatrix} K & h_{12} \\ o & h_{22} \end{pmatrix}$$

where entry h_{22} satisfies the contraction identities with respect to the restrictions of the face operators ε_i to the supplementation kernels, i.e. with respect to the maps $\bar{\varepsilon}_i$. Thus, we can switch to

$$h'_n = \begin{pmatrix} K & o \\ o & h_{22} \end{pmatrix}$$

which is also a matrix representation of a K-linear contraction

$$\begin{array}{ccc} \Lambda G^{n+1} & \xrightarrow{h'_n} & \Lambda G^{n+2} \\ K \oplus \overline{\Lambda G^{n+1}} & \xrightarrow[K \oplus \bar{h}_n]{} & K \oplus \overline{\Lambda G^{n+2}} \end{array}$$

where we have written $\bar{h}_n$ in place of h_{22}. This change can be made for all $n \geq -1$ so we get a K-contraction which is in diagonal form.

The next step is to find that the coproduct of two K-supplemented algebras can be written in a special form. Consider the direct sum

$$\begin{aligned} W = K + \bar{\Lambda} + \bar{\Gamma} + \bar{\Lambda} \otimes \bar{\Gamma} + \bar{\Gamma} \otimes \bar{\Lambda} \\ + \bar{\Lambda} \otimes \bar{\Gamma} \otimes \bar{\Lambda} + \dots \end{aligned}$$

of all words formed by tensoring $\bar{\Lambda}$ and $\bar{\Gamma}$ together with no repetitions allowed. There is an evident K-linear map

$$W \to \Lambda * \Gamma$$

given on the fifth summand above, for example, by

$$\bar{\Lambda} \otimes \bar{\Gamma} \otimes \bar{\Lambda} \to \Lambda \otimes \Gamma \otimes \Lambda \to (\Lambda \oplus \Gamma)F \to \Lambda * \Gamma .$$

$W \to \Lambda * \Gamma$ is one-one because its image in F does not intersect the ideal I $((\Lambda + \Gamma)F/I = \Lambda * \Gamma)$ and it is onto, clearly. Thus viewing $\Lambda * \Gamma$ as a K-module, we have

$$\Lambda * \Gamma = K + \bar{\Lambda} + \bar{\Gamma} + \bar{\Lambda} \otimes \bar{\Gamma} + \ldots .$$

Now let $\Lambda \xrightarrow{f} \Lambda_1$, $\Gamma \xrightarrow{g} \Gamma_1$ be K-<u>linear</u> maps which respect both units and supplementations, that is, writing $\Lambda = K \oplus \bar{\Lambda}$, and $\Lambda_1, \Gamma, \Gamma_1$ similarly,

$$K \oplus \bar{\Lambda} \xrightarrow{f = K \oplus \bar{f}} K \oplus \bar{\Lambda}_1 ,$$

$$K \oplus \bar{\Gamma} \xrightarrow{g = K \oplus \bar{g}} K \oplus \bar{\Gamma}_1 .$$

Then a K-linear map $\Lambda * \Gamma \to \Lambda_1 * \Gamma_1$, which we take the liberty of denoting by $f * g$, is induced:

$$\begin{array}{ccccccc} K + & \bar{\Lambda} & + & \bar{\Gamma} & + & \bar{\Lambda} \otimes \bar{\Gamma} & + \ldots \\ & \downarrow \bar{f} & & \downarrow \bar{g} & & \downarrow \bar{f} \otimes \bar{g} & \\ K + & \bar{\Lambda}_1 & + & \bar{\Gamma}_1 & + & \bar{\Lambda}_1 \otimes \bar{\Gamma}_1 & + \ldots \end{array}$$

If we are also given $f_1 : \Lambda_1 \to \Lambda_2$, $g_1 : \Gamma_1 \to \Gamma_2$, then $f_1 * g_1 : \Lambda_1 * \Gamma_1 \to \Lambda_2 * \Gamma_2$ and functoriality holds: $(f * g)(f_1 * g_1) = (ff_1 * gg_1)$ (because $\overline{ff_1} = \bar{f}\,\bar{f}_1$, and similarly for g, g_1).

To complete the argument, let

$$\Lambda \xrightarrow{K \oplus \bar{h}_{-1}} \Lambda G \xrightarrow{K \oplus \bar{h}_o} \Lambda G^2 \to \ldots \to \Lambda G^{n+1} \xrightarrow{K \oplus \bar{h}_n} \ldots$$

be the K-contraction with diagonal matrix constructed above. The $K \oplus h_n$ preserve both units and supplementations, so

$$\Lambda * \Gamma \xrightarrow{(K \oplus \pi_{-1}) * \Gamma} \Lambda G * \Gamma \to \cdots \to \Lambda G^{n+1} * \Gamma \xrightarrow{(K \oplus \pi_n) * \Gamma} \cdots$$

is a sequence of well defined K-linear maps which satisfies the contraction identities by virtue of the above functoriality.

__Alternative argument__.

The commutative diagram of adjoint pairs

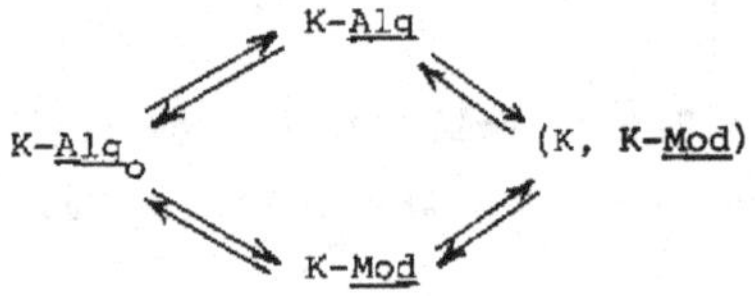

arises from a distributive law $TS \to ST$ in K-__Mod__; T is the triple $M \to K \oplus M$, whose algebras are __unitary__ K-modules (objects of the comma category (K, K-__Mod__)) and S is the triple $M \to M + M \otimes M + \ldots\ldots$, whose algebras are associative K-algebras without unit; this is the category denoted K-__Alg__$_o$. Let $\mathbb{G}$ be the cotriple in K-__Alg__ relative to K-__Mod__, and $\mathbb{G}_1$ that relative to unitary K-modules (K, K-__Mod__). By an easy extrapolation of [31,5.2], the cotriples $\mathbb{G}$, $\mathbb{G}_1$ operate in the full subcategory K-__Alg__' consisting of those $\Lambda \in$ K-__Alg__ whose underlying unitary K-modules are projective relative to K-modules, and $\mathbb{G}$, $\mathbb{G}_1$ restricted to this subcategory have the same projective objects. Now Λ, as a unitary K-module, is projective relative to K-modules $\Leftrightarrow$ there is a commutative diagram of K-linear maps

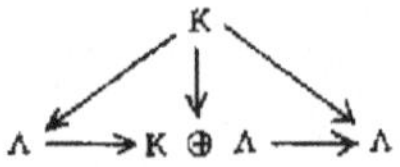

$\Leftrightarrow$ Λ has a K-linear supplementation. Thus if Λ is such an algebra, the standard resolutions

$$(\Lambda G^{n+1}) \ , \ (\Lambda G_1^{n+1})_{n \geq -1}$$

are chain equivalent in $Z(K\text{-}\underline{Alg})$. $(\ ) * \Gamma$ extends to an additive endofunctor of $Z(K\text{-}\underline{Alg})$.

$$(\Lambda G^{n+1} * \Gamma),\ (\Lambda G_1^{n+1} * \Gamma)_{n > -1}$$

are therefore also chain equivalent in $Z(K\text{-}\underline{Alg})$. Finally, $(\Lambda G_1^{n+1} * \Gamma)_{n > -1}$ has a K-linear contraction. This implies that

$$H_p(\Lambda G,\ (\Lambda G^{n+1} * \Gamma)_{n > -1}) = o\ ,\quad p > o\ ,$$

so $(\Lambda G^{n+1} * \Gamma)$ is $\mathbb{G}$-acyclic.

As to the last K-contraction, if Λ is any K-algebra for a moment, and Γ is K-linearly supplemented, then as a K-module $\Lambda * \Gamma$ can be viewed as a direct sum

$$\Lambda + \Gamma + \Lambda \otimes \overline{\Gamma} + \overline{\Gamma} \otimes \Lambda + \ldots.$$

modulo the relations $\gamma \otimes 1_\Lambda = 1_\Lambda \otimes \gamma = \gamma$, and the ideal generated by them, such as $\lambda \otimes \gamma \otimes 1_\Lambda = \lambda \otimes \gamma$,.... Thus if $f : \Lambda \to \Lambda_1$ is a <u>unitary</u> K-linear map, "$f * \Gamma$" $: \Lambda * \Gamma \to \Lambda_1 * \Gamma$ is induced, and functoriality holds: $ff_1 * \Gamma = (f * \Gamma)(f_1 * \Gamma)$ Now the resolution $(\Lambda G_1^{n+1})_{n > -1}$ has a <u>unitary</u> K-linear contraction (5.3). This contraction goes over into $(\Lambda G_1^{n+1} * \Gamma)_{n > -1}$ provided Γ is K-linearly supplemented.

7. <u>Homology coproduct theorems</u>.

Let $\mathbb{G}$ be a cotriple in $\underline{C}$ and let E be a coefficient functor $\underline{C} \to \mathcal{a}$. E <u>preserves coproducts</u> if the map induced by the coproduct injections $X,Y \to X * Y$ is an isomorphism for all $X,Y \in \underline{C}$:

$$XE \oplus YE \xrightarrow{\ \sim\ } (X * Y)E\ .$$

(In this § we assume $\underline{C}$ has coproducts). Particularly if E preserves coproducts, it is plausible that the similarly-defined natural map in homology is an isomorphism; if it is indeed the case that

$$H_n(X,E)_{\mathbb{G}} \oplus H_n(Y,E)_{\mathbb{G}} \longrightarrow H_n(X*Y,E)_{\mathbb{G}}$$

is an isomorphism, we say that <u>the homology coproduct theorem holds</u> (strictly speaking, for the objects X, Y, in dimension n ; it is characteristic of the theory to be developed that the coproduct theorem often holds only for objects X , Y with special properties) .

In this section we show that the homology coproduct theorem holds for the various categories and various cotriples considered in §6. As one gathers from the arguments resorted to in that § , there must be something the matter with the slick method of proving coproduct theorems sketched in [4 , §5]. First, to correct a slip, (5.4) in [4] should read $u : (X_1*X_2)GU \rightarrow (X_1G*X_2G)U$, that is, it is in the underlying category that u should be sought. However, even with that correction, such a natural u does not exist so far as we now know, in group theory (relative to sets) or in Hochschild theory, contrary to our earlier claims. The morphisms u which we had in mind in these cases turned out on closer inspection not to be natural, because of misbehavior of neutral elements of one kind or another in coproducts viewed at the underlying-category level . Only in "case 3" of [4 , §5], namely that of commutative algebras relative to K-modules, does the method of that paper work. However, we are able to retrieve most of the results claimed there although in the case of Hochschild theory we are forced to impose an additional linear-supplementation hypothesis.

Such tests as we possess for the coproduct theorem are contained in the next two propositions.

(7.1) <u>Proposition</u>. If X and Y possess $\mathbb{G}$-resolutions

$$X = X_{-1} \leftarrow X_o \Leftarrow X_1 \Lleftarrow \cdots$$

$$Y = Y_{-1} \leftarrow Y_o \Leftarrow Y_1 \Lleftarrow \cdots$$

such that the coproduct

$$X*Y \leftarrow X_o*Y_o \Leftarrow X_1*Y_1 \Lleftarrow \cdots$$

is a $\mathbb{G}$-resolution, then the coproduct theorem holds for X,Y and any coproduct-preserving coefficient functor. (The issue is $\mathbb{G}$-acyclicity.)

In particular, if each row and column of the double augmented simplicial object

$$(X_m * Y_n)_{m,n \geq -1}$$

is a $\mathbb{G}$-resolution, then the above diagonal object $(X_n * Y_n)_{n \geq -1}$ is a $\mathbb{G}$-resolution.

(7.2) Proposition. Suppose that the cotriple $\mathbb{G}$ factors through an adjoint pair

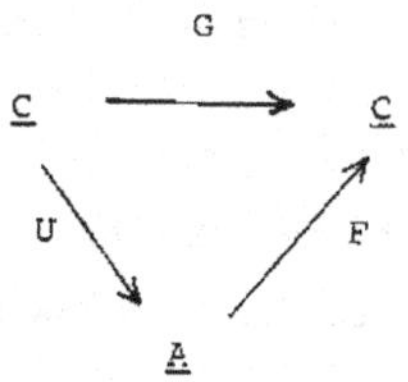

and that $(X*Y)U$ is naturally equivalent to $XU *_{\underline{A}} YU$ where $*_{\underline{A}} : \underline{A} \times \underline{A} \to \underline{A}$ is some bifunctor; in other words, the coproduct is definable at the underlying-category level. Then the homology coproduct theorem holds for any coproduct-preserving coefficient functor.

As to (7.1), coproducts of projectives being projective, we are left to consider the augmented double simplicial set

$$(AG, X_m * Y_n)_{m,n \geq -1}$$

As the rows and columns lack homology, so does the diagonal, by the Eilenberg-Zilber theorem [35]. For (7.2), identify

$$((XG^{n+1} * YG^{n+1})U)_{n \geq -1} \quad \text{with} \quad (XG^{n+1}U *_{\underline{A}} YG^{n+1}U) ,$$

of which

$$XU *_{\underline{A}} YU \xrightarrow{h_{-1} *_{\underline{A}} k_{-1}} XGU *_{\underline{A}} YGU \longrightarrow \cdots$$

is a contraction (see (5.3)).

In the following examples we use the fact that the coproduct in the category of objects over X, $(\underline{C},X)$, is "the same" as the coproduct in $\underline{C}$:

$$(X_1 \xrightarrow{P_1} X) * (X_2 \xrightarrow{P_2} X) = (X_1 * X_2 \xrightarrow{(P_1,P_2)} X)$$

(7.3) <u>Groups</u>.

$$H_n(\Pi_1,E) \oplus H_n(\Pi_2,E) \xrightarrow{\sim} H_n(\Pi_1 * \Pi_2, E)$$

for any coproduct-preserving functor $E : G \longrightarrow \mathcal{A}$, such as $\otimes M$ or $\mathrm{Hom}(\ ,M)$ where M is a fixed abelian group.

To deduce the usual coproduct theorems for homology and cohomology with coefficients in a module, we apply the complement to (5.3) to see that $(\Pi_1 G^{n+1})_{n \geq -1}$ is a $(\mathfrak{G},\Pi_1 * \Pi_2)$-resolution of Π_1 as a group over $\Pi_1 * \Pi_2$ (using the coproduct injection $\Pi_1 \to \Pi_1 * \Pi_2$) . By (6.1) and the complement to (5.3) again, $(\Pi_1 G^{n+1} * \Pi_2 G^{n+1})_{n \geq -1}$ is a $(\mathfrak{G},\Pi_1 * \Pi_2)$-resolution of $\Pi_1 * \Pi_2 \to \Pi_1 * \Pi_2$. If M is a $\Pi_1 * \Pi_2$-module, then M can be regarded both as a Π_1-module and as a Π_2-Module by means of $\Pi_1, \Pi_2 \to \Pi_1 * \Pi_2$. Thus in homology we have a chain equivalence between the complexes

$$(\mathrm{Diff}_{\Pi_1 * \Pi_2}(\Pi_1 G^{n+1}) \otimes M \oplus \mathrm{Diff}_{\Pi_1 * \Pi_2}(\Pi_2 G^{n+1}) \otimes M) ,$$

$$(\mathrm{Diff}_{\Pi_1 * \Pi_2}(\Pi_1 G^{n+1} * \Pi_2 G^{n+1}) \otimes M) , \qquad n \geq 0 ,$$

$\otimes$ being over $\Pi_1 * \Pi_2$. As a result,

$$H_n(\Pi_1,M) \oplus H_n(\Pi_2,M) \xrightarrow{\sim} H_n(\Pi_1 * \Pi_2, M) .$$

In cohomology, taking coefficients in $\mathrm{Hom}_{\Pi_1 * \Pi_2}(\ ,M)$,

$$H^n(\Pi_1 * \Pi_2,M) \xrightarrow{\sim} H^n(\Pi_1,M) \oplus H^n(\Pi_2,M) .$$

These isomorphisms, apparently known for some time, appear to have been first proved (correctly) in print in [25]. Similar isomorphisms hold for (co-)homology of $W_1 * W_2$ where $W_1 \to \Pi_1$, $W_2 \to \Pi_2$ are groups over Π_1, Π_2. (Earlier proof: [45].)

(7.4) <u>Commutative Algebras</u>. Let B, C be A-algebras over D, that is, $B, C \to D$, and M a D-module.

If H is homology relative to the "absolute" cotriple $\mathbb{G}$ coming from $(A, \underline{\text{Comm}}) \to$ <u>Sets</u>, we have

$$H_n(B,M) \oplus H_n(C,M) \xrightarrow{\ \sim\ } H_n(B \otimes_A C, M) ,$$

$$H^n(B \otimes_A C, M) \xrightarrow{\ \sim\ } H^n(B,M) \oplus H^n(C,M)$$

provided $\mathrm{Tor}_p^A(B,C) = 0$ for $p > 0$; this is because the coproduct of the standard resolutions,

$$(BG^{n+1} \otimes_A CG^{n+1})_{n \geq -1} ,$$

has $\mathrm{Tor}(B,C)$ as its homology (use the Eilenberg-Zilber theorem), which is the obstruction to a contraction in the underlying category of sets. The result is also proved in [19], [28], [1], [24] .

If H is the theory relative to A-modules, the isomorphisms hold without any con-- dition (6.2).

(7.5) <u>Associative K-algebras</u>. If $\Lambda_1, \Lambda_2 \to \Gamma$ are K-algebra maps and M is a two-sided Γ-module, and Λ_1, Λ_2 possess K-linear supplementations, then

$$H_n(\Lambda_1, M) \oplus H_n(\Lambda_2, M) \xrightarrow{\ \sim\ } H_n(\Lambda_1 * \Lambda_2, M) ,$$

$n > 0$; the same for cohomology with coefficients in M, or for any coproduct-preserving coefficient functor. The cotriple employed is that relative to K-modules; the proofs are from (6.3), (7.1) .

8. On the homology of a map.

In §2 we defined homology groups of a map so as to obtain an exact sequence

$$\cdots\cdot \to H_n(X,E) \to H_n(Y,E) \to H_n(X \to Y,E) \xrightarrow{\partial} H_{n-1}(X,E) \to \cdots\cdot$$

In fact, although we had to use a mapping cone instead of a quotient complex, the definition is the same as in algebraic topology. In this section we show (with a proviso) that these groups are the same as the cotriple groups

$$H_n(X \to Y,\ (X,E))_{(X,\mathbb{G})}\ , \qquad n > o\ ,$$

where $X \to Y$ is considered as an object under X, (X,E) is the extension to a functor $(X,\underline{C}) \to \mathcal{A}$ of a given coefficient functor $E : \underline{C} \to \mathcal{A}$ and $(X,\mathbb{G})$ is $\mathbb{G}$ lifted into the comma category as described before (1.2); the proviso is that a homology coproduct theorem should hold for the coproduct of any object with a free object.

The coefficient functor we use,

$$(X,\underline{C}) \xrightarrow{(X,E)} \mathcal{A}\ ,$$

is defined by $(X \to Y)(X,E) = \text{coker}\ XE \to YE$. Recalling the formulas for $(X,\mathbb{G})$, we have that the $H_n(X \to Y,(X,E))_{(X,\mathbb{G})}$ are the homology groups of the standard complex which in dimensions o and 1 reads:

$$o \longleftarrow \text{coker}\ (XE \to (X{*}YG)E) \xleftarrow{\partial_1} \text{coker}\ (XE \to (X{*}(X{*}YG)G)E) \xleftarrow{\partial_2} \cdots\cdot$$

(8.1) Theorem. There is a sequence of homology maps

$$H_n(X \to Y,E)_{\mathbb{G}} \xrightarrow{H_n(\varphi)} H_n(X \to Y,(X,E))_{(X,\mathbb{G})}\ , \qquad n > o\ ,$$

resulting from a natural chain transformation

$$C(X \to Y)_n \xrightarrow{\varphi_n} (X \to Y)(X,\mathbb{G})^{n+1}(X,E)\ .$$

($C(X \to Y)$ is the mapping cone defined in §2 and functoriality is with respect to maps of objects under X). The $H_n(\varphi)$ are <u>isomorphisms</u> if the following theorem holds: for all $X,Y \in \underline{C}$, the coproduct injections induce isomorphisms

$$H_n(X*YG,E)_{\mathbb{G}} \xleftarrow{\ \sim\ } \begin{cases} H_o(X,E)_{\mathbb{G}} \oplus YGE\ , & n = o\ , \\ H_n(X,E)_{\mathbb{G}}\ , & n > o\ , \end{cases}$$

that is, if E satisfies the homology coproduct theorem when one summand is $\mathbb{G}$-free.

<u>Proof</u>. We augment both complexes by attaching H_o as (-1)-dimensional term. We first define φ_o, φ_1 so as to obtain the commutative square $\varphi_1\partial_1 = \partial_1\varphi_o$, which induces a natural map φ_{-1} on the augmentation terms.

$$\begin{array}{ccccc}
YG^2E \oplus XGE & \xrightarrow{\ \varphi_1\ } & (X*(X*YG)G)E & \xrightarrow{\ \text{coker}\ } & (X \to Y)(X,G)^2(X,E) \\
\downarrow{\scriptstyle \partial_1} & & \downarrow{\scriptstyle \partial_1} & & \downarrow{\scriptstyle \partial_1} \\
YGE & \xrightarrow{\ \varphi_o\ } & (X*YG)E & \xrightarrow{\ \text{coker}\ } & (X \to Y)(X,G)(X,E) \\
\downarrow & & & & \downarrow \\
H_o(X \to Y,E)_{\mathbb{G}} & & \xrightarrow[\ \varphi_{-1}\ ]{} & & H_o(X \to Y,(X,E))_{(X,\mathbb{G})}
\end{array}$$

If we write $i : X \to Y * YG$, $j : YG \to X * YG$ for coproduct injections, then $\varphi_o = jE$, and φ_1 is determined by

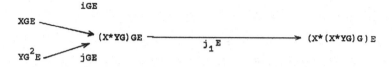

$$\begin{array}{ccccc}
XGE & \xrightarrow{\ iGE\ } & & & \\
 & & (X*YG)GE & \xrightarrow[\ j_1E\ ]{} & (X*(X*YG)G)E \\
YG^2E & \xrightarrow[\ jGE\ ]{} & & &
\end{array}$$

where j_1 is also a coproduct injection. That φ_o,φ_1 commute with ∂_1 is readily checked.

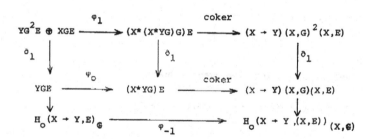

The higher φ_n could be written down similarly but we don't bother with that as they automatically fall out of the acyclic-models argument which we need for the isomorphism anyway. We use $(X,\mathbb{G})$ as the comparison cotriple. The cotriple complex $(X \to Y)(X,G)^{n+1}(X,E)$ is representable and contractible with respect to this cotriple, as always. Furthermore, $C(X \to Y)$ is $(X,\mathbb{G})$-representable via $\theta_n : C(X \to Y)_n \to C(X \to X{*}YG)_n$,

$$YG^{n+1}E \oplus XG^nE \xrightarrow{(Y\delta . jG)G^nE \,\oplus\, \mathrm{id.}} (X{*}YG)G^{n+1}E \oplus XG^nE \;,$$

if $n > o$, and $\theta_o = jE$. This proves φ_{-1} can be extended to a chain transformation defined in all dimensions. It happens that the extension produced by (11.1) agrees with the above φ_o, φ_1 in the lowest dimensions.

To conclude, if the homology coproduct assumption in (8.1) also holds, then

$$H_n(X \to X{*}YG,E)_{\mathbb{G}} \simeq \begin{cases} YGE \,, & n = o \,, \\ o \,, & n > o \,, \end{cases}$$

since this homology group H_n fits into the exact sequence

$$\begin{array}{l} \ldots \longrightarrow H_n(X,E) \longrightarrow H_n(X{*}YG,E) \longrightarrow H_n \\ \quad \overset{\partial}{\longrightarrow} \cdots \\ H_o(X,E) \longrightarrow H_o(X{*}YG) \longrightarrow H_o \to o \,. \end{array}$$

Thus the φ_n induce homology isomorphisms between the two theories ((11.3)), q.e.d.

(8.2) <u>Groups</u>. If

$$\begin{array}{ccc} \Pi_o & \xrightarrow{\;f\;} & \Pi_1 \\ & \searrow \quad \swarrow & \\ & \Pi & \end{array}$$

is a map in $(\mathcal{G},\Pi)$ and M is a Π-module we get an exact sequence

$$\ldots \to H_n(\Pi_o,M) \to H_n(\Pi_1,M) \to H_n(f,M) \to H_{n-1}(\Pi_o,M) \to \ldots$$

and a similar one in cohomology. The relative term arises either as in §2 or by viewing f as an object in the double comma category $(\Pi_o, \mathcal{G}, \Pi_1)$ and using this §. The equivalence results from the fact that the homology coproduct theorem holds for groups.

This sequence can be obtained topologically by considering the map of Eilenberg-MacLane spaces $K(\Pi_o,1) \to K(\Pi_1,1)$. It is also obtained in [44].

As a special case, if

$$\begin{array}{ccc} \Pi & \xrightarrow{f} & \Pi/N \\ & \searrow \quad \swarrow & \\ & 1 & \end{array}$$

is division by a normal subgroup and we take coefficients in Z as a 1-module, then $H_o(f) = o$ and $H_1(f) \cong N/[\Pi,N]$. Thus the Stallings-Stammbach sequence ([7],[8]) falls out:

$$H_1(\Pi) \to H_1(\Pi/N) \to N/[\Pi,N] \to H_o(\Pi) \to H_o(\Pi/N) \to o$$

(our dimensional indices). Doubtless many of the other sequences of this type given in [9] can be got similarly.

(8.3) <u>Commutative rings and algebras</u>. Given maps of commutative rings

$$\begin{array}{ccccc} A & \longrightarrow & B & \longrightarrow & C \\ & \searrow & \downarrow & \swarrow & \\ & & D & & \end{array}$$

we obtain exact sequences

$$\ldots \to H_n(A,M) \to H_n(B,M) \to H_n(A \to B,M) \xrightarrow{\partial} H_{n-1}(A,M) \to \ldots$$

$$\ldots \to H_n(A \to B,M) \to H_n(A \to C,M) \to H_n(B \to C,M) \xrightarrow{\partial} H_{n-1}(A \to B,M) \to \ldots$$

for a D-module M ; similar sequences are obtainable in cohomology. Taking B = C = D , and homology with respect to the couple $\mathbb{G}$ arising from

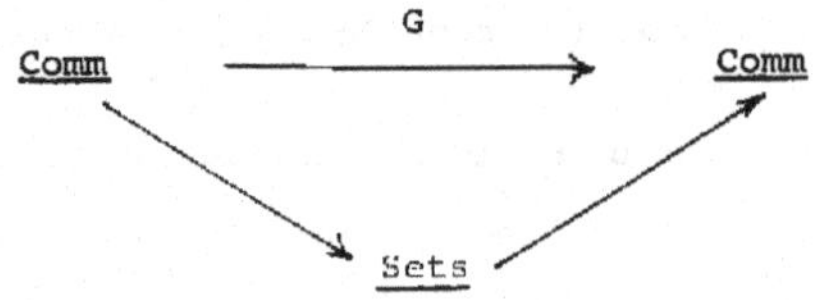

these sequences coincide with those of [1],[19],[28], as a result of the following facts:

(a) $(A,\mathbb{G})$ is the cotriple arising from

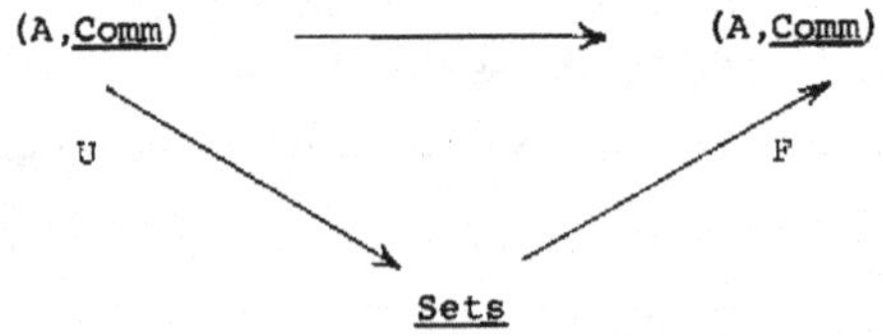

where $(A \to B)U = B$.

(b) If $E : (\underline{Comm},D) \to D\text{-}\underline{Mod}$ is $AE = Diff_D(A) \otimes_D M$, then $(A \to B)(A,E) = \Omega^1_{B/A} \otimes_B M$. If $E : (\underline{Comm},D)^* \to D\text{-}\underline{Mod}$ is $AE = Der\,(A,M)_D$, then $(A \to B)(A,E) = A\text{-}Der\,(B,M)_D$.

(c)

$$H_n(A \to B,E)_{\mathbb{G}} \xrightarrow{\ \sim\ } H_n(A \to B,(A,E))_{(A,\mathbb{G})}$$

for any coproduct preserving coefficient functor $E : (\underline{Comm},D) \to \mathcal{A}$ (writing A for $A \to D$) .

(a) has been noted in §1. For (b),

$$(A \to B)(A,E) = \text{coker}\,(Diff_D(A) \otimes_D M \to Diff_D(B) \otimes_D M)$$

$$= \text{coker}\,(Diff_D(A) \to Diff_D(B)) \otimes_D M$$

$$= (\Omega^1_{B/A} \underset{B}{\otimes} D) \otimes_D M$$

$$= \Omega^1_{B/A} \otimes_B M \; .$$

In the dual theory, it is appropriate to lift a functor $E : \underline{C}^* \to \mathcal{A}$ to a functor $(E,A) : (\underline{C}^*,A) \to \mathcal{A}$, by defining $(B \to A)(E,A) = \ker (BE \to AE)$. For E the contravariant functor in (b), we have then

$$(A \to B)(A,E) = \ker (\mathrm{Der}(B,M)_D \to \mathrm{Der}(A,M))$$

$$= A\text{-}\mathrm{Der}\,(B,M)_D \; .$$

Alternatively and of course equivalently, dualize the coefficient category of D-modules. Finally (c) follows from the fact that the coproduct theorem holds for homology in this category when one factor is free. Indeed, $C.(A,G)$ is the polynomial A-algebra $A[C]$ and is A-flat ; thus the coproduct $(\;) \otimes_A C.(A,G)$ preserves $(A,\mathbb{G})$-resolutions.

For the A-relative theory (1.2), the same exact sequences are available.

(8.4) Associative Algebras. Let $\mathbb{G}$ devote the cotriple on K-Alg arising out of the adjoint pair

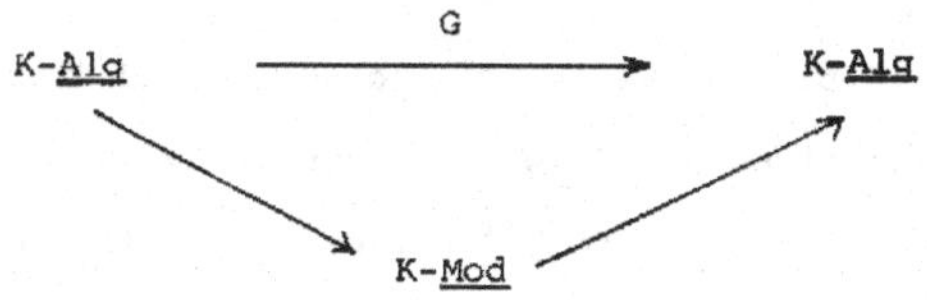

If $\Lambda \in$ K-Alg, $\Gamma \to \Lambda \in (\text{K-Alg},\Lambda)$ and M is a Λ-bimodule, we let $H_n(\Gamma,M)_{\mathbb{G}}$ and $H^n(\Gamma,M)_{\mathbb{G}}$ denote the derived functors with respect to $\mathbb{G}$ of $\mathrm{Diff}_\Lambda(\;) \otimes_{\Lambda^e} M$ and $\mathrm{Der}_\Lambda(\;,M)$ respectively. Let us drop Λ from the notation from now on. Hence $\Gamma \to \Gamma_1$ below really refers to $\Gamma \to \Gamma_1 \to \Lambda$ etc.

(8.5) Theorem.

$$H_n(\Gamma \to \Gamma_1,M)_{\mathbb{G}} \simeq H_n(\Gamma_1,M)_{(\Gamma,\mathbb{G})} ,$$

$$H^n(\Gamma \to \Gamma_1,M)_{\mathbb{G}} \simeq H^n(\Gamma_1,M)_{(\Gamma,\mathbb{G})} .$$

Proof. According to (8.1) this requires showing that for any Γ',

$$H_n(\Gamma * \Gamma'G,M)_{\mathbb{G}} \simeq \begin{cases} H_o(\Gamma,M)_{\mathbb{G}} \oplus \text{Der}\ (\Gamma'G,M), & n = o \\ H_n(\Gamma,M)_{\mathbb{G}} , & n > o , \end{cases}$$

and similarly for cohomology. Before doing this we require

(8.6) Proposition. Let $\mathbb{G}_1$ be the cotriple described in (6.3) above. Then

$$H_n(\Gamma,M)_{\mathbb{G}} \simeq H_n(\Gamma,M)_{\mathbb{G}_1} ,$$

$$H^n(\Gamma,M)_{\mathbb{G}} \simeq H^n(\Gamma,M)_{\mathbb{G}_1} .$$

The proof will be given at the end of this section.

Now observe that any $\mathbb{G}$-projective is $\mathbb{G}_1$-projective and also is supplemented. Now $(\Gamma G_1^{n+1})_{n > o}$ is a $\mathbb{G}_1$-resolution of Γ, which means it has a unitary K-linear contraction. As observed in (6.3) above, $(\Gamma G_1^{n+1} * \Gamma'G)_{n > o}$ also has a unitary K-linear contraction and it clearly consists of $\mathbb{G}_1$-projectives. Thus it is a $\mathbb{G}_1$-resolution of $\Gamma * \Gamma'G$. But then

$$\text{Diff}_\Lambda(\Gamma G_1^{n+1} * \Gamma'G)_{n > o} \simeq \text{Diff}_\Lambda(\Gamma G_1^{n+1})_{n > o} \oplus \text{Diff}_\Lambda \Gamma'G ,$$

the second summand being a constant simplicial object, and the result follows easily.

To prove (8.6) we use acyclic models in form (11.3) below with $\mathbb{G}_1$ as the comparison cotriple. First observe that there is a natural transformation $\varphi : G \to G_1$

which actually induces a morphism of cotriples (meaning it commutes with both comultiplication and counit). Actually $\mathbb{G}_1$ is presented as a quotient of $\mathbb{G}$ and φ is the natural projection. Now we prove the theorem for cohomology. The proof for homology is similar. For any $\Gamma' \to \Gamma$ and any Γ-bimodule M, let $\Gamma'E = \text{Der}(\Gamma', M)$, $\Gamma'E^n = \text{Der}(\Gamma' G^{n+1}, M)$. Let $\overline{\varphi} : \text{Der}(\Gamma', M) \to \Gamma'E$ be the identity and $\overline{\varphi}^n : \text{Der}(\Gamma' G_1{}^{n+1}, M) \to \Gamma'E^n$ be the map $\text{Der}(\Gamma'\varphi^{n+1}, M)$. Define $\theta^n : \Gamma' G_1 E^n \to \Gamma' E^n$ to be the composite

$$\text{Der}(\Gamma' G_1 G^{n+1}, M) \xrightarrow{\text{Der}(\Gamma'\varphi G^{n+1}, M)} \text{Der}(\Gamma' G^{n+2}, M) \xrightarrow{\text{Der}(\Gamma'\delta G^{n}, M)} \text{Der}(\Gamma' G^{n+1}, M) .$$

Then it is easily seen that $\text{Der}(\Gamma'\varepsilon_1 G^{n+1}, M) \cdot \theta^n$ is the identity. (Of course, everything is dualized for cohomology.) Thus the proof is finished by showing that the complex

$$\cdots \to \Gamma' G_1 E^n \to \Gamma' G_1 E^{n-1} \to \cdots \to \Gamma' G_1 E^1 \to \Gamma' G_1 E \to 0$$

is exact. But the homology of that complex is simply the Hochschild homology of $\Gamma'G$, (with the usual degree shift), which in turn is $\text{Ext}_{(\Gamma^e, K)}(\text{Diff}_\Gamma \Gamma' G_1, M)$. Hence we complete the proof by showing that $\text{Diff}_\Gamma \Gamma' G$, is a K-relative Γ^e- projective. But $\text{Der}(\Gamma' G_1, M)$ consists of those derivations of $\Gamma' G \to M$ which vanish on the ideal of $\Gamma' G$ generated by $1_{\Gamma'} - 1_K$ or, since all derivations vanish on 1_K, it simply consists of those derivations which vanish on $1_{\Gamma'}$. But $\text{Der}(\Gamma' G, M) \simeq \text{Hom}_K(\Gamma', M)$ and it is easily seen that $\text{Der}(\Gamma' G_1 M) \simeq \text{Hom}_K(\Gamma'/K, M)$ where Γ'/K denotes coker $(K \to \Gamma)$. This in turn is $\simeq \text{Hom}_{\Lambda^e}(\Lambda^e \otimes \Gamma'/K, M)$ and so $\text{Diff}_\Gamma \Gamma' G_1 \simeq \Lambda^e \otimes \Gamma'/K$ which is clearly a K-relative Λ^e- projective. This completes the proof.

9. Mayer-Vietoris theorems.

Using assumptions about the homology of coproducts, we shall deduce some theorems of Mayer-Vietoris type. We learned of such theorems from André's work [19]. In the case of commutative algebras we obtain slightly more comprehensive results (9.5). Mostly, however, we concentrate on the case of groups (9.4).

Let $E : \underline{C} \to \mathcal{A}$ be a coefficient functor.

(9.1) Theorem. Let

$$\begin{array}{ccc} X & \longrightarrow & X_1 \\ \downarrow & & \downarrow \\ X_2 & \longrightarrow & Y \end{array}$$

be a pushout diagram in $\underline{C}$ and suppose that the homology coproduct theorem holds for Y viewed as a coproduct in $(X,\underline{C})$:

$$H_n(X \to X_1,E) \oplus H_n(X \to X_2,E) \xrightarrow{\sim} H_n(X \to Y,E), \qquad n \geq o .$$

Then there is an exact sequence

$$\begin{array}{l} \cdots\cdots \to H_n(X,E) \to H_n(X_1,E) \oplus H_n(X_2,E) \to H_n(Y,E) \\ \xrightarrow{\ \partial\ } H_{n-1}(X,E) \to \cdots\cdots\cdots\cdots\cdots\cdots \to H_o(Y,E) \to o \end{array}$$

(The maps in the sequence are the usual Mayer-Vietoris maps $(\beta,-\gamma)$, (β_1,γ_1) transpose, if we momentarily write

$$\beta : H(X) \longrightarrow H(X_1) , \qquad \beta_1 : H(X_1) \longrightarrow H(Y) ,$$
$$\gamma : H(X) \longrightarrow H(X_2) , \qquad \gamma_1 : H(X_2) \longrightarrow H(Y) .)$$

(9.2) Theorem. Suppose that the natural map is an isomorphism

$$H_n(X,E) \oplus H_n(Y,E) \xrightarrow{\sim} H_n(X*Y,E) \;, \qquad r > o \;.$$

Then for any map $X * Y \to Z$ there is an exact sequence

$$\begin{array}{l} \cdots\cdots \to H_n(Z,E) \to H_n(X \to Z,E) \oplus H_n(Y \to Z,E) \to H_n(X * Y \to Z,E) \\ \xrightarrow{\partial} H_{n-1}(Z,E) \to \cdots\cdots\cdots\cdots\cdots \to H_o(X * Y \to Z,E) \to o \end{array}$$

For the proof of (9.1), write down the diagram

$$\begin{array}{ccccccc} H(X,E) & \to & H(X_1,E) & \longrightarrow & H(X \to X_1,E) & \xrightarrow{\partial} & H(X,E) \\ \| & & \downarrow & & \downarrow & & \| \\ H(X,E) & \to & H(Y,E) & \to & H(X \to X_1,E) \oplus H(X \to X_2,E) & \xrightarrow{\partial} & H(X,E) \\ \| & & \uparrow & & \uparrow & & \| \\ H(X,E) & \to & H(X_2,E) & \longrightarrow & H(X \to X_2,E) & \xrightarrow{\partial} & H(X,E) \end{array}$$

All three triangles are exact, the middle one by the coproduct theorem in $(X,\underline{C})$. Lemma (9.3) below then yields that

$$H(X,E) \to H(X_1,E) \oplus H(X_2,E) \to H(Y,E) \to H(X,E)$$

is an exact triangle. For (9.2), write

$$\begin{array}{ccccccc} H(Z,E) & \to & H(X \to Z,E) & \xrightarrow{\partial} & H(X,E) & \longrightarrow & H(Z,E) \\ \| & & \downarrow & & \downarrow & & \| \\ H(Z,E) & \to & H(X * Y \to Z,E) & \xrightarrow{\partial} & H(X,E) \oplus H(Y,E) & \to & H(Z,E) \\ \| & & \uparrow & & \uparrow & & \| \\ H(Z,E) & \to & H(Y \to Z,E) & \xrightarrow{\partial} & H(Y,E) & \longrightarrow & H(Z,E) \end{array}$$

and again apply

(9.3) <u>Lemma</u>. In an abelian category

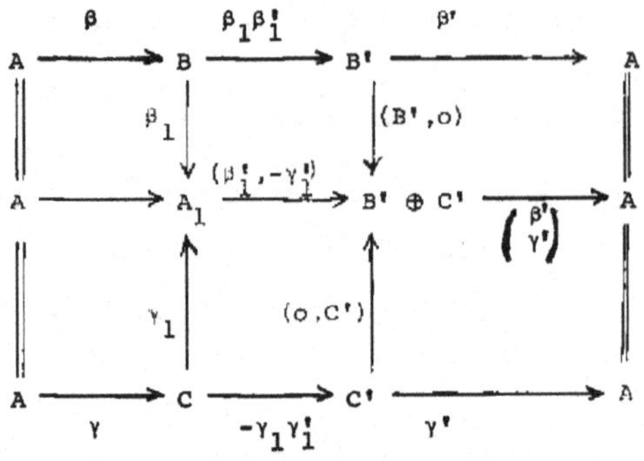

is commutative with exact triangles for rows $\Longleftrightarrow$

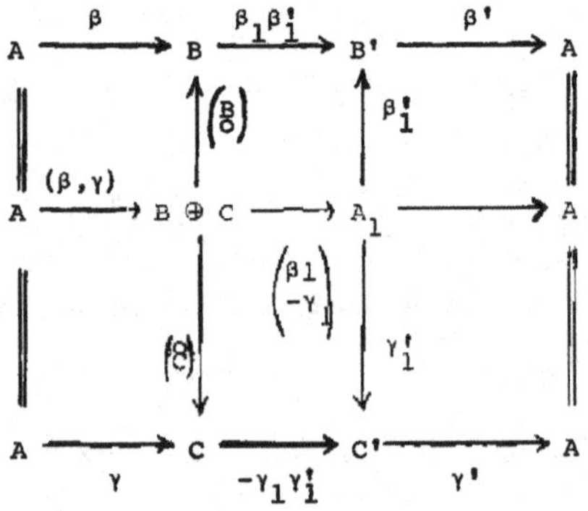

is commutative and has exact triangle for rows.

This lemma is dual to its converse and needn't be proved.

(9.4) Groups. Theorem (9.2) holds without restriction. Because of the validity of the homology coproduct theorem (7.3), if $\Pi_o * \Pi_1 \to \Pi$ we get an exact sequence

$$\ldots\ldots \to H_n(\Pi,M) \to H_n(\Pi_o \to \Pi,M) \oplus H_n(\Pi_1 \to \Pi,M) \to H_n(\Pi_o * \Pi_1 \to \Pi,M)$$

$$\to H_{n-1}(\Pi,M) \to \ldots\ldots$$

if M is a Π-module; similar sequences hold in cohomology, or in homology with coefficients in any coproduct-preserving functor.

As to (9.1), its applicability is a little more restricted. Suppose that Π_o is a subgroup of Π_1 and Π_2 and that Π is the pushout or amalgamated coproduct:

$$\begin{array}{ccc} \Pi_o & \longrightarrow & \Pi_1 \\ \downarrow & & \downarrow \\ \Pi_2 & \longrightarrow & \Pi \end{array}$$

It will be shown that if M is a Π-module, then the Π_o-coproduct theorem holds for homology:

$$H_n(\Pi_o \to \Pi_1, M) \oplus H_n(\Pi_o \to \Pi_2, M) \xrightarrow{\sim} H_n(\Pi_o \to \Pi, M) \ .$$

Thus the Mayer-Vietoris sequence

$$\cdots\cdot \to H_n(\Pi_o, M) \to H_n(\Pi_1, M) \oplus H_n(\Pi_2, M) \to H_n(\Pi, M)$$
$$\to H_{n-1}(\Pi_o, M) \to \cdots\cdot\cdot$$

is exact. There is a similar exact sequence for cohomology with coefficients in M. As our argument will involve Diff_Π, we cannot claim this for arbitrary coefficient functors but only for those that are a composition of $\mathrm{Diff}_\Pi : (\mathcal{G}, \Pi) \to \Pi\text{-}\underline{\mathrm{Mod}}$ and an additive functor $E : \Pi\text{-}\underline{\mathrm{Mod}} \to \mathcal{A}$. This theorem also has a topological proof using Eilenberg-Mac Lane spaces. Similar results have been obtained by Ribes [43]. We now launch into the algebraic details:

The free group cotriple preserves monomorphisms; let Y_n be the pushout or amalgamated coproduct

$$\begin{array}{ccc} \Pi_o G^{n+1} & \longrightarrow & \Pi_1 G^{n+1} \\ \downarrow & & \downarrow \\ \Pi_2 G^{n+1} & \longrightarrow & Y_n \end{array} \qquad (n > -1)$$

Thus (Y_n) is an augmented simplicial group, with $Y_{-1} = \Pi$. Moreover Y_n is a free group when $n > o$ as $Y_n = SF$ where S is the settheoretic pushout

$$\begin{array}{ccc} \Pi_o G^n U & \longrightarrow & \Pi_1 G^n U \\ \downarrow & & \downarrow \\ \Pi_2 G^n U & \longrightarrow & S \end{array}$$

and F is the free group functor $\mathcal{S} \to \mathcal{G}$, which as a left adjoint preserves pushouts. Applying Diff_Π, we get a square

$$\text{(a)} \qquad \begin{array}{ccc} \mathrm{Diff}_\Pi(\Pi_o G^{n+1}) & \longrightarrow & \mathrm{Diff}_\Pi(\Pi_1 G^{n+1}) \\ \downarrow & & \downarrow \\ \mathrm{Diff}_\Pi(\Pi_2 G^{n+1}) & \longrightarrow & \mathrm{Diff}_\Pi(Y_n) \end{array}$$

which is exact, i.e., simultaneously a pushout and a pullback. We will prove this later, as also fact (b) arrayed below. For the rest of this § we write

$$\mathrm{Diff} = \mathrm{Diff}_\Pi .$$

Now using the usual Mayer-Vietoris maps we get an exact sequence of chain complexes

$$o \to (\mathrm{Diff}(\Pi_o G^{n+1})) \to (\mathrm{Diff}(\Pi_1 G^{n+1}) \oplus \mathrm{Diff}(\Pi_2 G^{n+1}))$$
$$\to (\mathrm{Diff}(Y_n)) \to o , \qquad n \geq o$$

whence the homology sequence

$$\text{(b)} \qquad \begin{array}{ccccccc} \cdots\cdots & \longrightarrow & o & \longrightarrow & o & \longrightarrow & H_p(\mathrm{Diff}(Y_*)) \\ & & o & \longrightarrow & \cdots & \longrightarrow & H_1(\mathrm{Diff}(Y_*)) \\ & & & & \partial & & \\ \mathrm{Diff}(\Pi_o) & \longrightarrow & \multicolumn{3}{c}{\mathrm{Diff}(\Pi_1) \oplus \mathrm{Diff}(\Pi_2)} & \longrightarrow & H_o(\mathrm{Diff}(Y_*)) \longrightarrow o \end{array}$$

(The p illustrated is ≥ 2); in addition, the map ∂ is zero. This yields the conclusion that $(\mathrm{Diff}(Y_n))$, $n \geq -1$, is a Π-free resolution of $\mathrm{Diff}(\Pi) = I\Pi$ in the category of Π-modules.

Let E be any additive functor π-Mod $\to \mathcal{A}$. The first two columns of the following commutative diagramm are exact, hence the third column which consists of the mapping cones of the horizontal maps is also exact.

$$\begin{array}{ccccc}
0 & & 0 & & 0 \\
\downarrow & & \downarrow & & \downarrow \\
\mathrm{Diff}(\Pi_o G^{n+1})E & \longrightarrow & \mathrm{Diff}(\Pi_o G^{n+1})E & \longrightarrow & C' \\
\downarrow & & \downarrow & & \downarrow \\
\mathrm{Diff}(\Pi_o G^{n+1})E \oplus \mathrm{Diff}(\Pi_o G^{n+1})E & \rightarrow & \mathrm{Diff}(\Pi_1 G^{n+1})E \oplus \mathrm{Diff}(\Pi_2 G^{n+1})E & \rightarrow & C \\
\downarrow & & \downarrow & & \downarrow \\
\mathrm{Diff}(\Pi_o G^{n+1})E & \longrightarrow & (\mathrm{Diff}\ Y_*)E & \longrightarrow & C'' \\
\downarrow & & \downarrow & & \downarrow \\
0 & & 0 & & 0
\end{array}$$

$(n > o)$

C' is acyclic as it is the mapping cone of an identity. Clearly,

$$H_n(C) = H_n(\Pi_o \to \Pi_1, E) \oplus H_n(\Pi_o \to \Pi_2, E) \quad ,$$

$$H_n(C'') = H_n(\Pi_o \to \Pi, E) \quad .$$

The homology sequence of $o \to C' \to C \to C'' \to 0$ then proves the coproduct theorem for groups under Π_o. This completes the proof that (9.1) applies to amalgamated coproduct diagrams in $\mathcal{G}$, modulo going back and proving (a), (b).

Square (a) is obviously a pushout since $\mathrm{Diff}_\Pi : (\mathcal{G},\Pi) \to \Pi$-<u>Mod</u> is a left adjoint and preserves pushouts. The hard part is proving that it is a pullback. For that it is enough to show that the top map $\mathrm{Diff}_\Pi(\Pi_o G^{n+1}) \to \mathrm{Diff}_\Pi(\Pi_1 G^{n+1})$ is a monomorphism, in view of:

<u>Lemma</u>. In abelian category.

$$\begin{array}{ccc}
A_o & \xrightarrow{\alpha_1} & A_1 \\
{\scriptstyle \alpha_2}\downarrow & & \downarrow{\scriptstyle \beta_1} \\
A_2 & \xrightarrow[\beta_2]{} & A_3
\end{array}$$

is a pushout $\longleftrightarrow$

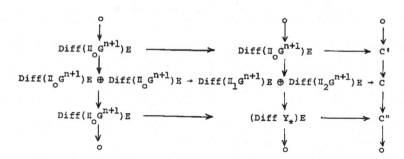

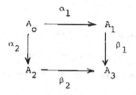

$$A_o \xrightarrow{(\alpha_1,\alpha_2)} A_1 \oplus A_2 \xrightarrow{\begin{pmatrix}\beta_1\\ -\beta_2\end{pmatrix}} A_3 \longrightarrow o$$

is exact, and dually, is a pullback $\Longleftrightarrow$

$$o \longrightarrow A_o \xrightarrow{(\alpha_1,\alpha_2)} A_1 \oplus A_2 \xrightarrow{\begin{pmatrix}\beta_1\\ -\beta_2\end{pmatrix}} A_3$$

is exact.

This is standard. Thus, we are reduced to:

Lemma. If $\Pi_o \to \Pi$ is a subgroup, then $\mathrm{Diff}_\Pi(\Pi_o) \to \mathrm{Diff}_\Pi(\Pi)$ is a monomorphism of Π-modules. If

$$\begin{array}{ccc} \Pi_o & \longrightarrow & \Pi_1 \\ & \searrow \quad \swarrow & \\ & \Pi & \end{array}$$

is a diagram of subgroups, then

$$\begin{array}{ccc} \mathrm{Diff}_\Pi(\Pi_o) & \longrightarrow & \mathrm{Diff}_\Pi(\Pi_1) \\ & \searrow \quad \swarrow & \\ & \mathrm{Diff}_\Pi(\Pi) & \end{array}$$

commutes, hence $\mathrm{Diff}_\Pi(\Pi_o) \to \mathrm{Diff}_\Pi(\Pi_1)$ is a monomorphism of Π-modules.

Proof.[1] We write $x \in \Pi_o$, $y \in \Pi$ and present an isomorphism

$$\begin{array}{ccc} \mathrm{Diff}_\Pi(\Pi_o) = Z\Pi \underset{\Pi_o}{\otimes} I\Pi_o & \xrightarrow{f} & D \\ & \searrow \quad \swarrow & \\ & I\Pi & \end{array}$$

where D is the Π-submodule generated by all $x-1$. f and f^{-1} are the Π-linear maps determined by the correspndence $1 \otimes (x-1) \longleftrightarrow x-1$. f is more-or-less obviously well-defined. As for f^{-1} , it is deduced from the exact sequence of Π-modules

[1] There is a simple exact-sequences argument.

$$\begin{array}{ccccccc} R & \xrightarrow{\partial_1} & F & \xrightarrow{\partial_o} & D & \longrightarrow & o \\ & & \Big\downarrow{\scriptstyle f_o} & \swarrow{\scriptstyle f^{-1}} & & & \\ & & Z\Pi \otimes_{\Pi_o} I\Pi_o & & & & \end{array}$$

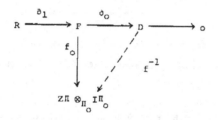

where F is the free Π-module on generators $[x]$, $[x]\partial_o = x-1$, and ∂_1 is the submodule generated by all elements of the form

$$y[x] + y_1[x_1] - y_1[y_1^{-1}yx]$$

where $y = y_1x_1$. f_o is defined by $[x]f_o = 1 \otimes (x-1)$ and annihilates ∂_1, q.e.d.

For the proof of the statements around (b), we know that

$$H_p(\mathrm{Diff}_{\Pi_o}(\Pi_o G^{n+1})_{n > o}) = \begin{cases} \mathrm{Diff}_{\Pi_o}(\Pi_o) & , \ p = o \\ o & , \ p > o \end{cases}$$

by (1.2). After tensoring over Π_o with $Z\Pi$, which is Π_o-projective since $\Pi_o \to \Pi$ is a subgroup, we find that the homology becomes $\mathrm{Diff}_\Pi(\Pi_o)$ in dimension o and o in dimensions $> o$. This accounts for the two columns of o's in (b). The fact that $\partial = o$ results from exactness and the above Lemma, which implies that $\mathrm{Diff}_\Pi(\Pi_o) \to \mathrm{Diff}_\Pi(\Pi_1) \oplus \mathrm{Diff}_\Pi(\Pi_2)$ is monomorphic.
This completes the proof.

(9.5) <u>Commutative algebras</u>. If

$$\begin{array}{ccc} A & \longrightarrow & B \\ \Big\downarrow & & \Big\downarrow \\ C & \longrightarrow & B\otimes_A C \end{array}$$

is a pushout in the category of commutative K-algebras, where K is a commutative ring and M is a $B \otimes_A C$-module, then

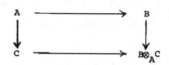

$$H_n(A \to B, M) \oplus H_n(A \to C, M) \xrightarrow{\sim} H_n(A \to B \otimes_A C, M)$$

for $n \geq o$ if $\mathrm{Tor}_p^A(B,C) = o$ for $p > o$ (homology with respect to the absolute cotriple in the category of commutative K-algebras (cf. (7.4)). In this case (9.1) gives an exact sequence

$$\ldots . \to H_n(A,M) \to H_n(B,M) \oplus H_n(C,M) \to H_n(B \otimes_A C, M)$$
$$\to H_{n-1}(A,M) \to \ldots .$$

A similar sequence holds in cohomology under the same Tor assumption. If $K = A$, this coincides with the homology coproduct theorem.

If $A \otimes_K B \to C$ is a K-algebra map and $\mathrm{Tor}_p^K(A,B) = o$ for $p > o$, then the homology assumption in (9.2) is satisfied and we get the sequence

$$\ldots . \to H_n(C,M) \to H_n(A \to C,M) \oplus H_n(B \to C,M)$$
$$\to H_n(A \otimes_K B \to C,M) \to H_{n-1}(C,M) \to \ldots .$$

if M is a C-module; similarly in cohomology. This is the same sequence as in [19, 19.5], but the assumption $\mathrm{Tor}_p^K(C,C) = o$, $p > o$, employed there is seen to be superfluous.

10. Cotriples and models.

For our purposes it is sufficient to consider a category with models to be a functor $\underline{M} \to \underline{C}$ where $\underline{M}$ is discrete. The objects of $\underline{M}$ are known as the models. Many cotriples can be constructed in the following manner.

(10.1.) Model-induced cotriple. If $X \in \underline{C}$, let

$$XG = \mathop{*}_{\substack{M \to X \\ M \in \underline{M}}} M ,$$

the coproduct indexed by all maps of model objects $M \to X$.
We assume that such coproducts exist in $\underline{C}$, and write $\underline{M} \to X$ instead of $MI \to X$ in order to avoid having to name $I : \underline{M} \to \underline{C}$

Let $\langle x \rangle : M \to XG$ denote the canonical map of the cofactor indexed by a map $x : M \to X$. Then

$$XG \xrightarrow{X\varepsilon} X$$

is the map such that $\langle x \rangle X\varepsilon = x$ for all $x : M \to X$, $M \in \underline{M}$.

$$XG \xrightarrow{X\delta} XGG$$

is the map such that $\langle x \rangle X\delta = \langle\langle x \rangle\rangle$ for all such x. (Since $\langle x \rangle : M \to XG$, $\langle\langle x \rangle\rangle : M \to (XG)G$.) Both ε and δ are natural transformations, and as

$$\langle x \rangle X\delta . XG\delta = \langle\langle\langle x \rangle\rangle\rangle = \langle x \rangle X\delta . X\delta G ,$$

$$\langle x \rangle X\delta . XG\varepsilon = \langle\langle x \rangle\rangle XG\varepsilon = \langle x \rangle = \langle\langle x \rangle\rangle X\varepsilon G = \langle x \rangle X\delta . X\varepsilon G ,$$

we have that $\mathfrak{G} = (\mathfrak{G}, \varepsilon, \delta)$ is a cotriple in $\underline{C}$, which we call <u>model-induced</u>. (This special case is dual to the "triple structure" which Linton discusses in [47] ; see also Appelgate-Tierney [13].)

If M is a model, then M viewed as an object in $\underline{C}$ is $\mathfrak{G}$-projective (even a $\mathfrak{G}$-coalgebra):

$$M \xrightarrow{\langle M \rangle} MG \xrightarrow{M\varepsilon} M .$$

Some other relations between model concepts and cotriple concepts are: a simplicial object X_* has zero homotopy relative to $\mathfrak{G}$ (every (AG, X_*) has zero homotopy) $\Longleftrightarrow$ every simplicial set (M, X_*) has zero homotopy. In the additive case, $\mathfrak{G}$-acyclicity is equivalent to acyclicity relative to all of the objects $M \in \underline{M}$.

(10.2) <u>Examples of model-induced cotriples.</u> (a) Let $1 \to R$-<u>Mod</u> be the functor whose value is R. Then $AG = \oplus R$, over all elements $R \to A$, is the free R-module cotriple (1.1).

More generally, if $\underline{C}$ is tripleable over sets and $1 \to \underline{C}$ has value $1F$, the free object on 1 generator, then the modelinduced cotriple $\mathbf{G}$ is the free cotriple in $\underline{C}$, e.g., $\underline{C}$ = K-Alg, $1F = K[x]$, $\underline{C}$ = Groups, $1F = Z$.

(b) Let $1 \to$ Ab have value Q/Z (rationals mod one). Let T be the model-induced triple in Ab

$$AT = \prod_{A \to Q/Z} Q/Z.$$

$(AT^{n+1})_{n \geq -1}$ is an injective resolution of A. The composition

$$R\text{-}\underline{Mod} \to \underline{Ab} \xrightarrow{T} \underline{Ab} \xrightarrow{Hom_Z(R,\ )} R\text{-}\underline{Mod}$$

is the Eckmann-Schopf triple T_R in R-Mod. $(AT_R^{n+1})_{n \geq -1}$ is an R-injective resolution of an R-module A.

(c) Let $\underline{M} \to$ R-Mod be the subset of cyclic R-modules. The model induced cotriple is the pure cotriple

$$CG = \bigoplus_{\substack{R/I \to C \\ I \subset R \\ I \neq 0}} R/I .$$

The $\mathbf{G}$-homology and cohomology of $C \in$ R-Mod with coefficients in $A \otimes_R (\ )$, resp. $Hom_R(\ ,A)$, are Harrison's $Ptor_n^R(A,C)$, $Pext_R^n(C,A)$; Pext classifies pure extensions of R-modules [34] . This example is one of the original motivations for relative homological algebra.

(d) Let $\Delta \to$ Top be the discrete subcategory whose objects are the standard Euclidean simplices Δ_p, $p \geq 0$. Then

$$XG = \bigcup_{\substack{\Delta_p \to X \\ p \geq 0}} \Delta_p$$

If $\underline{Top} \xrightarrow{E} \mathcal{a}$ is $H_0(\ ,M)_{sing}$, the o-th singular homology group of X with coefficients in M, then

$$H_n(X,H_o(\ ,M)_{sing})_{\mathbb{G}} \cong H_n(X,M)_{sing} .$$

This is proved by a simple acyclic-models argument (11.2) or equivalently by collapsing of a spectral sequence like that in (10.5) . Singular cohomology is similarly captured.

(e) Let $\underline{\Delta} \to$ Simp be the discrete subcategory of all Δ_p , $p \geq o$, where Simp is the category of simplicial spaces. The model-induced cotriple is $XG = \cup \Delta_p$ over all simplicial maps $\Delta_p \to X$, $p \geq o$. The $\mathbb{G}$-homology is simplicial homology.

(10.3) Homology of a category. In [38] , [39] , Roos and André defined a homology theory $H_n(\mathcal{K},E)$ of a category $\mathcal{K}$ with coefficients in a functor $E : \mathcal{K} \to \mathcal{A}$. The homology theory arises from a complex

$$C_n(\mathcal{K},E) = \sum_{M_o \xrightarrow{\alpha_o} M_1 \to \cdots \xrightarrow{\alpha_{n-1}} M_n} M_oE , \qquad n \geq o .$$

Using the $\langle\rangle$ notation for the coproduct injections $M_oE \to C_n(\mathcal{K},E)$, the face operators $\varepsilon_i : C_n \to C_{n-1}$, $o \leq i \leq n$, are

$$\langle\alpha_o,\dots,\alpha_{n-1}\rangle\ \varepsilon_i = \begin{cases} \alpha_oE.\langle\alpha_1,\dots,\alpha_{n-1}\rangle , & i = o , \\ \langle\alpha_o,\dots,\alpha_{i-1}\alpha_i,\dots,\alpha_{n-1}\rangle & o < i < n , \\ \langle\alpha_o,\dots,\alpha_{n-2}\rangle , & i = n ; \end{cases}$$

it is understood that $\langle\alpha_o\rangle\ \varepsilon_o = \alpha_oE.\langle M_1\rangle$, $\langle\alpha_o\rangle\ \varepsilon_1 = \langle M_o\rangle$, and $C_o(\mathcal{K},E) = \Sigma ME$ over all $M \in \mathcal{K}$. The homology groups of this complex, with respect to the boundary operator $\partial = \Sigma(-1)^i\varepsilon_i$, are denoted by $H_n(\mathcal{K},E)$.

Clearly, $H_o(\mathcal{K}\ E) = \varinjlim E$, and Roos proves that if $\mathcal{A}$ has exact direct sums (AB4) , then $H_n(\mathcal{K},E) = (L_n \varinjlim)(E)$, the left satellite of the direct limit functor $(\mathcal{K},\mathcal{A}) \to \mathcal{A}$, for $n \geq o$.

If there is a terminal object $1 \in \mathcal{K}$, then $H_n(\mathcal{K},E) = o$ for $n > o$. This follows from the existence of homotopy operators

$$C_o \xrightarrow{h_o} C_1 \longrightarrow \cdots \longrightarrow C_n \xrightarrow{h_n} \cdots$$

defined by $\langle\alpha_o,\ldots,\alpha_n\rangle h_n = \langle\alpha_o,\ldots,\alpha_n,(\ )\rangle$, where $(\ )$ is the unique map of the appropriate object into 1 . This is also obvious from the fact that $\varinjlim E = 1E$, so that $\varinjlim$ is an exact functor (assuming $\mathcal{A}$ is AB4) .

More generally, if $\mathcal{X}$ is directed and $\mathcal{A}$ is AB5 , then $H_n(\mathcal{X},E) = o$ for $n > o$. "Directed" means that if $X_o, X_1 \in \mathcal{X}$, then there exist an object $X \in \mathcal{X}$ and maps $X_o \to X \leftarrow X_1$, and if $x,y : X_1 \to X_o$, then there exists a map $z : X_o \to X$ such that $xz = yz$. AB5 is equivalent to exactness of direct limits over directed index categories.

(10.4) <u>André-Appelgate homology</u>. In a models situation, let $\mathrm{Im}\underline{M}$ be the full subcategory of $\underline{C}$ generated by the image of $\underline{M} \to \underline{C}$. If $X \in \underline{C}$, $(\mathrm{Im}\underline{M},X)$ is the category whose objects are maps of models $M \to X$ and whose maps are triangles $X \leftarrow M_o \to M_1 \to X$. If $E_o : \mathrm{Im}\underline{M} \to \mathcal{A}$ is a coefficient functor, E_o can be construed as a functor $(\mathrm{Im}\underline{M},X) \to \mathcal{A}$ by $(M \to X)E_o = ME_o$.

The <u>André-Appelgate homology</u> of X <u>with coefficients in</u> E_o (relative to the models $\underline{M} \to \underline{C}$) is

$$A_n(X,E_o) = H_n[(\mathrm{Im}\underline{M},X),E_o] ,$$

where on the right we have the Roos-André homology of the comma category. Explicitly, the chain complex which gives rise to this homology theory has

$$C_n(X,E_o) = \sum_{M_o \xrightarrow{\alpha_o} M_1 \to \cdots \to M_n \xrightarrow{x} X} M_oE_o$$

with boundary operator as in [19,§1]. We note that H.Appelgate [18] developed this homology theory in a different way. He viewed the above complex as being generated by its o-chains acting as a cotriple in the functor category $(\underline{C},\mathcal{A})$.

A basic property of this theory is that if M is a model, then

$$A_n(M,E_o) \simeq \begin{cases} ME_o , & n = o , \\ o , & n > o , \end{cases}$$

for any functor $E_o : \mathrm{Im}\underline{M} \to \mathcal{A}$. The category $(\mathrm{Im}\underline{M},M)$ has M as final object, and the contracting homotopy in (10.3) in available [19 ,1.1] .

In general,

$$A_o(X,E_o) = X.E_J(E_o)$$

where $E_J : (\mathrm{Im}\underline{M},\mathcal{A}) \to (\underline{C},\mathcal{A})$, the Kan extension , is left adjoint to the restriction functor $(\mathrm{Im}\underline{M},\mathcal{A}) \leftarrow (\underline{C},\mathcal{A})$. As the Kan extension can also be written as $\varinjlim(E_o : (\mathrm{Im}\underline{M},X) \to \mathcal{A})$, Roos's result implies that

$$A_n(X,E_o) = X.(L_nE_J)(E_o) , \qquad n > o .$$

provided that $\mathcal{A}$ is AB4 . (For further information about Kan extension, see Ulmer's paper in this volume.)

The theory $A_n(X,E)$ is also defined when $E : \underline{C} \to \mathcal{A}$, by restricting E to $\mathrm{Im}\underline{M}$. It can always be assumed that the coefficient functor is defined on all of $\underline{C}$. If not, take the Kan extension. The restriction of $E_J(E_o)$ to $\mathrm{Im}\underline{M}$ is equivalent to the given E_o since $J : \mathrm{Im}\underline{M} \to \underline{C}$ is full.

Now suppose we have both a models situation $\underline{M} \to \underline{C}$ and a cotriple $\mathbb{G}$ in $\underline{C}$. To compare the homology theories $A_n(X,E)$ and $H_n(X,E) = H_n(X,E)_{\mathbb{G}}$, we use:

(10.5) Spectral sequence. Suppose that all models $M \in \underline{M}$ are $\mathbb{G}$-projective. Then there is a spectral sequence

$$H_p(X,A_q(\ \ ,E)) \Rightarrow A_{p+q}(X,E)$$

where the total homology is filtered by levels $\leq p$.

Proof. For each $M \in \underline{M}$ choose a map $M\sigma : M \to MG$ such that $M\sigma . M\varepsilon = M$. Define $\theta_q : C_q(X,E) \to C_q(XG,E)$ by the identity map from the $\langle \alpha_o, \ldots, \alpha_{q-1}, x\rangle$-th summand to the $\langle \alpha_o, \ldots, \alpha_{q-1} . M_q \sigma, xG\rangle$-th . This makes $C_*(X,E)$ $\mathfrak{G}$-representable, and the result follows from (11.3) .

(10.6) Proposition. If each category $(Im\underline{M},XG)$ is directed and the coefficient category $\mathcal{A}$ is AB 5 , then the above spectral sequence collapses and gives edge isomorphisms

$$H_n(X,A_o(\ ,E)) \xrightarrow{\ \sim\ } A_n(X,E) , \qquad n > o .$$

The André-Appelgate theory has a natural augmentation $A_o(\ ,E) \to E$, which is induced by the following cokernel diagram and map e such that $\langle x\rangle e = xE$:

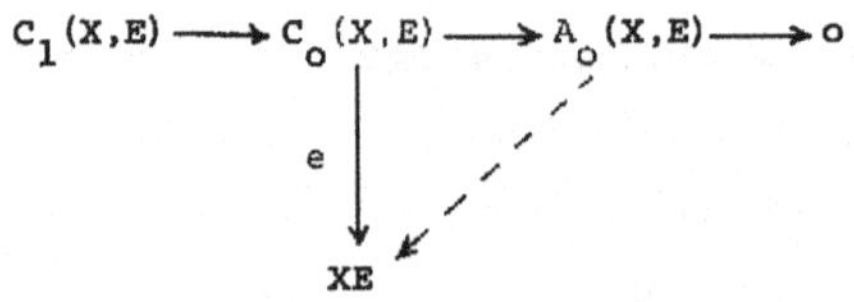

We obtain isomorphisms $H_n \cong A_n$ from (10.6) when the augmentation is an isomorphism:

(10.7) Proposition. Equivalent are:

(1) $A_o(\ ,E) \to E$ is an isomorphism,

(2) $E = E_J(E_o)$, where $E_o : Im\underline{M} \to \mathcal{A}$

(3) $E = E_J(E)$, the Kan extension of E restricted to $Im\underline{M}$.

Finally, (1)(2)(3) are implied by:

(4) E commutes with direct limits and $Im\underline{M} \to \underline{C}$ is adequate (Isbell [40])/dense (Ulmer [29]).

The equivalences are trivial in view of fullness of $J : Im\underline{M} \to \underline{C}$. As to (4), this results from the fact that J is adequate/dense $\Longleftrightarrow \varinjlim[(Im\underline{M},X) \to \underline{C}] = X$ for all $X \in \underline{C}$

(10.8) <u>Examples in which the models are $\mathfrak{G}$-projective.</u> (a) Let the models be the values of the cotriple, that is, all $X\mathfrak{G}$, $X \in \underline{C}$. The comma category $(\mathrm{Im}\underline{M}, X\mathfrak{G})$ has $X\mathfrak{G}$ as terminal object, hence is directed. Thus $A_q(X\mathfrak{G},E) = 0$ for $q > 0$ and any E, and (10.6) gives an isomorphism

$$H_n(X, A_0(\ ,E)) \xrightarrow{\ \sim\ } A_n(X,E) \ , \qquad n > 0.$$

(10.7) is inapplicable in general.

A stronger result follows directly from acyclic models (11.2). The complex $C_*(X,E)$ is $\mathfrak{G}$-representable (10.5) and is $\mathfrak{G}$-acyclic since each $X\mathfrak{G}$ is a model. Thus $H_n(X,E) \xrightarrow{\sim} A_n(X,E)$.

(b) Another convenient set of models with the same properties is that of all $\mathfrak{G}$-projectives.

Here and above, existence of the André-Appelgate complex raises some difficulties. The sets of models are too large. However, for coefficient functors with values in AB 5 categories with generators, the problem can be avoided. Such categories are <u>Ab</u>-topos (Roos), realizable as categories of abelian sheaves on suitable <u>sites</u>, and it suffices to pass to models of abelian groups in a larger universe. (See the discussion of this point in [19] as well.)

(c) Let $\mathfrak{G}$ be the free R-module cotriple and let $\underline{M} \to R\text{-}\underline{\mathrm{Mod}}$ be the set of finitely generated free R-modules. The categories $(\mathrm{Im}\underline{M}, X\mathfrak{G})$ are directed, since any $M \to X\mathfrak{G}$ can be factored

$$\begin{array}{ccc} M & \longrightarrow & X\mathfrak{G} = XUF \\ & \searrow & \swarrow \scriptstyle{iF} \\ & SF & \end{array}$$

where $i : S \to XU$ is a finite subset of the free basis XU. (U is the underlying set functor.) Moreover, $\mathrm{Im}\underline{M} \to R\text{-}\underline{\mathrm{Mod}}$ is adequate. Thus if E is any cocontinuous coefficient functor with values in an AB 5 category, then $H_n(X,E) \cong A_n(X,E)$ for all R-modules X.

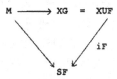

(d) More generally, if $\underline{C}$ is tripleable over sets, $\aleph$ is a rank of the triple [12] and $\underline{M}$ is the set of free algebras on fewer than $\aleph$ generators, then $(\mathrm{Im}\underline{M}, XG)$ can be proved directed in the same way. Thus homology relative to the models agrees with the cotriple homology (for cocontinuous AB 5 - category-valued coefficient functors; $\mathbb{G}$ is the free algebra cotriple relative to sets).

In these examples, adequacy/denseness of $\mathrm{Im}\underline{M}$ is well known or easily verified. In the following case adequacy fails. Let $1 \to R\text{-}\underline{\mathrm{Mod}}$ have R as value. $A_o(X, \mathrm{id.}) \to X$ is non-isomorphic (coefficients are in the identity functor $R\text{-}\underline{\mathrm{Mod}} \to R\text{-}\underline{\mathrm{Mod}}$). In fact $A_o(X, \mathrm{id.}) = R(X)/I$, the free R-module on X modulo the submodule generated by all $r(x)-(rx)$. Of course, $H_o(X, \mathrm{id.}) \simeq X$ (homology with respect to the absolute cotriple, which is induced by the above model).

(e) Cohomology. Let $E : \underline{C} \to \mathcal{A}^*$ be a "contravariant" coefficient functor. Isomorphisms $A^n(X,E) \cong H^n(X,E)$ follow purely formally in cases (a), (b) above. Cases (c), (d) offer the difficulty that the coefficient category $\mathcal{A}^*$ cannot be assumed to be AB 5, since in practice it is usually dual to a category of modules and therefore AB 5*. Assuming that the rank $\aleph$ of the triple $\mathbb{T}$ is $\aleph_o$, however, one can proceed as follows.

If X is a $\mathbb{T}$-algebra, the category of X-modules is abelian, AB 5, has a projective generator and is complete and cocomplete. Thus injective resolutions can be constructed, in the abelian category sense. Moreover, the free abelian group functor $\mathrm{Diff}_X : (\underline{C}, X) \to X\text{-}\underline{\mathrm{Mod}}$ exists. Consider the André-Appelgate complex with values in $X\text{-}\underline{\mathrm{Mod}}$: $(C_p(X)) = (C_p(X, \mathrm{Diff}_X)_{p \geq o})$. Its homology, written $A_p(X)$, measures the failure of the André-Appelgate theory to be a derived functor on the category $X\text{-}\underline{\mathrm{Mod}}$. If $Y \to X$ is an X-module, $(\mathrm{Hom}_X(C_pX, Y)_{p \geq o})$ has $A^p(X,Y)$ as its cohomology. Let $(Y^q)_{q \geq o}$ be an injective resolution of Y. We get a double complex $(\mathrm{Hom}_X(C_pX, Y^q)_{p,q \geq o})$, hence a universal-coefficients spectral sequence

$$\mathrm{Ext}^q_X(A_p(X), Y) \Longrightarrow A^{p+q}(X,Y) ,$$

where the total cohomology is filtered by q. (Use the fact that the complex $C_p(X)$ consists of projective X-modules.)

For example, in the case of commutative A-algebras over B , one obtains

$$\mathrm{Ext}^q_B(H_p(A,B,B),M) \longrightarrow H^{p+q}(A,B,M)$$

in the notation of [19 , §16].

Similarly, in the cotriple theory, there is a spectral sequence $\mathrm{Ext}^q_{\mathbf{X}}(H_p(X),Y) \longrightarrow H^{p+q}(X,Y)$.

Now, by the assumption that the rank of the triple is $\aleph_o$, the free $\mathbf{T}$-algebra $XG \to X$ is a filtered direct limit of free $\mathbf{T}$-algebras of finite type, that is, of models . Since the homology $A_p(\ )$ commutes with filtered limits, $A_p(XG) = o$ for $p > o$, $A_o(XG) = \mathrm{Diff}_X(XG)$. Thus the above spectral sequence yields $A^n(XG,Y) = o$ for $n > o$, $A^o(XG,Y) = \mathrm{Hom}_X(XG,Y)$. Acyclic models (11.2) now yields isomorphisms

$$A^n(X,Y) \xrightarrow{\ \sim\ } H^n(X,Y) .$$

A case in which this comparison technique runs into difficulty is the following. Let $\underline{M} \to$ K-$\underline{\mathrm{Alg}}$ be the set of tensor algebras of finitely generated K-modules, and let $\mathbf{G}$ be the cotriple in K-$\underline{\mathrm{Alg}}$ relative to K-modules. Homology isomorphisms $H_n \tilde{\to} A_n$ are easily obtained, as in (d). But the above derivation of the universal-coefficients spectral sequence does not work, because one seems to need to resolve the module variable both $\mathcal{G}$-relatively and K-relatively at the same time.

11. Appendix on acyclic models.

Let $o \leftarrow C_{-1} \leftarrow C_o \leftarrow C_1 \leftarrow \cdots$ be a chain complex of functors $\underline{C} \to \mathcal{A}$. (C_n) is $\mathbf{G}$-<u>representable</u>, where $\mathbf{G}$ is a cotriple in $\underline{C}$, if there are natural transformations $\theta_n : C_n \to GC_n$ such that $\theta_n \cdot \varepsilon C_n = C_n$ for all $n > o$. (C_n) is $\mathbf{G}$-<u>contractible</u> if the complex $(GC_n)_{n > -1}$ has a contracting homotopy (by natural transformations).

(11.1) <u>Proposition</u> [4]. Suppose that (C_n) is $\mathfrak{G}$-representable, (K_n) is $\mathfrak{G}$-contractible, and $\varphi_{-1} : C_{-1} \to K_{-1}$ is a given natural transformation. Then φ_{-1} can be extended to a natural chain transformation $(\varphi_n) : (C_n) \to (K_n)_{n \geq -1}$ by the inductive formula

$$\begin{array}{ccccccc} C_n & \xrightarrow{\theta_n} & GC_n & & GK_n & \xrightarrow{\varepsilon K_n} & K_n \\ & & \downarrow G\partial & & \uparrow h & & \\ & & GC_{n-1} & \xrightarrow{G\varphi_{n-1}} & GK_{n-1} & & \end{array}$$

Any two extensions of φ_{-1} are naturally chain homotopic (we omit the formula).

In particular, if $C_{-1} = K_{-1}$, then there are natural chain equivalences $(C_n) \overset{\sim}{\longleftrightarrow} (K_n)$.

If $E : \underline{C} \to \mathcal{A}$ is a functor with values in an additive category, then the standard chain complex

$$o \longleftarrow E \longleftarrow GE \longleftarrow G^2E \longleftarrow \cdots$$

is $\mathfrak{G}$-representable and $\mathfrak{G}$-contractible by virtue of $\delta G^n E : G^{n+1}E \to G^{n+2}E$. Thus if

$$o \longleftarrow E \longleftarrow E_o \longleftarrow E_1 \longleftarrow \cdots$$

is any $\mathfrak{G}$-representable chain complex of functors $\underline{C} \to \mathcal{A}$, there exists a unique natural chain transformation $(\varphi_n) : (E_n) \to (G^{n+1}E)$ such that $\varphi_{-1} = E$ (up to homotopy).

The proof is more or less contained in the statement. The term "$\mathfrak{G}$-contractible" was not used in [4], the term "$\mathfrak{G}$-acyclic" used there is reintroduced below with a different meaning.

The conclusions of (11.1) in practice are often too hard to establish and too strong to be relevant. At present all we need is homology isomorphism - a conclusion which is much weaker than chain equivalence. Thus it is convenient and reasonably satisfying to have the following weaker result available (as M. André has pointed out to us - see also [19]), that one can conclude a homology isomorphism $H(XE_*) \to H(X,E)_{\mathfrak{G}}$ from the information that the complex E_* is $\mathfrak{G}$-representable as above and $\mathfrak{G}$-<u>acyclic</u> merely in the sense that $H_n(XGE_*) = o$ if $n > o$, and $= XGE$ if $n = o$. This observation greatly simplifies proofs of agreement between homology theories arising from standard complexes, such as those of [4].

(11.2) <u>Proposition</u>. Let

$$o \longleftarrow E \longleftarrow E_o \longleftarrow E_1 \longleftarrow \cdots$$

be a complex of functors $\underline{C} \to \mathcal{A}$ such that

$$H_n(XGE_*) = \begin{cases} XGE, & n = o, \\ o, & n > o, \end{cases}$$

and the $\mathfrak{G}$-homology groups

$$H_p(X,E_q)_{\mathfrak{G}} = \begin{cases} XE_q, & p = o, \\ o, & p > o, \end{cases}$$

for all $X \in \underline{C}$, $q \geq o$. Then the spectral sequences obtained from the double complex

$$(XG^{p+1}E_q)_{p,q \geq o}$$

by filtering by levels $\leq p$ and $\leq q$ both collapse, giving edge isomorphisms

$$H_n(XE_*) \xrightarrow{\simeq} \text{total } H_n \qquad \text{(p filtration)}$$

$$H_n(X,E)_{\mathfrak{G}} \xrightarrow{\simeq} \text{total } H_n \qquad \text{(q filtration)}$$

for all $n \geq o$, hence natural isomorphisms $H_n(XE_*) \cong H_n(X,E)_{\mathfrak{G}}$.

In particular, $\mathfrak{G}$-representability of the complex (E_n) guarantees the second acyclicity condition, since the E_q are then retracts of the $\mathfrak{G}$-acyclic functors GE_q, $q \geq o$.

There is an obvious overlap between these two propositicns which we encountered in Theorem (8.1):

(11.3) Proposition. Let $0 \leftarrow E \leftarrow E_0 \leftarrow \dots$ be a $\mathbb{G}$-representable chain complex of functors $\underline{C} \to \mathcal{A}$ and $(\varphi_n) : (E_n) \to (G^{n+1}E)$, $n \geq -1$, a chain transformation such that $\varphi_{-1} = E$ (see (11.1)). By $\mathbb{G}$-representability, the acyclicity hypothesis

$$H_p(X,E_q) = \begin{cases} XE_q, & p = 0, \\ 0, & p > 0, \end{cases}$$

is satisfied and the rows of the double complex $(XG^{p+1}E_q)$ have homology zero. We obtain a spectral sequence

$$H_p(X,H_q(\ ,E_*))_G \Longrightarrow H_{p+q}(XE_*) ,$$

where the total homology is filtered by levels $\leq p$. The edge homomorphisms are

$$H_0(X,H_n(\ ,E_*))_{\mathbb{G}} \xrightarrow{\lambda_{\mathbb{G}}} H_n(XE_*)$$

and the top map in the commutative diagram

$$\begin{array}{ccc} H_n(XE_*) & \longrightarrow & H_n(X,H_0(\ ,E_*))_{\mathbb{G}} \\ & \searrow^{H_n(X\varphi_*)} & \downarrow \\ & & H_n(X,E)_{\mathbb{G}} \end{array}$$

Finally suppose that

$$H_n(XGE_*) = \begin{cases} XGE, & n = 0, \\ 0, & n > 0. \end{cases}$$

The spectral sequence collapses, as $H_p(X,H_q(\ ,E_*))_{\mathbb{G}} = 0$ if $q > 0$. The edge homomorphism $\lambda_{\mathbb{G}}$ is zero. The second edge homomorphism and the vertical map in the above triangle both become isomorphisms. Thus the homology isomorphism produced by (11.2) is actually induced by the chain map $\varphi_* : E_* \to (G^{n+1}E)_{n \geq 0}$.

The proof is left to the reader.

BIBLIOGRAPHY

[1] S. Lichtenbaum-M.Schlesinger, The cotangent complex of a morphism, Trans. AMS, 128 (1967), 41 - 70.

[2] A. Grothendieck - J. Dieudonné, Eléments de Géometrie algébrique, IV, I.H.E.S.

[3] M. Barr, Cohomology in tensored categories, COCA (La Jolla), Springer 1966, 344 - 354.

[4] M.Barr, J.Beck, Acyclic models and triples, COCA (La Jolla), Springer Verlag 1966, 336 - 343.

[5] P. Freyd, Abelian Categories, Harper and Row, New York, 1964.

[6] R. Hartshorne, Residues and Duality, Lecture Notes in Mathematics 20, Springer Verlag, 1966.

[7] J. Stallings, Homology and central series of groups, J. Alg., 2 (1965), 170 - 181.

[8] U. Stammbach, Anwendungen der Homologietheorie der Gruppen ..., Math.Z. 94 (1966), 157 - 177.

[9] B. Eckmann - U. Stammbach, Homologie et différentielles. Basses dimensions, cas spéciaux. Comptes Rendus, 265 (1967), 46 - 48.

[10] H. Cartan - S. Eilenberg, Homological Algebra, Princeton 1956.

[11] S. Mac Lane, Homology, Springer 1963 .

[12] F.E.J. Linton, Some aspects of equational categories, COCA (La Jolla), Springer 1966, 84 - 94.

[13] H. Appelgate - M. Tierney, Categories with models, this volume

[14] P. Huber, Homotopy Theory in general categories,
Math. Ann., 144 (1961), 361 - 385.

[15] P. Huber, Standard constructions in abelian categories,
Math. Ann., 146 (1962), 321 - 325.

[16] S. Eilenberg - J.C. Moore, Foundations of Relative Homological Algebra,
AMS Memoir No. 55, 1965.

[17] ______________________, Adjoint Functors and triples,
Ill J.Math. 9 (1965), p. 381 - 398.

[18] H. Appelgate, Acyclic models and resolvent functors, Dissertation,
Columbia University, 1965.

[19] M. André, Méthode simpliciale en algèbre homologique et algèbre
commutative, Lecture Notes in Mathematics No. 32, Springer 1967.

[20] J. Beck, Triples, algebras and cohomology,
Dissertation, Columbia, 1964 - 67.

[21] J.C. Moore, Seminar on algebraic homotopy theory,
Princeton 1956 (mimeographed).

[22] U. Shukla, Cohomologie des algèbres associatives, Ann.Sci.
Éc.Norm. Sup. 78 (1961), 163 - 209.

[23] M. Barr, Shukla cohomology and triples,
J. Alg., 5 (1967) 222 - 231

[24] D.K. Harrison, Commutative Algebras and cohomology,
Trans. AMS 104 (1962), 191 - 204.

[25] M. Barr - G. Rinehart, Cohomology as the derived functor of derivations,

Trans. AMS, 122 (1966), 416 - 426

[26] J. Beck, untitled mimeograph, Cornell 1966

[27] F.E.J. Linton, Applied functorial semantics, II, this volume.

[28] D.G. Quillen, Notes on homology of commutative algebras (mimeographed) M.I.T. 1967.

[29] F. Ulmer, Properties of dense and relative adjoint functors, J.Alg. 8 (1968), 77 - 95.

[30] H. Kleisli, Résolutions dans les catégories avec multiplication, C.R. 264 (1967).

[31] M. Barr, Composite cotriples and derived functors, this volume.

[32] J. Beck, Distributive laws, this volume.

[33] M. Barr, Harrison cohomology, Hochschild cohomology and triples, J. Alg. 8 (1968), 314 - 323.

[34] D.K. Harrison, Infinite abelian groups and homological methods, Ann. of Math. 69 (1959), 366 - 391.

[35] S. Eilenberg - J.A. Zilber, On products of complexes Ann. of Math. 75 (1953), 200 - 204.

[36] S. Eilenberg - N. Steenrod, Foundations of Algebraic Topology, Princeton, 1952.

[37] C.T.C. Wall, On the exactness of interlocking sequences, L'Enseignement mathém. 12 (1966), 95 - 100.

[38] J.-E. Roos, Sur les dérivés du foncteur $\varprojlim$, C.R. 252 (1961), 3702 - 3704.

[39] M. André, Limites et fibrés, C.R. 260 (1965), 756 - 759

[40] J.R. Isbell, Subobjects, adequacy, completeness and categories of algebras, Rozprawy Mat., 36 (1964), 3 - 32.

[41] F. Ulmer, mimeographed notes, ETH, Zürich.

[42] H. Röhrl, Satelliten halbexakter Funktoren, Math. Z. 79 (1962), 193 - 223.

[43] L. Ribes, A cohomology theory for pairs of groups, Ph. D. Thesis, The University of Rochester, 1967.

[44] S. Takasu, Relative homology and relative cohomology of groups, J. Fac. Sci. Tokyo, Sec.I, 8 (1959-60), 75 - 110.

[45] H.T. Trotter, Homology of group systems with applications to knot theory, Ann. of Math. 76 (1962), 464 - 498.

[46] G. Hochschild, Relative homological algebra, Trans. AMS 82 (1956), 246-269.

[47] F.E.J. Linton, An outline of functorial semantics, this volume.

COMPOSITE COTRIPLES AND DERIVED FUNCTORS

by Michael Barr 1)

Introduction

The main result of [2] is that the cohomology of an algebra with respect to the free associative algebra cotriple can be described by the resolution given by U. Shukla in [8]. That looks like a composite resolution; first an algebra is resolved by means of free modules (over the ground ring) and then this resolution is given the structure of a DG-algebra and resolved by the categorical bar resolution. This suggests that similar results might be obtained for all categories of objects with "two structures". Not surprisingly this turns out to involve a coherence condition between the structures which, for ordinary algebras, turns out to reduce to the distributive law. It was suggested in this connection by J. Beck and H. Appelgate.

If α and β are two morphisms in some category whose composite is defined we let $\alpha.\beta$ denote that composite. If S and T are two functors whose composite is defined we let ST denote that composite; we let $\alpha\beta = \alpha T'.S\beta = S'\beta.\alpha T : ST \to S'T'$ denote the natural transformation induced by $\alpha : S \to S'$ and $\beta : T \to T'$. We let $\alpha X : SX \to S'X$ denote the X component of α. We let the symbol used for an object, category or functor denote also its identity morphism, functor or natural transformation respectively. Throughout we let $\mathfrak{M}$ denote a fixed category and $\mathfrak{A}$ a fixed abelian category. $\mathfrak{N}$ will denote the category of simplicial $\mathfrak{M}$ objects (see 1.3. below) and $\mathfrak{B}$ the category of cochain complexes over $\mathfrak{A}$.

1. Preliminaries

In this section we give some basic definitions that we will need. More details on cotriples may be found in [3], [5] and [7]. More details on simplicial complexes and their relevance to derived functors may be found in [7] and [8].

1) This research has been partially supported by the NSF under grant GP-5478.

<u>Definition 1.1</u>. A cotriple $\mathfrak{G} = (G,\varepsilon,\delta)$ on $\mathfrak{A}$ consists of a functor $G : \mathfrak{A} \to \mathfrak{A}$ and natural transformations $\varepsilon : G \to \mathfrak{A}$ and $\delta : G \to G^2$ $(= GG)$ satisfying the identities $\varepsilon G.\delta = G\varepsilon.\delta = G$ and $G\delta.\delta = \delta G.\delta$. From our notational conventions $\varepsilon^n : G^n \to \mathfrak{A}$ is given the obvious definition and we also define $\delta^n : G \to G^{n+1}$ as any composite of δ's . The "coassociative" law guarantees that they are all equal.

<u>Proposition 1.2</u>. For any integers $n,m \geq 0$,

(1) $\quad \varepsilon^n.G^i\varepsilon^m G^{n-i} = \varepsilon^{n+m}$, $\qquad$ for $0 \leq i \leq n$

(2) $\quad G^i\delta^m G^{n-i}.\delta^n = \delta^{m+n}$, $\qquad$ for $0 \leq i \leq n$

(3) $\quad G^{n-i+1}\varepsilon^m G^i.\delta^{n+m} = \delta^n$, $\qquad$ for $0 \leq i \leq n+1$

(4) $\quad \varepsilon^{n+m}.G^i\delta G^{n-i-1} = \varepsilon^n$, $\qquad$ for $0 \leq i \leq n-1$.

The proof is given in the appendix (A.1) .

<u>Definition 1.3</u>. A simplicial $\mathfrak{A}$ object $X = \{X_n, d_n^iX, s_n^iX\}$ consists of objects X_n , $n \geq 0$ of $\mathfrak{A}$ together with morphisms $d^i = d_n^iX : X_n \to X_{n-1}$ for $0 \leq i \leq n$ called face operators and morphisms $s^i = s_n^iX : X_n \to X_{n+1}$ for $0 \leq i \leq n$ called degeneracies subject to the usual commutation identities (see, for example [7]). A morphism $\alpha : X \to Y$ of simplicial objects consists of a sequence $\alpha_n : X_n \to Y_n$ of morphisms commuting in the obvious way with all faces and degeneracies. A homotopy $h : \alpha \sim \beta$ of such morphisms consists of morphisms $h^i = h_n^i : X_n \to Y_{n+1}$ for $0 \leq i \leq n$ for each $n \geq 0$ satisfying $d^0h_n^0 = \alpha_n$, $d^{n+1}h_n^n = \beta_n$ and five additional identities tabulated in [7].

From now on we will imagine $\mathfrak{A}$ embedded in $\mathfrak{S}$ as the subcategory of constant simplicial objects, those $X = \{X_n, d_n^i, s_n^i\}$ for which $X_n = C$, $d_n^i = s_n^i = C$ for all n and all $0 \leq i \leq n$.

<u>Definition 1.4</u>. Given a cotriple $\mathfrak{G} = (G,\varepsilon,\delta)$ on $\mathfrak{A}$ we define a functor $G^* : \mathfrak{S} \to \mathfrak{S}$ by letting $X = \{X_n, d_n^iX, s_n^iX\}$ and $G^*X = Y = \{Y_n, d_n^iY, s_n^iY\}$ where $Y_n = G^{n+1}X_n$, $d_n^iY = G^i\varepsilon G^{n-i}(d_n^iX)$ and $s_n^iY = G^i\delta G^{n-i}(s_n^iX)$.

Theorem 1.5. If $h : \alpha \sim \beta$ where $\alpha, \beta : X \to Y$, then $G^* h : G^* \alpha \sim G^* \beta$ where $(G_* h)^i_n = G^i \delta G^{n-i} h^i_n$.

The proof is given in the appendix (A.2).

Theorem 1.6. Suppose $\mathfrak{N}$ is any subcategory of $\mathfrak{M}$ containing all the terms and all the faces and degeneracies of an object X of $\mathfrak{N}$. Suppose there is a natural transformation $\vartheta : \mathfrak{N} \to G|\mathfrak{N}$ such that $\varepsilon . \vartheta = \mathfrak{N}$. Then there are maps $\alpha : G^* X \to X$ and $\beta : X \to G^* X$ such that $\alpha . \beta = X$ and $G^* X \sim \beta . \alpha$.

The proof is given in the appendix (A.3).

2. The distributive law

The definitions 2.1 and Theorem 2.2 were first discovered by H. Appelgate and J. Beck (unpublished).

Definition 2.1. Given cotriples $\mathfrak{G}_1 = (G_1, \varepsilon_1, \delta_1)$ and $\mathfrak{G}_2 = (G_2, \varepsilon_2, \delta_2)$ on $\mathfrak{M}$, a natural transformation $\lambda : G_1 G_2 \to G_2 G_1$ is called a distributive law of $\mathfrak{G}_1$ over $\mathfrak{G}_2$ provided the following diagrams commute

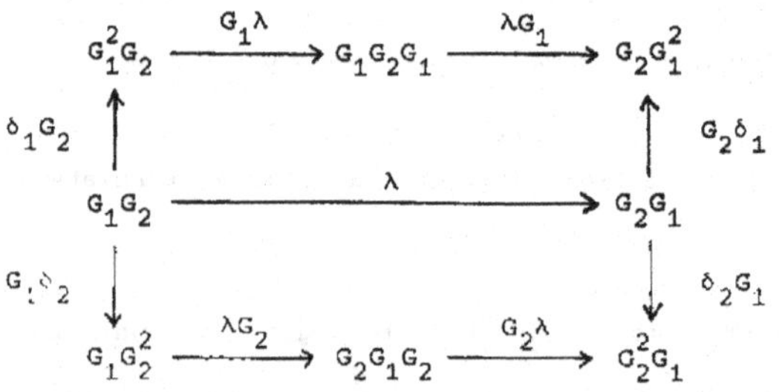

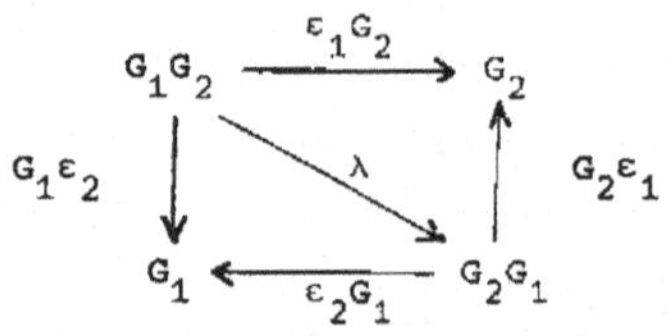

Theorem 2.2. Suppose $\lambda : G_1 G_2 \to G_2 G_1$ is a distributive law of $\mathfrak{G}_1$ over $\mathfrak{G}_2$. Let $G = G_1 G_2$, $\varepsilon = \varepsilon_1 \varepsilon_2$ and $\delta = G_1 \lambda G_2 . \delta_1 \delta_2$. Then $\mathfrak{G} = (G, \varepsilon, \delta)$ is a cotriple. We write $\mathfrak{G} = \mathfrak{G}_1 \circ_\lambda \mathfrak{G}_2$.

The proof is given in the appendix (A.4).

<u>Definition 2.3</u>. For $n \geq 0$ we define $\lambda^n : G_1^n G_2 \to G_2 G_1^n$ by $\lambda^o = G_2$ and $\lambda^n = G_1^{n-1}\lambda . \lambda^{n-1} G_1$. Also $\lambda_n : G_1^{n+1} G_2^{n+1} \to G^{n+1}$ is defined by $\lambda_o = G$ and $\lambda_n = \lambda_{n-1} G_2 . G_1^n \lambda^n G_2$. Let $\lambda^* : G_1^* G_2^* \to G^*$ be the natural transformation whose n-th component is λ_n .

<u>Proposition 2.4</u>.

(1) $G_2^n \varepsilon_1 . \lambda^n = \varepsilon_1 G_1^n$, for $n \geq 0$

(2) $G_2^n \delta_1 . \lambda^n = \lambda^n G_1 . G_1 \lambda^n . \delta G_2^n$, for $n \geq 0$

(3) $G_2^i \varepsilon_2 G_2^{n-i} G_1 . \lambda^{n+1} = \lambda^n . G_1 G_2^i \varepsilon_2 G_2^{n-i}$, for $0 \leq i \leq n$

(4) $G_2^i \delta_2 G_2^{n-i} G_1 . \lambda^{n+1} = \lambda^{n+2} . G_1 G_2^i \delta_2 G_2^{n-i}$, for $0 \leq i \leq n$.

The proof is given in the appendix (A.5).

3. Derived Functors

<u>Definition 3.1</u>. Given a functor $E : \mathfrak{M} \to \mathfrak{A}$ we define $E_c : \mathfrak{N} \to \mathfrak{B}$ by letting $E_c X$ where $X = \{X_n, d_n^i, s_n^i\}$ be the complex with EX_n in degree n and boundary $\sum_{i=0}^n (-1)^i Ed_n^i : EX_n \to EX_{n-1}$.

The following proposition is well known and its proof is left to the reader.

<u>Proposition 3.2</u>. If $\alpha, \beta : X \to Y$ are morphisms in $\mathfrak{N}$ and $h : \alpha \sim \beta$ and we let $E_c h : E_c X_n \to E_c Y_{n+1}$ be $\sum_{i=0}^n (-1)^i Eh_n^i$ then $E_c h : E_c \alpha \sim E_c \beta$.

<u>Definition 3.3</u>. If $E : \mathfrak{M} \to \mathfrak{A}$ is given, the derived functors of E with respect to the cotriple $\mathfrak{G}$, denoted by $H(\mathfrak{G};-,E)$, are the homology groups of the chain complex $E_c G^* X$ (where X is thought of as a constant simplicial object).

<u>Theorem 3.4</u>. If $\mathfrak{G} = \mathfrak{G}_1 o_\lambda \mathfrak{G}_2$ then for any $E : \mathfrak{M} \to \mathfrak{A}$, $E_c \lambda^* : E_c G_1^* G_2^* \to E_c G^*$ is a chain equivalence.

Proof. The proof uses the method of acyclic models described (in dual form) in [3].

We let V and W be the chain complexes $E_cG_1^*G_2^*$ and E_cG^* respectively. Then we show that $E_c\lambda^*$ induces an isomorphism of O-homology, that both V_n and W_n are $\mathfrak{G}$-retracts (in the sense given below - we use this term in place of $\mathfrak{G}$-representable to avoid conflict with the more common use of that term) and that each becomes naturally contractible when composed with $\mathfrak{G}$. For W , being the $\mathfrak{G}$-chain complex, these properties are automatic (see [3]).

Proposition 3.5. $E_c\lambda^*$ induces an isomorphism of O-homology.

Proof. Consider the commutative diagram with exact rows

$$\begin{array}{ccccccc}
EG_1^2G_2^2 & \xrightarrow{d} & EG_1G_2 & \xrightarrow{p} & H_oV & \longrightarrow & 0 \\
\downarrow{\scriptstyle E\lambda_1} & & \downarrow{\scriptstyle EG} & & \downarrow{\scriptstyle \zeta} & & \\
EG_2 & \xrightarrow{\partial} & EG & \xrightarrow{\pi} & H_oW & \longrightarrow & 0
\end{array}$$

where $d = E\varepsilon_1G_1\varepsilon_2G_2 - EG_1\varepsilon_1G_2\varepsilon_2$, $\partial = E\varepsilon G - EG\varepsilon$, $p = \text{coker } d$, $\pi = \text{coker } \partial$ and ζ is induced by $EG : EG_1G_2 \to EG$ since $\pi.d = \pi.\partial.E\lambda_1 = 0$. To show ζ is an isomorphism we first show that $p.\partial = 0$. In fact, $p.E\varepsilon G = p.E\varepsilon_1\varepsilon_2G_1G_2 = p.E\varepsilon_1G_1G_2.EG_1\varepsilon_2G_1G_2$ $= p.E\varepsilon_1G_1\varepsilon_2G_2 . EG_1\varepsilon_2G_1\delta_2 = p.EG_1\varepsilon_1G_2\varepsilon_2.EG_1\varepsilon_2G_1\delta_2 = p.EG_1\varepsilon_1\varepsilon_2G_2$. In a similar way this is also equal to $p.EG\varepsilon$ and so $p.\partial = 0$. But then there is a $\xi : H_oW \to H_oV$ such that $\xi.\pi = p$. But then $\xi.\zeta.p = \xi.\pi = p$ from which, since p is an epimorphism we conclude $\xi.\zeta = H_oV$. Similarly $\zeta.\xi = H_oW$.

Now we return to the proof of 3.4. To say that V_n is a $\mathfrak{G}$-retract means that there are maps $\vartheta_n : V_n \to V_nG$ such that $V_n\varepsilon.\vartheta_n = V_n$. Let $\vartheta_n = E_cG_1^n(G_1\lambda^{n+1}G_2.\delta_1G_2^n\delta_2)$. Then $V_n\varepsilon.\vartheta_n = E_cG_1^{n+1}G_2^{n+1}\varepsilon_1\varepsilon_2.E_cG_1^n(G_1\lambda^{n+1}G_2.\delta_1G_2^n\delta_2) =$ $E_cG_1^n(G_1G_2^{n+1}\varepsilon_1\varepsilon_2.G_1\lambda^{n+1}G_2.\delta_1G_2^n\delta_2) = E_cG_1^n(G_1\varepsilon_1G_2^{n+1}.\delta_1G_2^n\delta_2) = EG_1^n(G_1G_2^{n+1}) = V_n$.

To see that the augmented complex $VG \to H_oVG \to 0$ has a natural contracting homotopy, observe that for any X the constant simplicial object GX satisfies Theorem 2.

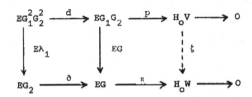

with respect to the cotriples $\mathbb{G}_1$ and $\mathbb{G}_2$ taking $\mathfrak{R}$ to be the full subcategory generated by the image of G. In fact $\delta_1 G_2 X : GX \to G_1 GX$ and $\lambda G_2 X . G_1 \delta_2 X : GX \to G_2 GX$ are natural maps whose composite with $\varepsilon_1 GX$ and $\varepsilon_2 GX$, respectively, is the identity. This means, for $i = 1,2$, that the natural map $\alpha_i X : G_i^* GX \to GX$ whose n-th component is $\varepsilon_i^{n+1} GX$ has a homotopy inverse $\beta_i X : GX \to G_i^* GX$ with $\alpha_i . \beta_i = G$. Let $h_i : G_i^* G \sim \beta_i . \alpha_i$ denote the natural homotopy. Then if $\alpha = \alpha_1 . G_1^* \alpha_2$, $\beta = G_1^* \beta_2 . \beta_1$ we have $E_c \alpha : E_c G_1^* G_2^* G \to E_c G$ and $E_c \beta : E_c G \to E_c G_1^* G_2^* G$. Moreover, noting that the boundary operator in $E_c G$ simply alternates 0 and EG it is obvious that the identity map of degree 1 denoted by h_3 is a contracting homotopy. Then if

$$h = E_c G_1^* h_2 + E_c (G_1^* \beta_2 . h_1 . G_1^* \alpha_2) + E_c (\beta . h_3 . \alpha) ,$$

$$\begin{aligned} d.h + h.d &= d.E_c G_1^* h_2 + d.E_c (G_1^* \beta_2 . h_1 . G_1^* \alpha_2) + d.E_c (\beta . h_3 . \alpha) + E_c G_1^* h_2 . d + E_c (G_1^* \beta_2 . h_1 . G_1^* \alpha_2) . d \\ &\quad + E_c (\beta . h_3 . \alpha) . d \\ &= E_c (G_1^* G_2^* G - G_1^* (\beta_2 . \alpha_2)) + E_c G_1^* \beta_2 . E_c (dh_1 + h_1 d) . E_c G_1^* \alpha_2 + E_c \beta . E_c (dh_3 + h_3 d) . E_c \alpha \\ &= VG - E_c G_1^* (\beta_2 . \alpha_2) + E_c G_1^* \beta_2 . E_c (G_1^* G - \beta_1 . \alpha_1) . E_c G_1^* \alpha_2 + E_c (\beta . \alpha) \\ &= VG - E_c G_1^* (\beta_2 . \alpha_2) + E_c G_1^* (\beta_2 . \alpha_2) - E_c (G_1^* \beta_2 . \beta_1 . \alpha_1 . G_1^* \alpha_2) + E_c (\beta . \alpha) \\ &= VG . \end{aligned}$$

This completes the proof.

4. Simplicial Algebras

In this section we generalize from the category of associative k-algebras to the category of simplicial associative k-algebras the theorem of [3] which states that the triple cohomology with respect to the underlying category of k-modules is equivalent to a "suspension" of the Hochschild cohomology. The theorem we prove will be easily seen to reduce to the usual one for a constant simplicial object.

Let Λ be an ordinary algebra. We let $\mathfrak{M}$ be the category of k-algebras over Λ. More precisely, an object of $\mathfrak{M}$ is a $\Gamma \to \Lambda$ and a morphism of $\mathfrak{M}$ is a commutative triangle $\Lambda \leftarrow \Gamma \to \Gamma' \to \Lambda$. In what follows we will normally drop any explicit reference to Λ. As before we let $\mathfrak{N}$ denote the category of simplicial $\mathfrak{M}$ objects.

Let $\mathbb{G}_t$ denote the tensor algebra cotriple on $\mathfrak{M}$ lifted to $\mathfrak{N}$ in the obvious way: $G_t\{X_n,d^i,s^i\} = \{G_tX_n,G_td^i,G_ts^i\}$. Let G_p denote the functor on $\mathfrak{N}$ described by $G_p\{X_n,d_n^i,s_n^i\} = \{X_{n+1},d_{n+1}^{i+1},s_{n+1}^{i+1}\}$. This means that the n-th term is X_{n+1} and the i-th face and degeneracy are d^{i+1} and s^{i+1} respectively. Let $\varepsilon_p : G_tX \to X$ be the map whose n-th component is d_{n+1}^o and $\delta_p : G_tX \to G_t^2X$ be the map whose n-th component is s_{n+1}^o.

Proposition 4.1.

(1) $\mathbb{G}_p = (G_p,\varepsilon_p,\delta_p)$ is a cotriple; in particular ε_p and δ_p are simplicial maps.

(2) If $\mathbb{G}$ is any cotriple "lifted" from a cotriple on $\mathfrak{M}$, then the equality $GG_p = G_pG$ is a distributive law.

(3) The natural transformations α and β where $\alpha X : G_pX \to X_o$ whose n-th component is $d^1.d^1. \ldots d^1$ and $\beta X : X_o \to G_pX$ whose n-th component is $s^o.s^o. \ldots s^o$ are maps between G_pX and the constant object X_o such that $\alpha.\beta = X_o$. There is a natural homotopy $h : G_pX \sim \beta.\alpha$.

Proof. (1) The simplicial identity $d^od^{i+1} = d^id^o$, $i > 0$, says that d^o commutes with the face maps. The identity $d^os^{i+1} = s^id^o$, $i > 0$, does the same for the degeneracies and so ε_p is simplicial. For δ_p we have $s^od^{i+1} = d^{i+2}s^o$ and $s^os^{i+1} = s^{i+2}s^o$ for $i > 0$, so it is simplicial. $G_p\delta_p$ has n-th component s_{n+2}^o and δ_pG_p has n-th component s_{n+2}^1, and so $\delta_pG_p.\delta_p = s_{n+2}^1.s_{n+1}^o = s_{n+2}^o.s_{n+1}^o = G_p\delta_p.\delta_p$, which is the coassociative law. Finally, $\varepsilon_pG_p.\delta_p = d_{n+2}^1.s_{n+1}^o = X_{n+1} = d_{n+2}^o.s_{n+1}^o = G_p\varepsilon_p.\delta_p$.

(2) This is completely trivial.

(3) This is proved in the appendix (A.6).

We note that under the equivalence between simplicial sets and simplicial topological spaces the "same" functor G_p is analogous to the topological path space.

From this we have the cotriple $\mathbb{G} = \mathbb{G}_t o \mathbb{G}_p$ where the distributive law is the identity map. If we take as functor the contravariant functor E, whose value at X is $\mathrm{Der}(\pi_oX,M)$ where M is a Λ-bimodule, the $\mathbb{G}$-derived functors are given by the

homology of the cochain complex $0 \to \mathrm{Der}(\pi_o GX,M) \to \dots \to \mathrm{Der}(\pi_o G^{n+1}X,M) \to \dots$.
$\pi_o X$ is most easily described as the coequalizer of $X_1 \rightrightarrows X_o$. Let $d^o = d^o_o : X_o \to \pi_o X$ be the coequalizer map. But by the above, $\pi_o GX \simeq G_t X_o$ and $G_t X = \varepsilon_t d^o$. Then $\pi_o G^{n+1}X = G_t^{n+1}X_n$ and the i-th face is $G_t^i \varepsilon_t G_t^{n-i} d^i$. Thus $H(\mathfrak{G};X,E)$ is just the homology of KX , the cochain complex whose n-th term is $\mathrm{Der}(G_t^{n+1}X_n,M)$. When X is the constant object Γ , this reduces to the cotriple cohomology of Γ with respect to $\mathfrak{G}_t$.

If X is in $\mathfrak{R}$, the normalized chain complex NX given by $N_n X = \bigcap_{i=1}^{n} \ker d_n^i$ naturally bears the structure of a DG-algebra. In fact, if $NX \otimes NX$ is the tensor product in the category of DG modules over k given by $(NX \otimes NX)_n = \sum N_i X \otimes N_{n-i} X$ and $X \otimes X$ is the tensor product in the category of simplicial k-modules given by $(X \otimes X)_n = X_n \otimes X_n$, then the Eilenberg-Zilber map $g : NX \otimes NX \to N(X \otimes X)$ is known to be associative in the sense that $g.(NX \otimes g) = g.(g \otimes NX)$. From this it follows easily that if $\mu : X \otimes X \to X$ is the multiplication map in X , then $N\mu.g$ makes NX into a DG-algebra. Actually it can be shown that the Dold-Puppe equivalence ([6]) between the categories of simplicial k-modules and DG-modules (chain complexes) induces an analogous equivalence between the categories of simplicial algebras and DG-algebras. Given a DG-algebra $V \overset{\alpha}{\to} \Lambda$, we let $\tilde{B}V$ be the chain complex given by $\tilde{B}_n V = \sum \Lambda \otimes V_{i_1} \otimes \dots \otimes V_{i_m} \otimes \Lambda$, the sum taken over all sets of indices for which $i_1 + \dots + i_m + m = n$. The boundary $\partial = \partial\tilde{B}$ is given by $\partial = \partial' + \partial''$ where ∂' is the Hochschild boundary and ∂'' arises out of the boundary in V . Let $\lambda[v_1, \dots, v_m]\lambda'$ denote the chain $\lambda \otimes v_1 \otimes \dots \otimes v_m \otimes \lambda'$, $\deg[v_1,\dots,v_m]$ denote the total degree of $[v_1,\dots,v_m]$, and $\exp q$ denote $(-1)^q$ for an integer q . Then

$$\partial'[v_1,\dots,v_m] = \alpha v_1[v_2,\dots,v_m] + \sum \exp(\deg[v_1,\dots,v_i])[v_1,\dots,v_i v_{i+1},\dots,v_m] + $$
$$+ \exp(\deg[v_1,\dots,v_{n-1}])[v_1,\dots,v_{n-1}]\alpha v_n .$$

$\partial''[v_1,\dots,v_m] = \sum \exp(\deg[v_1,\dots,v_{i-1}])[v_1,\dots,dv_i,\dots,v_m]$ where d is the boundary in V . Then it may easily be seen that $\partial'\partial'' + \partial''\partial' = 0$, and so $\partial\tilde{B} = \partial' + \partial''$ is a boundary operator. It is clear that $\tilde{B}$ reduces to the usual Hochschild complex when V is concentrated in degree zero.

BV is defined by letting $B_nV = \tilde{B}_{n+1}V$ and $\partial B = -\partial\tilde{B}$. This is where the degree shift in the comparison theorems between triple cohomology and the classical theories comes in. Then we define for a simplicial algebra over Λ and M a Λ-bimodule

$$LX = \mathrm{Hom}_{\Lambda-\Lambda}(BNX,M).$$

Theorem 4.2. The cochain complexes K and L are homotopy equivalent.

Proof. We apply the theorem of acyclic models of [3] with respect to $\mathfrak{G}$. As usual, the complex K, being the cotriple resolution, automatically satisfies both hypotheses of that theorem. Let $\vartheta^n : L^nG \to L^n$ (where L^n is the n-th term of L) be the map described as follows. We have for each $n \geq 0$ a k-linear map $\varphi_nX : X_n \to (GX)_n$ given by the composite $X_n \xrightarrow{s_0} X_{n+1} = G_pX_n \to (G_tG_pX)_n$ where the second is the isomorphism of an algebra with the terms of degree 1 in its tensor algebra. Also it is clear that $\varepsilon X.\varphi_nX = X_n$. Thus we have k-linear maps $\tilde{\varphi}_n : N_n \to N_nG$ with $N_n\varepsilon.\tilde{\varphi}_n = N_n$. This comes about because N is defined on the level of the underlying modules and extends to algebras. Then the Λ-bilinear map

$$\Lambda \otimes \tilde{\varphi}_{i_1} \otimes \dots \otimes \tilde{\varphi}_{i_m} \otimes \Lambda : \Lambda \otimes N_{i_1} \otimes \dots \otimes N_{i_m} \otimes \Lambda \to \Lambda \otimes N_{i_1}G \otimes \dots \otimes N_{i_m}G \otimes \Lambda \qquad (*)$$

is a map whose composite with the map induced by ε is the identity. Then forming the direct sum of all those maps (*) for which $i_1 + i_2 + \dots + i_m + m = n+1$ we have the map of $B_n \to B_nG$ whose composite with $B_n\varepsilon$ is B_n. Let $\vartheta^n : \mathrm{Hom}_{\Lambda-\Lambda}(B_nG,M) \to$ $\to \mathrm{Hom}_{\Lambda-\Lambda}(B_n,M)$ be the map induced. Clearly $\vartheta^n.K^n\varepsilon = K^n$.

Now we wish to show that the augmented complex $L^+GX = LGX \leftarrow H^0(LGX) \leftarrow 0$ is naturally contractible. First note that by proposition 4.1 (3) there are natural maps $\alpha = \alpha X : GX = G_pG_tX \to G_tX_0$ and $\beta = \beta X : G_tX_0 \to GX$ with $\alpha.\beta = G_tX_0$, and there is a natural homotopy $h : GX \sim \beta.\alpha$. Then we have $L^+\alpha : L^+GX \to L^+G_tX_0$ and $L^+\beta : L^+G_tX_0 \to L^+GX$ such that $L^+\alpha.L^+\beta = L^+G_tX_0$ and $L^+h : L^+GX \sim L^+\beta.L^+\alpha$. If we can find a contracting homotopy t in $L^+G_tX_0$, then $s = h + L^+\beta.t.L^+\alpha$ will satisfy $ds + sd = dh + hd + L^+\beta.(dt + td).L^+\alpha = L^+GX - L^+\beta.L^+\alpha + L^+\beta.L^+\alpha = L^+GX$. But NG_tX_0 is just the normalized complex associated with a constant. For $n > 0$, $\bigcap_{i>0} \ker d^i_n = 0$, since each $d^i_n = G_tX_0$. Thus NG_tX_0 is the DG-algebra consisting of G_tX_0 concentrate in degree zero. But then LG_tX_o is simply the Hochschild complex with degree lowered by one. I.e. LG_tX_o is the complex $\ldots \to (G_tX_o)^{(4)} \to (G_tX_o)^{(3)} \to 0$ with the usual boundary operator. But this complex was shown to be naturally contractible in [1]. In fact this was the proof that the Hochschild cohomology was essentially the triple cohomology with respect to $\mathbb{G}_t$. What remains in order to finish the proof of theorem 4.2 is to show

<u>Proposition 4.3</u>. $H^o(K) \cong H^o(L) \cong \mathrm{Der}(\pi_oX,M)$.

An auxiliary proposition will be needed. It is proved in the appendix (A.7).

<u>Proposition 4.4</u>. If X is as above, then $\varepsilon_td^o : G_tX_o \to \pi X_o$ is the coequalizer of $\varepsilon_tG_td^o$ and $G_t\varepsilon_td^1$ from $G_t^2X_1$ to G_tX_o .

Proof of Proposition 4.3. From Proposition 4.4 it follows that for any Γ , $\mathfrak{M}(\pi_oX,\Gamma)$ is the equalizer of $\mathfrak{M}(G_tX_o,\Gamma) \rightrightarrows \mathfrak{M}(G_t^2X_1,\Gamma)$. But by letting Γ be the split extension $\Lambda \times M$ and using the well-known fact that $\mathrm{Der}(Y,M) \cong \mathfrak{M}(Y,\Lambda\times M)$ for any Y of $\mathfrak{M}$, we have that $\mathrm{Der}(\pi_oX,M)$ is the equalizer of $\mathrm{Der}(G_tX_o,\Gamma) \rightrightarrows \mathrm{Der}(G_t^2X_1,\Gamma)$ or simply the kernel of the difference of the two maps. I.e. $\mathrm{Der}(\pi_oX,M)$ is the kernel of $K^oX \to K^1X$ and thus is isomorphic to H^oKX .

To compute H^oL , it suffices to show that $H_o(BNX) = \mathrm{Diff}\ \pi_oX$ where, for an algebra $\varphi : \Gamma \to \Lambda$, $\mathrm{Diff}\ \Gamma$ represents $\mathrm{Der}(\Gamma,-)$ on the category of Λ-modules. Explicitly, $\mathrm{Diff}\ \Gamma$ is the cokernel of $\Lambda\otimes\Gamma\otimes\Gamma\otimes\Lambda \to \Lambda\otimes\Gamma\otimes\Lambda$ where the map is the Hochschild boundary operator $\partial(\lambda\otimes\gamma\otimes\gamma'\otimes\lambda') = \lambda\cdot\varphi\gamma\otimes\gamma'\otimes\lambda' - \lambda\otimes\gamma\gamma'\otimes\lambda' + \lambda\otimes\gamma\otimes\varphi\gamma'\cdot\lambda'$. If for convenience we denote the cokernel of an $f : A \to B$ by B/A , we have $\pi_oX = N_oX/N_1X$, and then

$$H_o(BNX) = \frac{\Lambda\otimes N_oX\otimes\Lambda}{\Lambda\otimes N_1X\otimes\Lambda + \Lambda\otimes N_oX\otimes N_oX\otimes\Lambda} \cong \frac{\Lambda\otimes\pi_oX\otimes\Lambda}{\Lambda\otimes N_oX\otimes N_oX\otimes\Lambda} \cong \frac{\Lambda\otimes\pi_oX\otimes\Lambda}{\Lambda\otimes\pi_oX\otimes\pi_oX\otimes\Lambda} \cong$$

$$\cong \mathrm{Diff}\ \pi_oX .$$

The next to last isomorphism comes from the fact that $\Lambda\otimes N_oX\otimes N_oX\otimes\Lambda \to \Lambda\otimes\pi_oX\otimes\Lambda$

factors through the surjection $\Lambda \otimes N_oX \otimes N_oX \otimes \Lambda \to \Lambda \otimes \pi_oX \otimes \pi_oX \otimes \Lambda$. This argument is given by element chasing in [2], proposition 3.1.

We now recover the main theorem 1.1 of [2] as follows.

<u>Definition 4.5</u>. Given a k-algebra $\Gamma \to \Lambda$ we define $G_k\Gamma \to \Lambda$ by letting $G_k\Gamma$ be the free k-module on the elements of Γ made into an algebra by letting the multiplication in Γ define the multiplication on the basis. That is, if $\gamma_1, \gamma_2 \in \Gamma$ and if $[\gamma_i]$ denotes the basis element of $G_k\Gamma$ corresponding to γ_i , $i = 1,2$, then
$[\gamma_1][\gamma_2] = [\gamma_1\gamma_2]$.

<u>Theorem 4.6</u>. There are natural transformations ε_k and δ_k such that $\mathbb{G}_k = (G_k, \varepsilon_k, \delta_k)$ is a cotriple. Also there is a natural $\lambda : G_tG_k \to G_kG_t$ which is a distributive law.

Proof. $\varepsilon_k : G_k\Gamma \to \Gamma$ takes $[\gamma]$ to γ and δ_k takes $[\gamma]$ to $[[\gamma]]$ for $\gamma \in \Gamma$. G_k is made into a functor by $G_kf[\gamma] = [f\gamma]$ for $f : \Gamma \to \Gamma'$ and $\gamma \in \Gamma$. Then
$G_k\delta_k.\delta_k[\gamma] = G_k\delta_k[[\gamma]] = [\delta_k[\gamma]] = [[[\gamma]]] = \delta_kG_k[[\gamma]] = \delta_kG_k.\delta_k[\gamma]$. Also
$G_k\varepsilon_k.\delta_k[\gamma] = G_k\varepsilon_k[[\gamma]] = [\varepsilon_k[\gamma]] = [\gamma] = \varepsilon_kG_k[[\gamma]] = \varepsilon_kG_k.\delta_k[\gamma]$. To define λ we note that $G_tG_k\Gamma$ is the free algebra on the set underlying Γ . In fact, any algebra homomorphism $G_tG_k\Gamma \to \Gamma'$ is, by adjointness of the tensor product with the underlying k-module functor, determined by its value on the k-module underlying $G_k\Gamma$. As a k-module this is simply free on the set underlying Γ . Thus an algebra homomorphism $G_tG_k\Gamma \to G_kG_t\Gamma$ is prescribed by a set map of $\Gamma \to G_kG_t\Gamma$. Let $\langle\gamma\rangle$ denote the element of $G_t\Lambda$ corresponding to $\gamma \in \Gamma$. Then $\lambda\langle[\gamma]\rangle = [\langle\gamma\rangle]$ is the required map. In this form the laws that must be verified become completely transparent. For example,
$\lambda G_t.G_t\lambda.\delta_tG_k\langle[\gamma]\rangle = \lambda G_t.G_t\lambda\langle\langle[\gamma]\rangle\rangle = \lambda G_t.\langle\lambda\langle[\gamma]\rangle\rangle = \lambda G_t\langle[\langle\gamma\rangle]\rangle = [\langle\langle\gamma\rangle\rangle] = [\delta_t\langle\gamma\rangle] =$
$= G_k\delta_t[\langle\gamma\rangle] = G_k\delta_t.\lambda\langle[\gamma]\rangle$. The remaining identities are just as easy. It is, however, instructive to discuss somewhat more explicitly what λ does to a more general element of $G_tG_k\Gamma$.

A general element of $G_tG_k\Gamma$ is a formal (tensor) product of elements which are formal k-linear combinations of elements of Γ. We are required to produce from this an element of $G_kG_t\Gamma$ which is a formal k-linear combination of formal products of elements of Γ. Clearly the ordinary distributive law is exactly that: a prescription for turning a product of sums into a sum of products. For example $\lambda(\langle[\gamma]\rangle \otimes (\langle\alpha_1[\gamma_1] + \ldots + \alpha_n[\gamma_n]\rangle)) = \alpha_1[\langle\gamma\rangle \otimes \langle\gamma_1\rangle] + \ldots + \alpha_n[\langle\gamma\rangle \otimes \langle\gamma_n\rangle]$. The general form is practically impossible to write down but the idea should be clear. It is from this example that the term "distributive law" comes.

Now $G_k^*\Gamma$ is, for any $\Gamma \to \Lambda$, an object of $\mathfrak{R}$. Its cohomology with respect to $G = G_pG_t$ is with coefficients in the Λ-module M, as we have seen, the cohomology of $0 \to \mathrm{Der}(G_t\, G_k\Gamma,M) \to \ldots \to \mathrm{Der}(G_t^{n+1}G_k^{n+1}\Gamma,M) \to \ldots$ which by theorem 3.4 is chain equivalent to $0 \to \mathrm{Der}(G_tG_k\Gamma,M) \to \ldots \to \mathrm{Der}((G_tG_k)^{n+1}\Gamma,M) \to \ldots$, in other words the cohomology of Γ with respect to the free algebra cotriple G_tG_k. On the other hand, $NG_k\Gamma$ is a DG-algebra, acyclic and k-projective in each degree. Thus $BNG_k\Gamma$ is, except for the dimension shift, exactly Shukla's complex. Thus if $\mathrm{Shuk}^n(\Gamma,M)$ denotes the Shukla cohomology groups as given in [9], the above, together with proposition 4.3, shows

<u>Theorem 4.7</u>. There are natural isomorphisms

$$H^n(\mathbb{G}_t\circ_\lambda\mathbb{G}_k;\Gamma,M) \simeq \begin{cases} \mathrm{Der}(\Gamma,M), & n = 0 \\ \mathrm{Shuk}^{n+1}(\Gamma,M), & n > 0 \end{cases}.$$

5. Other applications

In this section we apply the theory to get two theorems about derived functors, each previously known in cohomology on other grounds.

<u>Theorem 5.1</u>. Let $\mathbb{G}_f$ and $\mathbb{G}_{bf}$ denote the cotriples on the category of groups for which G_fX is the free group on the elements of X and $G_{bf}X$ is the free group on the elements of X different from 1. Then the $\mathbb{G}_f$ and $\mathbb{G}_{bf}$ derived functors are equivalent.

<u>Theorem 5.2</u>. Let $\mathfrak{M}$ be the category of k-algebras whose underlying k-modules are k-projective. Then if $\mathbb{G}_t$, $\mathbb{G}_k$ and λ are as above (§4), the $\mathbb{G}_t$ and $\mathbb{G}_t \circ_\lambda \mathbb{G}_k$ derived functors are equivalent.

Before beginning the proofs we need the following:

<u>Definition 5.3</u>. If $\mathbb{G}$ is a cotriple on $\mathfrak{M}$, then an object X of $\mathfrak{M}$ is said to be $\mathbb{G}$-projective if there is a sequence $X \xrightarrow{\alpha} GY \xrightarrow{\beta} X$ with $\beta . \alpha = X$. We let $P(\mathbb{G})$ denote the class of all $\mathbb{G}$-projectives.

The following theorem is shown in [4].

<u>Theorem 5.4</u>. If $\mathbb{G}_1$ and $\mathbb{G}_2$ are cotriples on $\mathfrak{M}$ with $P(\mathbb{G}_1) = P(\mathbb{G}_2)$, then the $\mathbb{G}_1$ and $\mathbb{G}_2$ derived functors are naturally equivalent.

<u>Proposition 5.5</u>. Suppose $\mathbb{G}_1$ and $\mathbb{G}_2$ are cotriples on $\mathfrak{M}$, $\lambda : G_1G_2 \to G_2G_1$ is a distributive law, and $\mathbb{G} = \mathbb{G}_1 \circ_\lambda \mathbb{G}_2$. Then $P(\mathbb{G}) = P(\mathbb{G}_1) \cap P(\mathbb{G}_2)$.

Proof. If X is $\mathbb{G}$-projective, it is clearly $\mathbb{G}_1$-projective. If $X \xrightarrow{\alpha} G_1G_2Y \xrightarrow{\beta} X$ is a sequence with $\beta . \alpha = X$, then $X \xrightarrow{\alpha} G_1G_2Y \xrightarrow{G_1\delta_2} G_1G_2^2Y \xrightarrow{\lambda G_2 Y} G_2G_1G_2Y \xrightarrow{\varepsilon_2\beta} X$ is a sequence whose composite is X . If X is both $\mathbb{G}_1$- and $\mathbb{G}_2$-projective, find $X \xrightarrow{\alpha_i} G_iY_i \xrightarrow{\beta_i} X$, for $i = 1,2$ with $\beta_i . \alpha_i = X$; then $X \xrightarrow{\alpha_1} G_1Y_1 \xrightarrow{\delta_1 Y_1} G_1^2Y_1 \xrightarrow{G_1\beta_1} G_1X \xrightarrow{G_1\alpha_2} G_1G_2Y_2 \xrightarrow{\varepsilon_1 G_2 Y} G_2Y_2 \xrightarrow{\beta_2} X$ is a sequence for which

$$\beta_2 . \varepsilon_1 G_2 Y . G_1\alpha_2 . G_1\beta_1 . \delta_1 Y_1 . \alpha_1 = \varepsilon_1 X . G_1\beta_2 . G_1\alpha_2 . G_1\beta_1 . \delta_1 Y_1 . \alpha_1 = \varepsilon_1 X . G_1\beta_1 . \delta_1 Y_1 . \alpha_1 =$$

$= \beta_1 . \varepsilon_1 G_1 Y_1 . \delta_1 Y_1 . \alpha_1 = \beta_1 . \alpha_1 = X$, and thus exhibits X as a retract of GY_2 .

<u>Theorem 5.6</u>. Suppose $\mathbb{G}_1$, $\mathbb{G}_2$, λ , $\mathbb{G}$ are as above. If $P(\mathbb{G}_1) \subset P(\mathbb{G}_2)$, then the $\mathbb{G}_1$-derived functors and the $\mathbb{G}$-derived functors are equivalent; if $P(\mathbb{G}_2) \subset P(\mathbb{G}_1)$, then the $\mathbb{G}_2$-derived functors and the $\mathbb{G}$-derived functors are equivalent.

Proof. The first condition implies that $P(\mathbb{G}) = P(\mathbb{G}_1)$, while the second that $P(\mathbb{G}) = P(\mathbb{G}_2)$.

Proof of theorem 5.1. Let $\mathbb{G}_z$ denote the cotriple on the category of groups for which $G_z X = Z + X$ where Z is the group of integers and $+$ is the coproduct (free product). The augmentation and comultiplication are induced by the trivial map $Z \to 1$ and the "diagonal" map $Z \to Z + Z$ respectively. By the "diagonal" map $Z \to Z + Z$ is meant the map taking the generator of Z to the product of the two generators of $Z + Z$. Map $Z \to G_{bf}Z$ by the map which takes the generator of Z to the generator of $G_{bf}Z$ corresponding to it. For any X, map $G_{bf}Z \to G_{bf}(Z+X)$ by applying G_{bf} to the coproduct inclusion. Also map $G_{bf}X \to G_{bf}(Z + X)$ by applying G_{bf} to the other coproduct inclusion. Putting these together we have a map which is natural in X, $\lambda X : Z + G_{bf}X \to G_{bf}(Z + X)$, which can easily be seen to satisfy the data of a distributive law. Also it is clear that $Z + G_{bf}X \simeq G_f X$, since the latter is free on exactly one more generator than $G_{bf}X$. Thus the theorem follows as soon as we observe that $P(\mathbb{G}_z) \supset P(\mathbb{G}_{bf})$. In fact, the coordinate injection $\alpha : X \to Z + X$ is a map with $\varepsilon_Z . \alpha = X$, and thus $P(\mathbb{G}_z)$ is the class of all objects.

Proof of theorem 5.2. It suffices to show that on $\mathfrak{M}$, $P(\mathbb{G}_t) \subset P(\mathbb{G}_t \circ_\lambda \mathbb{G}_k)$. To do this, we factor $G_t = F_t U_t$ where $U_t : \mathfrak{M} \to \mathfrak{R}$, the category of k-projective k-modules, and F_t is its coadjoint (the tensor algebra). For any Y, the map $U_t \varepsilon_k Y : U_t G_k Y \to U_t Y$ is easily seen to be onto, and since $U_t Y$ is k-projective, it splits, that is, there is a map $\gamma : U_t Y \to U_t G_k Y$ such that $U_t \varepsilon_k Y . \gamma = U_t Y$. Then $G_t Y \xrightarrow{F_t \gamma} G_t G_k Y \xrightarrow{G_t \varepsilon_k Y} G_t Y$ presents any $G_t Y$ as a retract of $G_t G_k Y$. Clearly any retract of $G_t Y$ enjoys the same property.

The applicability of these results to other situations analogous to those of theorems 5.1 and 5.2 should be clear to the reader.

Appendix

In this appendix we give some of the more computational - and generally unen - lightening - proofs so as to avoid interrupting the exposition in the body of the paper.

A.1. Proof of proposition 1.2.

(1) When $n = i = 0$ there is nothing to prove. If $i = 0$ and $n > 0$, we have by induction on n, $\varepsilon^{n}\cdot\varepsilon^{m}G^{n} = \varepsilon.\varepsilon^{n-1}G.\varepsilon^{m}G^{n} = \varepsilon.(\varepsilon^{n-1}\cdot\varepsilon^{m}G^{n-1})G = \varepsilon.\varepsilon^{n+m-1}G = \varepsilon^{n+m}$. If $i = n > 0$, then we have by induction $\varepsilon^{n}\cdot G^{n}\varepsilon^{m} = \varepsilon.G\varepsilon^{n-1}\cdot G^{n}\varepsilon^{m} = \varepsilon.G(\varepsilon^{n-1}\cdot G^{n-1}\varepsilon^{m}) =$ $= \varepsilon.G\varepsilon^{n+m-1} = \varepsilon^{n+m}$. Finally, we have for $0 < i < n$, again by induction, $\varepsilon^{n}.G^{i}\varepsilon^{m}G^{n-i} = \varepsilon^{i}\cdot G^{i}\varepsilon^{n-i}\cdot G^{i}\varepsilon^{m}G^{n-i} = \varepsilon^{i}\cdot G^{i}\varepsilon^{n+m-i} = \varepsilon^{n+m}$.

(2) This proof follows the same pattern as in (1) and is left to the reader.

(3) When $n = 0$ and $m = 1$ these are the unitary laws. Then for $n = 0$, we have, by induction on m, $G\varepsilon^{m}\cdot\delta^{m} = G(\varepsilon.\varepsilon^{m-1}G).\delta^{m} = G\varepsilon.G\varepsilon^{m-1}G.\delta^{m-1}G.\delta = G\varepsilon.(G\varepsilon^{m-1}\cdot\delta^{m-1})G.\delta =$ $= G\varepsilon.\delta = G = \delta^{0}$ and similarly $\varepsilon^{m}G.\delta^{m} = \delta^{0}$. Then for $n > 0$, we have for $i < n+1$, $G^{n-i+1}\varepsilon^{m}G^{i}\cdot\delta^{n+m} = G^{n-i+1}\varepsilon^{m}G^{i}\cdot G^{n-i}\delta^{m}G^{i}\cdot\delta^{n} = G^{n-i}(G\varepsilon^{m}\cdot\delta^{m})G^{i}\cdot\delta^{n} = \delta^{n}$. Finally, for $i = n+1$, $\varepsilon^{m}G^{n+1}\cdot\delta^{n+m} = \varepsilon^{m}G^{n+1}\cdot\delta^{m}G^{n}\cdot\delta^{n} = (\varepsilon^{m}G.\delta^{m})G.\delta^{n} = \delta^{n}$.

(4) The proof follows the same pattern as in (3) and is left to the reader.

A.2. Proof of theorem 1.5.

We must verify the seven identities which are to be satisfied by a simplicial homotopy. In what follows we drop all lower indices.

(1) $\varepsilon G^{n+1}d^{0}.\delta G^{n}h^{0} = G^{n+1}(d^{0}\cdot h^{0}) = G^{n+1}\alpha_{n}$.

(2) $G^{n+1}\varepsilon d^{n+1}\cdot G^{n}\delta h^{n} = G^{n+1}(d^{n+1}\cdot h^{n}) = G^{n+1}\beta_{n}$.

(3) For $i < j$, $G^{i}\varepsilon G^{n+1-i}d^{i}\cdot G^{j}\delta G^{n-j}h^{j} = G^{i}(\varepsilon G^{n+1-i}d^{i}\cdot G^{j-i}\delta G^{n-j}h^{j}) =$

$= G^{i}(G^{j-i-1}\delta G^{n-j}h^{j-1}\cdot\varepsilon G^{n-i}d^{i}) = G^{j-1}\delta G^{n-j}h^{j-1}\cdot G^{i}\varepsilon G^{n-i}d^{i}$.

(4) For $0 < i = j < n+1$, $G^{i}\varepsilon G^{n+1-i}d^{i}\cdot G^{i}\delta G^{n-i}h^{i} = G^{n+1}(d^{i}\cdot h^{i}) = G^{n+1}(d^{i}\cdot h^{i-1}) =$

$= G^{i}\varepsilon G^{n+1-i}d^{i}\cdot G^{i-1}\delta G^{n-i+1}h^{i-1}$.

(5) For $i > j+1$, $G^{i}\varepsilon G^{n+1-i}d^{i}\cdot G^{j}\delta G^{n-j}h^{j} = G^{j}(G^{i-j}\varepsilon G^{n+1-i}d^{i}\cdot\delta G^{n-j}h^{j}) =$

$= G^{j}(\delta G^{n-j-1}h^{j}\cdot G^{i-j-1}\varepsilon G^{n+1-i}d^{i-1}) = G^{j}\delta G^{n-j-1}h^{j}\cdot G^{i-1}\varepsilon G^{n+1-i}d^{i-1}$.

(6) For $i < j$, $G^{i}\delta G^{n+1-i}s^{i}\cdot G^{j}\delta G^{n-j}h^{j} = G^{i}(\delta G^{n+1-i}s^{i}\cdot G^{j-i}\delta G^{n-j}h^{j}) =$

$= G^{i}(G^{j-i+1}\delta G^{n-j}h^{j+1}\cdot\delta G^{n-i}s^{i}) = G^{j+1}\delta G^{n-j}h^{j+1}\cdot G^{i}\delta G^{n-i}s^{i}$.

(7) For $i > j$, $G^i\delta G^{n+1-i}s^i.G^j\delta G^{n-j}h^j = G^j(G^{i-j}\delta G^{n+1-i}s^i.\delta G^{n-j}h^j) =$

$= G^j(\delta G^{n+1-j}h^j.G^{i-1-j}\delta G^{n+1-i}s^{i-1}) = G^j\delta G^{n+1-j}h^j.G^{i-1}\delta G^{n+1-i}s^{i-1}$.

A.3. Proof of theorem 1.6.

We define $\alpha_n = \varepsilon^{n+1}X_n : G^{n+1}X_n \to X_n$ and $\beta_n = \delta^n X_n.\vartheta X_n : X_n \to G^{n+1}X_n$. First we show that these are simplicial. We have $d^i.\alpha_n = d^i.\varepsilon^{n+1}X_n = \varepsilon^{n+1}X_{n-1}.d^i =$ $= \varepsilon^n X_n.G^i\varepsilon G^{n-i}X_{n-1}.G^{n+1}d^i = \alpha_n.G^i\varepsilon G^{n-i}d^i$. Similarly, $s^i.\alpha_n = s^i.\varepsilon^{n+1}X_n = \varepsilon^{n+1}X_{n+1}.s^i =$ $= \varepsilon^{n+2}X_{n+1}.G^i\delta G^{n-i}X_{n+1}.s^i = \alpha_{n+1}.G^i\delta G^{n-i}s^i$.

$G^i\varepsilon G^{n-i}d^i.\beta_n = G^i\varepsilon G^{n-i}d^i.\delta^n X_n.\vartheta X_n = G^n d^i.\delta^{n-i}X_n.\vartheta X_n = \delta^{n-1}X_{n-1}.\vartheta X_{n-1}.d^i = \beta_{n-1}.d^i$. Similarly, $G^i\delta G^{n-i}s^i.\beta_n = G^i\delta G^{n-i}s^i.\delta^n X_n.\vartheta X_n = \delta^{n+1}s^i.\vartheta X_n = \delta^{n+1}X_{n+1}.Gs^i.\vartheta X_n =$ $= \delta^{n+1}X_{n+1}.\vartheta X_{n+1}.s^i = \beta_{n+1}.s^i$. Moreover, $\alpha_n.\beta_n = \varepsilon^{n+1}X_n.\delta^n X_n.\vartheta X_n = \varepsilon X_n.\vartheta X_n = X_n$.

Let $h^i_n = G^{i+1}(\delta^{n-i}s^i_n.\vartheta X_n.\varepsilon^{n-i}X_n) : G^{n+1}X_n \to G^{n+2}X_{n+1}$ for $0 \le i \le n$. Then we will verify the identities which imply that $h : \beta.\alpha \sim G^*X$. At most places in the computation below we will omit both lower indices and the name of the objects under consideration.

(1) $\varepsilon G^{n+1}d^0.h^0_n = \varepsilon G^{n+1}d^0.G(\delta^n s^0.\vartheta.\varepsilon^n) = \delta^n(d^0.s^0).\vartheta.\varepsilon^n.\varepsilon G^n = \delta^n.\vartheta.\varepsilon^{n+1} = \beta_n.\alpha_n$.

(2) $G^{n+1}\varepsilon d^{n+1}.h^n_n = G^{n+1}\varepsilon d^{n+1}.G^{n+1}(Gs^n.\vartheta) = G^{n+1}(\varepsilon d^{n+1}.Gs^n.\vartheta) = G^{n+1}(\varepsilon.\vartheta) = G^{n+1}X_n$.

(3) For $i < j$, $G^i\varepsilon G^{n+1-i}d^i.h^j_n = G^i\varepsilon G^{n+1-i}d^i.G^{j+1}(\delta^{n-j}s^j.\vartheta.\varepsilon^{n-j}) =$

$= G^i(\varepsilon G^{n+1-i}d^i.G^{j-i+1}(\delta^{n-j}s^j.\vartheta.\varepsilon^{n-j})) = G^i(G^{j-i}(\delta^{n-j}s^{j-1}.\vartheta.\varepsilon^{n-j}).\varepsilon G^{n-i}d^i) =$

$= G^j(\delta^{n-j}s^{j-1}.\vartheta.\varepsilon^{n-j}).G^i\varepsilon G^{n-i}d^i = h^{j-1}_{n-1}.G^i\varepsilon G^{n-i}d^i$

(4) For $0 < i = j < n+1$, $G^i\varepsilon G^{n+1-i}d^i.h^i_n = G^i\varepsilon G^{n+1-i}d^i.G^{i+1}(\delta^{n-i}s^i.\vartheta.\varepsilon^{n-i}) =$

$= G^i(\varepsilon G^{n+1-i}d^i.G(\delta^{n-i}s^i.\vartheta.\varepsilon^{n-i})) = G^i(\delta^{n-i}(d^i.s^i).\vartheta.\varepsilon^{n-i}.\varepsilon G^{n-i}) =$

$= G^i(\delta^{n-i}.\vartheta.\varepsilon^{n+1-i}) = G^i(\delta^{n-i}(d^i.s^{i-1}).\vartheta.\varepsilon^{n-i+1}) = G^i(\varepsilon G^{n+1-i}d^i.\delta^{n-i+1}s^{i-1}.\vartheta.\varepsilon^{n-i+1})$

$= G^i\varepsilon G^{n+1-i}d^i.G^i(\delta^{n-i+1}s^{i-1}.\vartheta.\varepsilon^{n-i+1}) = G^i\varepsilon G^{n+1-i}d^i.h^{i-1}_n$.

(5) For $i > j+1$, $G^i\varepsilon G^{n+1-i}d^i.h^j_n = G^i\varepsilon G^{n+1-i}d^i.G^{j+1}(\delta^{n-j}s^j.\vartheta.\varepsilon^{n-j}) =$

$= G^{j+1}(G^{i-j-1}\varepsilon G^{n+1-i}d^i.\delta^{n-j}s^j.\vartheta.\varepsilon^{n-j}) = G^{j+1}(\delta^{n-j-1}(d^i.s^j).\vartheta.\varepsilon^{n-j}) =$

$= G^{j+1}(\delta^{n-j-1}(s^j.d^{i-1}).\vartheta.\varepsilon^{n-j}) = G^{j+1}(\delta^{n-j-1}s^j.\vartheta.\varepsilon^{n-j-1}.G^{i-1}\varepsilon G^{n-i+1}d^{i-1}) =$

$= h^j_{n-1}.G^{i-1}\varepsilon G^{n-i+1}d^{i-1}.$

(6) For $i < j$, $G^i\delta G^{n+1-i}s^i.h^j_n = G^i\delta G^{n+1-i}s^i.G^{j+1}(\delta^{n-j}s^j.\vartheta.\varepsilon^{n-j}) =$

$= G^i(\delta G^{n+1-i}s^i.G^{j-i+1}(\delta^{n-j}s^j.\vartheta.\varepsilon^{n-j})) = G^i(G^{j-i+2}(\delta^{n-j}s^{j+1}.\vartheta.\varepsilon^{n-j}).\delta G^{n-i}s^i) =$

$= G^{j+2}(\delta^{n-j}s^{j+1}.\vartheta.\varepsilon^{n-j}).G^i\delta G^{n-i}s^i = h^{j+1}_{n+1}.s^i.$

(7) For $i > j$, $G^i\delta G^{n+1-i}s^i.h^j_n = G^i\delta G^{n+1-i}s^i.G^{j+1}(\delta^{n-j}s^j.\vartheta.\varepsilon^{n-j}) =$

$= G^j(G^{i-j}\delta G^{n+1-i}s^i.G\delta^{n-j}s^j.G\vartheta.G\varepsilon^{n-j}) = G^j(G\delta^{n-j+1}(s^i.s^j).G\vartheta.G\varepsilon^{n-j}) =$

$= G^j(G\delta^{n-j+1}(s^j.s^{i-1}).G\vartheta.G\varepsilon^{n-j}) = G^j(G\delta^{n-j+1}s^j.G\vartheta.G\varepsilon^{n-j}.G^{i-j-1}(G\varepsilon.\delta)G^{n+1-i}s^{i-1}) =$

$= G^j(G\delta^{n-j+1}s^j.G\vartheta.G\varepsilon^{n-j}.G^{i-j}\varepsilon G^{n+1-i}.G^{i-j-1}\delta G^{n+1-i}s^{i-1}) =$

$= G^j(G\delta^{n-j+1}s^j.G\vartheta.G\varepsilon^{n-j+1}.G^{i-j-1}\delta G^{n+1-i}s^{i-1}) = G^{j+1}(\delta^{n-j+1}s^j.\vartheta.\varepsilon^{n-j+1}).G^{i-1}\delta G^{n+1-i}s^{i-1} =$

$= h^j_{n+1}.G^{i-1}\delta G^{n+1-i}s^{i-1}.$

This proof is adapted from the proof of Theorem 4.5 of [O].

A.4. Proof of theorem 2.2.

We must verify the three identities satisfied by a cotriple.

(1) $G\varepsilon.\delta = G_1G_2\varepsilon_1\varepsilon_2.G_1\lambda G_2.\delta_1\delta_2 = G_1\varepsilon_1G_2\varepsilon_2.\delta_1\delta_2 = (G_1\varepsilon_1.\delta_1)(G_2\varepsilon_2.\delta_2) = G_1G_2 = G$.

(2) $\varepsilon G.\delta = \varepsilon_1\varepsilon_2G_1G_2.G_1\lambda G_2.\delta_1\delta_2 = \varepsilon_1G_1\varepsilon_2G_2.\delta_1\delta_2 = (\varepsilon_1G_1.\delta_1)(\varepsilon_2G_2.\delta_2) = G_1G_2 = G$.

(3) $G\delta.\delta = G_1G_2G_1\lambda G_2.G_1G_2\delta_1\delta_2.G_1\lambda G_2.\delta_1\delta_2 = G_1G_2G_1\lambda G_2.G_1\lambda G_1G_2^2.G_1^2\lambda G_2^2.G_1\delta_1G_2\delta_2.\delta_1\delta_2 =$

$= \lambda_2.\delta_1^2\delta_2^2$ and by symmetry this latter is equal to $\delta G.\delta$.

A.5. Proof of proposition 2.4.

(1) For $n = 0$ this is vacuous and for $n = 1$ it is an axiom. For $n > 1$, we have by induction $G_2^n\varepsilon_1.\lambda^n = G_2^n\varepsilon_1.G_2\lambda^{n-1}.\lambda G_2^{n-1} = G_2(G_2^{n-1}\varepsilon_1.\lambda^{n-1}).\lambda G_2^{n-1} =$

$= G_2(\varepsilon_1 G_2^{n-1}).\lambda G_2^{n-1} = (G_2\varepsilon_1.\lambda)G_2{}^{n-1} = (\varepsilon_1 G_2)G_2^{n-1} = \varepsilon_1 G_2^n.$

(2) For $n = 0$ this is vacuous and for $n = 1$ it is an axiom. For $n > 1$, we have by induction $G_2^n\delta_1.\lambda^n = G_2^n\delta_1.G_2\lambda^{n-1}.\lambda G_2^{n-1} = G_2(G_2^{n-1}\delta_1.G_2\lambda^{n-1}).\lambda G_2^{n-1} =$

$= G_2(\lambda^{n-1}G_1.G_1\lambda^{n-1}.\delta_1 G_2^{n-1}).\lambda G_2^{n-1} = G_2\lambda^{n-1}G_1.G_2G_1\lambda^{n-1}.(G_2\delta_1.\lambda)G_2^{n-1} =$

$= G_2\lambda^{n-1}G_1.G_2G_1\lambda^{n-1}.(\lambda G_1.G_1\lambda.\delta_1 G_2)G_2^{n-1} = G_2\lambda^{n-1}G_1.G_2G_1\lambda^{n-1}.\lambda G_1G_2^{n-1}.G_1\lambda G_2^{n-1}.\delta_1 G_2^n =$

$= G_2\lambda^{n-1}G_1.\lambda G_2^{n-1}G_1.G_1G_2\lambda^{n-1}.G_1\lambda G_2^{n-1}.\delta_1 G_2^n = \lambda^n G_1.G_1\lambda^n.\delta_1 G_2^n$

(3) For $n = 0$ this is an axiom. For $n > 0$, first assume that $i = 0$. Then we have by induction $\varepsilon_2 G_2^n G_1.\lambda^{n+1} = \varepsilon_2 G_2^n G_1.G_2^n\lambda.\lambda^n G_2 = G_2^{n-1}\lambda.\varepsilon_2 G_2^{n-1}G_1G_2.\lambda^n G_2 =$

$= G_2^{n-1}\lambda.(\varepsilon_2 G_2^{n-1}G_1.\lambda^n)G_2 = G_2^{n-1}\lambda.(\lambda^{n-1}.G_1\varepsilon_2 G_2^{n-1})G_2 = G_2^{n-1}\lambda.\lambda^{n-1}G_2.G_1\varepsilon_2 G_2^n =$

$= \lambda^n.G_1\varepsilon_2 G_2^n.$

For $i > 0$ we have, again by induction,

$G_2^i\varepsilon_2 G_2^{n-i}G_1.\lambda^{n+1} = G_2^i\varepsilon_2 G_2^{n-i}G_1.G_2\lambda^n.\lambda G_2^n = G_2(G_2^{i-1}\varepsilon_2 G_2^{n-i}G_1.\lambda^n).\lambda G_2^n =$

$= G_2(\lambda^{n-1}.G_1G_2^{i-1}\varepsilon_2 G_2^{n-i}).\lambda G_2^n = G_2\lambda^{n-1}.G_2G_1G_2^{i-1}\varepsilon_2 G_2^{n-i}.\lambda G_2^n = G_2\lambda^{n-1}.\lambda G_2^{n-1}.G_1G_2^i\varepsilon_2 G_2^{n-i} =$

$= \lambda^n.G_1G_2^i\varepsilon_2 G_2^{n-i}.$

(4) When $n = 0$, this is an axiom. For $i = 0$, we have by induction

$\delta_2 G_2^n G_1.\lambda^{n+1} = \delta_2 G_2^n G_1.G_2^n\lambda.\lambda^n G_2 = G_2^{n+1}\lambda.\delta_2 G_2^{n-1}G_1G_2.\lambda^n G_2 = G_2^{n+1}\lambda.(\delta_2 G_2^{n-1}G_1.\lambda^n)G_2 =$

$= G_2^{n+1}\lambda.(\lambda^{n+1}.G_1\delta_2 G_2^{n-1})G_2 = G_2^{n+1}\lambda.\lambda^{n+1}G_2.G_1\delta_2 G_2^n = \lambda^{n+2}.G_1\delta_2 G_2^n.$

For $i > 0$ we have, again by induction,

$G_2^i\delta_2 G_2^{n-i}G_1.\lambda^{n+1} = G_2^i\delta_2 G_2^{n-i}G_1.G_2\lambda^n.\lambda G_2^n = G_2(G_2^{i-1}\delta_2 G_2^{n-i}G_1.\lambda^n).\lambda G_2^n =$

$= G_2(\lambda^{n+1}.G_1G_2^{i-1}\delta_2 G_2^{n-i}).\lambda G_2^n = G_2\lambda^{n+1}.G_2G_1G_2^{i-1}\delta_2 G_2^{n-i}.\lambda G_2^n = G_2\lambda^{n+1}.\lambda G_2^{n+1}.G_1G_2^i\delta_2 G_2^{n-i} =$

$= \lambda^{n+2}.G_1G_2^i\delta_2 G_2^{n-i}.$

A.6. Proof of proposition 4.1(3).

In the following we let d^i and s^i stand for d^iX and s^iX respectively. If $Y = G_pX$, then $Y_n = X_{n+1}$, $d^iY = d^{i+1}$ and $s^iY = s^{i+1}.$ $\alpha_n = (d^1)^{n+1} : Y_n \to X_o$

and $\beta_n = (s^0)^{n+1} : X_0 \to Y_n$. Then $\alpha_n \cdot \beta_n = (d^1)^{n+1} \cdot (s^0)^{n+1} = Y_n$. Let

$h_n^i = (s^0)^{i+1}(d^1)^i : Y_n \to Y_{n+1}$ for $0 \leq i \leq n$.

(1) $d^0 Y \cdot h^0 = d^1 \cdot s^0 = Y_n$.

(2) $d^{n+1} Y \cdot h^0 = d^{n+2} \cdot (s^0)^{n+1} \cdot (d^1)^n = (s^0)^{n+1} \cdot d^1 \cdot (d^1)^n = \beta_n \cdot \alpha_n$.

(3) For $i < j$, $d^i Y \cdot h^j = d^{i+1} \cdot (s^0)^{j+1} \cdot (d^1)^j = (s^0)^j \cdot (d^1)^j = (s^0)^j \cdot (d^1)^{j-1} \cdot d^{i+1} = h^{j-1} \cdot d^i Y$.

(4) $d^i Y \cdot h^i = d^{i+1} \cdot (s^0)^{i+1} \cdot (d^1)^i = (s^0)^i \cdot (d^1)^i = (s^0)^i \cdot d^1 \cdot (d^1)^{i-1} = d^{i+1} \cdot (s^0)^i \cdot (d^1)^{i-1} =$

$= d^i Y \cdot h^{i-1}$.

(5) For $i > j+1$, $d^i Y \cdot h^j = d^{i+1} \cdot (s^0)^{j+1} \cdot (d^1)^j = (s^0)^{j+1} \cdot d^{i-j} \cdot (d^1)^j =$

$= (s^0)^{j+1} \cdot (d^1)^j \cdot d^i = h^j \cdot d^{i-1} Y$.

(6) For $i < j$, $s^i Y \cdot h^j = s^{i+1} \cdot (s^0)^{j+1} \cdot (d^1)^j = (s^0)^{j+2} \cdot (d^1)^j =$

$= (s^0)^{j+2} \cdot (d^1)^{j+1} \cdot s^{i+1} = h^{j+1} \cdot s^i Y$.

(7) For $i > j$, $s^i Y \cdot h^j = s^{i+1} \cdot (s^0)^{j+1} \cdot (d^1)^j = (s^0)^{j+1} \cdot s^{i-j} \cdot (d^1)^j =$

$= (s^0)^{j+1} \cdot (d^1)^j \cdot s^i = h^j \cdot s^{i-1} Y$.

A.7. Proof of proposition 4.4.

Form the double simplicial object $E = \{E_{ij} = G_t^{i+1} X_j\}$ with the maps gotten by applying G to the faces and degeneracies of X in one direction and the cotriple faces and degeneracies in the other. Let $D = \{D_i = G_t^{i+1} X_i\}$ be the diagonal complex. We are trying to show that $\pi_0 D \cong \pi_0 X$. But the Dold-Puppe theorem asserts that $\pi_0 D \cong H_0 ND$ and the Eilenberg-Zilber theorem asserts that $H_0 ND$ is H_0 of the total complex associated with E . But we may compute the zero homology of

$$\begin{array}{ccccc}
G_t^2 X_1 & \rightrightarrows & G_t^2 X_0 & \longrightarrow & 0 \\
\Downarrow & & \Downarrow & & \\
G_t X_1 & \rightrightarrows & G_t X_0 & \longrightarrow & 0 \\
\downarrow & & \downarrow & & \\
0 & & 0 & &
\end{array}$$

by first computing the O homology vertically which gives, by another application of the Dold-Puppe theorem,

$$\pi_o(G_t^* X_1) \Longrightarrow \pi_o(G_t^* X_o) \longrightarrow 0 \quad .$$

But G_t^* is readily shown to be right exact (i.e. it preserves coequalizers) and so this is $X_1 \Longrightarrow X_o \longrightarrow 0$. Another application of the Dold-Puppe theorem gives that H_o of this is $\pi_o X$.

REFERENCES

1. M. Barr, Cohomology in tensored categories, "Proceedings of the La Jolla Conference on Categorical Algebra". Springer, Berlin (1966), 345-355.

2. ______, Shukla cohomology and triples, J. Algebra 5 (1967), 222-231.

3. ______, and J. Beck, Acyclic models and triples, "Proceedings of the La Jolla Conference on Categorical Algebra". Springer, Berlin (1966), 336-344.

4. _____, _____, Homology and standard constructions, (this volume).

5. J. Beck, "Triples, Algebras and Cohomology", dissertation, Columbia University, 1967.

6. A. Dold and D. Puppe, Homologie nicht-additiver Funktoren, Anwendungen, Ann. Inst. Fourier II (1961), 201-312.

7. P.J. Huber, Homotopy theory in general categories, Math. Ann. 144 (1961), 361-385.

8. S. Mac Jone, Homology, Springer, Berlin, 1963.

9. U. Shukla, Cohomologie des algèbres associatives, Ann. Scient. Éc. Norm. Supér., Sér. III 78 (1961), 163-209.

0. H. Appelgate, Acyclic models and resolvent functors, thesis, Columbia Univ. 1965.

COHOMOLOGY AND OBSTRUCTIONS: COMMUTATIVE ALGEBRAS

by Michael Barr

this work was partially supported by NSF grant GP-5478

Introduction

Associated with each of the classical cohomology theories in algebra has been a theory relating H^2 (H^3 as classically numbered) to obstructions to non-singular extensions and H^1 with coefficients in a "center" to the non-singular extension theory (see [EM] , [Hoch] , [Hoch'], [Mac], [Shuk], [Harr]). In this paper we carry out the entire process using triple cohomology. Because of the special constructions which arise, we do not know how to do this in any generality. Here we restrict attention to the category of commutative (associative) algebras. It will be clear how to make the theory work for groups, associative algebras and Lie algebras. My student, Grace Orzech, is studying more general situations at present. I would like to thank her for her careful reading of the first draft of this paper.

The triple cohomology is described at length elsewhere in this volume [B-B]. We use the adjoint pair

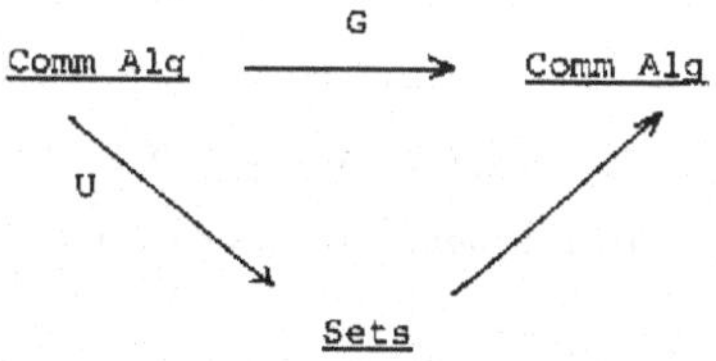

for our cotriple $\mathbb{G} = (\mathbb{G},\varepsilon,\delta)$. We let

$$\varepsilon^i = G^i \varepsilon G^{n-i} : G^{n+1} \to G^n ,$$
$$\delta^i = G^i \delta G^{n-i} : G^{n+1} \to G^{n+2} \quad \text{and}$$
$$\varepsilon = \Sigma(-1)^i \varepsilon^i : G^{n+1} \to G^n .$$

It is shown in [B-B] that the associated chain complex

$$\cdots \xrightarrow{\varepsilon} G^{n+1}R \xrightarrow{\varepsilon} \cdots \xrightarrow{\varepsilon} G^2R \xrightarrow{\varepsilon} GR \xrightarrow{\varepsilon} R \longrightarrow 0$$

is exact. This fact will be needed below.

More generally we will have occasion to consider simplicial objects (or at least the first few terms thereof)

$$X : \cdots \longrightarrow X_n \longrightarrow \cdots \substack{\longrightarrow\\\longrightarrow\\\longrightarrow\\\longrightarrow} X_2 \substack{\longrightarrow\\\longrightarrow\\\longrightarrow} X_1 \substack{\longrightarrow\\\longrightarrow} X_o$$

with face maps $d^i : X_n \to X_{n-1}$, $o \leq i \leq n$, and degeneracies $s^i : X_{n-1} \to X_n$, $o \leq i \leq n-1$ subject to the usual identities (see [Hu]). The simplicial normalization theorem*, which we will have occasion to use many times, states that the three complexes K_*X, T_*X and N_*X defined by

$$K_nX = \cap_{i=1}^n \operatorname{Ker}(d^i : X_n \to X_{n-1})$$

with boundary d induced by d^o,

$$T_nX = X_n ,$$

with boundary $d = \Sigma_{i=o}^n (-1)^i d^i$, and

$$N_nX = X_n/\Sigma_{i=o}^{n-1} \quad \operatorname{Im}(^i : X_{n-1} \to X_n)$$

with boundary d induced by $\Sigma_{i=o}^n (-1)^i d^i$ are all homotopic and in fact the natural inclusion $K_*X \subset T_*X$ and projection $T_*X \to N_*X$ have homotopy inverses . In our context the X_n will be algebras and d^i will be algebra homomorphisms, but of course d is merely an additive map.

We deliberately refrain from saying whether or not the algebras are required to have a unit. The algebras Z, A, Z(T,A), ZA below are proper ideals (notation $A \lhd T$) of other algebras and the theory becomes vacuous if they are required to be unitary. On the other hand the algebras labeled B, E, M, P, R, T can be required or not required to have a unit, as the reader desires. There is no effect on the cohomology (although

* (see [D-P])

G changes slightly, being in one case the polynominal algebra cotriple and in the other case the subalgebra of polynomials with o constant term). The reader may choose for himself between having a unit or having all the algebras considered in the same category. Adjunction of an identity is an exact functor which takes the one projective class on to the other (see [B-B], theorem 5.2 for the significance of that remark). (Also see [B], §3.)

Underlying everything is a commutative ring which everything is assumed to be an algebra over. It plays no role once it has been used to define G . By specializing it to the ring of integers we recover a theory for commutative rings.

1. The class $\mathbf{E}$.

Let A be a commutative algebra. If $A \prec T$, let $Z(A,T) = \{t \in T \mid tA = 0\}$. Then $Z(A,T)$ is an ideal of T . In particular $ZA = Z(A,A)$ is an ideal of A . Note that Z is not functorial in A (although $Z(A,-)$ is functorial on the category of algebras under A). It is clear that $ZA = A \cap Z(A,T)$. Let $\mathbf{E} = \mathbf{E}A$ denote the equivalence classes of exact sequences of algebras

$$0 \longrightarrow ZA \longrightarrow A \longrightarrow T/Z(A,T) \longrightarrow T/A+Z(A,T) \longrightarrow 0$$

for $A \prec T$. Equivalence is by isomorphisms which fix ZA and A . (*A priori* it is not a set: this possibility will disappear below.)

Let $\mathbf{E}'$ denote the set of $\lambda : A \to E$ where E is a subalgebra of $\mathrm{Hom}_A(A,A)$ which contains all multiplications $\lambda a : A \to A$, given by $(\lambda a)(a') = aa'$.

<u>Proposition 1.1</u>. There is a natural 1-1 correspondence $\mathbf{E} \cong \mathbf{E}'$.

Proof. Given $A \prec T$, let E be the algebra of multiplications on A by elements of T . There is a natural map $T \to E$ and its kernel is evidently $Z(A,T)$. If $T \succ A \prec T'$, then T and T' induce the same endomorphisms of A if and only if

$T/Z(A,T) \simeq T'/Z(A,T')$ by an isomorphism which fixes A and induces $T/A + Z(A,T) \simeq T'/A + Z(A,T')$.

To go the other way, given $\lambda : A \to E \in \mathbf{E}'$, let P be the algebra whose module structure is $E \times A$ and multiplication is given by $(e,a)(e',a') = (ee', ea'+e'a+aa')$. (ea is defined as the value of the endomorphism e). $A \to P$ is the coordinate mapping and embeds A as an ideal of P with $Z(A,P) = \{(-\lambda a, a) \mid a \in A\}$. The associated sequence is easily seen to be

$$o \longrightarrow ZA \longrightarrow A \xrightarrow{\lambda} E \xrightarrow{\pi} M \longrightarrow o$$

where π is coker λ .

From now on we will identify $\mathbf{E}$ with $\mathbf{E}'$ and call it $\mathbf{E}$.

Notice that we have constructed a natural representative $P = PE$ in each class of $\mathbf{E}$. It oames equipped with maps $d^o, d^1 : P \to E$ where $d^o(e,a) = e + \lambda a$ and $d^1(e,a) = e$. Note that $A = \text{Ker } d^1$ and $Z(A,P) = \text{Ker } d^o$. In particular $\text{Ker } d^o \cdot \text{Ker } d^1 = o$.

$P = P(T/Z(A,T))$ can be described directly as follows. Let $K \rightrightarrows T$ be the kernel pair of $T \to T/A$. This means that

$$\begin{array}{ccc} K & \longrightarrow & T \\ \downarrow & & \downarrow \\ T & \longrightarrow & T/A \end{array}$$

is a pullback. Equivalently $K = \{(t,t') \in T \times T \mid t + A = t' + A\}$, the two maps being the restrictions of the coordinate projections. It is easily seen that $\Delta_Z = \{(z,z) \mid z \in Z(A,T)\} \subset K$ and that $K/\Delta_Z \simeq P$.

Let $d^o, d^1, d^2 : B \to P$ be the kernel triple of $d^o, d^1 : P \to A$. This means $d^od^o = d^od^1$, $d^1d^1 = d^1d^2$, $d^od^2 = d^1d^o$, and B is universal with respect to these identities. Explicitly B is the set of all triples $(p,p',p'') \in P \times P \times P$ with $d^op = d^op'$, $d^1p' = d^1p''$, $d^op'' = d^1p$, the maps being the coordinate projections.

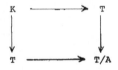

<u>Proposition 1.2.</u> The "truncated simplicial algebra",

$$o \longrightarrow B \rightrightarrows P \rightrightarrows E \longrightarrow M \longrightarrow o$$

is exact in the sense that the associated (normalized) chain complex

$$o \longrightarrow \ker d^1 \cap \ker d^2 \longrightarrow \ker d^1 \longrightarrow E \longrightarrow M \longrightarrow o$$

is exact. (The maps are those induced by restricting d^o as in K_* .)

The proof is an elementary computation and is omitted.

Note that we are thinking of this as a simplicial algebra even though the degeneracies have not been described. They easily can be, but we have need only for $s^o : E \to P$, which is the coordinate injection, $s^o e = (e,o)$. Recall that $d : B \to P$ is the additive map $d^o - d^1 + d^2$. The simplicial identities imply that $d^o d = d^o(d^o - d^1 + d^2)$ $= d^o d^2 = d^1 d^o = d^1(d^o - d^1 + d^2) = d^1 d$.

Finally note that ZA is a module over M , since it is an E-module on which the image of λ acts trivially. This implies that it is a module over B, P and E and that each face operator preserves the structure.

<u>Proposition 1.3.</u> There is a derivation $\partial : B \to ZA$ given by the formula $\partial x = (1 - s^o d^o) dx = (1 - s^o d^1) dx$.

Proof. First we see that $\partial x \in ZA = \ker d^o \cap \ker d^1$, since $d^i \partial x = d^i(1 - s^o d^i) dx (d^i - d^i) dx$ $= o$ for $i = o,1$. To show that it is a derivation, first recall that $\ker d^o \cdot \ker d^1$ $= Z(A,P) \cdot A = o$. Then for $b_1, b_2 \in B$,

$$\partial b_1 \cdot b_2 + b_1 \cdot \partial b_2 = (1 - s^o d^o) db_1 \cdot d^o b_2 + d^1 b_1 \cdot (1 - s^o d^o) db_2$$

$$= d^o b_1 \cdot d^o b_2 - d^1 b_1 \cdot d^o b_2 + d^2 b_1 \cdot d^o b_2 - s^o d^o d^2 b_1 \cdot d^o b_2$$

$$+ d^1 b_1 \cdot d^o b_2 - d^1 b_1 \cdot d^1 b_2 + d^1 b_1 \cdot d^2 b_2 - d^1 b_1 \cdot s^o d^o d^2 b_2 .$$

To this we add $(d^2 b_1 - d^1 b_1)(d^2 b_2 - s^o d^o d^2 b_2)$ and $(s^o d^o d^2 b_1 - d^2 b_1)(d^o b_2 - s^o d^o d^2 b_2)$, each easily seen to be in $\ker d^o \cdot \ker d^1 = o$, and get

$$d^ob_1 \cdot d^ob_2 - d^1b_1 \cdot d^1b_2 + d^2b_1 \cdot d^2b_2 - s^od^od^2b_1 \cdot s^od^od^2b_2$$

$$= d^o(b_1b_2) - d^1(b_1b_2) + d^2(b_1b_2) - s^od^od^2(b_1 \cdot b_2)$$

$$= (1 - s^od^o)\ d\ (b_1b_2) = \partial(b_1b_2)\ .$$

2. <u>The obstruction to a morphism</u>.

We consider an algebra R and are interested in extensions

$$o \longrightarrow A \longrightarrow T \longrightarrow R \longrightarrow o\ .$$

In the singular case, $A^2 = o$, such an extension leads to an R-module structure on A . This comes about from a surjection $T \to E$ where $E \in \mathbf{E}$, and, since A is annihilated, we get a surjection $R \to E$ by which R operates on A . In general we can only map $R \to M$. Obstruction theory is concerned with the following question. Given a surjection $p : R \to M$, classify all extensions which induce the given map. The first problem is to discover whether or not there are any. (Note: in a general category, "surjection" should probably be used to describe a map which has a kernel pair and is the coequalizer of them.) Since GR is projective in the category, we can find $p_o : GR \to E$ with $\pi p_o = p\varepsilon$. If $\tilde{d}^o, \tilde{d}^1 : \tilde{P} \to E$ is the kernel pair of π , then there is an induced map $u : P \to \tilde{P}$ such that $d^iu = \tilde{d}^i$, $i = o,1$ which is easily seen to be onto. The universal property of $\tilde{P}$ guarantees the existence of a map $\tilde{p}_1 : G^2R \to \tilde{P}$ with $\tilde{d}^i\tilde{p}_1 = p_o\varepsilon^i$, $i = o,1$. Projectivity of G^2R and the fact that u is onto guarantee the existence of $p_1 : G^2R \to P$ with $up_1 = \tilde{p}_1$, and then $d^ip_1 = d^iu\tilde{p}_1 = \tilde{d}^i\tilde{p}_1 = p_o\varepsilon^i$, $i = o,1$. Finally, the universal property of B implies the existence of $p_2 : G^3R \to E$ with $d^ip_2 = p_1\varepsilon^i$, $i = o,1,2$. Then $\partial . p_2 : G^3R \to ZA$ is a derivation and $\partial p_2\varepsilon = (1-s^od^o)dp_2\varepsilon = (1-s^od^o)p_1\varepsilon\varepsilon = o$. Thus ∂p_2 is a cocycle in $Der(G^3R,ZA)$.

<u>Proposition 2.1.</u> The homology class of ∂p_2 in $Der(G^3R,ZA)$ does not depend on the choices of p_o, p_1 and p_2 made . (p_2 actually is not an arbitrary choice.)

Proof. $\partial p_2 = (1-s^od^1)p_1\varepsilon$ and so doesn't depend on p_2 at all. Now let σ_o, σ_1 be new choices of p_o, p_1 . Since $\pi p_o = \varepsilon p = \pi p_1$, there is an $\tilde{h}^o : GR \to \tilde{p}$ with

$\tilde{d}^o\tilde{h}^o = p_o$, $\tilde{d}^1\tilde{h}^o = \sigma_o$. Again, since u is onto, there exists $h^o : GR \to P$ with $uh^o = \tilde{h}^o$, and then $d^oh^o = p_o$, $d^1h^o = \sigma_o$. Also $\pi d^o p_1 = \pi p_o \varepsilon^o = \pi\sigma_o\varepsilon^o = \pi d^o\sigma_1 =$ $= \pi d^1\sigma_1$ and by a similar argument we can find $v : G^2R \to P$ with $d^ov = d^op_1$ and $d^1v = d^1\sigma_1$. Now consider the three maps $p_1, v, h^o\varepsilon^1 : G^2R \to P$. $d^op_1 = d^ov$, $d^1v = d^1\sigma_1 = \sigma_o\varepsilon^1 = d^1h^o\varepsilon^1$ and $d^oh^o\varepsilon^1 = p_o\varepsilon^1 = d^1p_1$, so by the universal mapping property of B , there is $h^o : G^2R \to B$ with $d^oh^o = p_1$, $d^1h^o = v$, $d^2h^o = h^o\varepsilon^1$. By a similar consideration of $h^o\varepsilon^o, v, \sigma_1 : G^2R \to P$ we deduce the existence of $h^1 : G^2R \to B$ such that $d^oh^1 = h^od^o$, $d^1h^1 = v$, $d^2h^1 = \sigma_1$. The reader will recognize the construction of a simplicial homotopy between the p_i and the σ_i . We have

$(\partial h^o - \partial h^1)\varepsilon = (1-s^od^o)\ d\ (h^o - h^1)\varepsilon$

$= (1 - s^od^o)(d^oh^o - d^1h^o + d^2h^o - d^oh^1 + d^1h^1 - d^2h^1)\varepsilon$

$= (1 - s^od^o)(d^oh^o - d^2h^1 + h^o\varepsilon^1 - h^o\varepsilon^o)\varepsilon$

$= (1 - s^od^o)(p_1 - \sigma_1 + h^o\varepsilon)\varepsilon = (1 - s^od^o)(p_1 - \sigma_1)\varepsilon$

$= (1 - s^od^o)\ d\ (p_2 - \sigma_2) = \partial p_2 - \partial\sigma_2$. This shows that ∂p_2 and $\partial\sigma_2$ are in the same cohomology class in $\mathrm{Der}(G^3R, ZA)$, which class we denote by [p] and which is called the <u>obstruction</u> of p . We say p is <u>unobstructed</u> provided $[p] = o$.

<u>Theorem 2.2.</u> A surjection $p : R \to M$ arises from an extension if and only if p is unobstructed.

Proof. Suppose p arises from

$$o \longrightarrow A \longrightarrow T \longrightarrow R \longrightarrow o .$$

Then we have a commutative diagram

$$\begin{array}{ccccccccccc} & & o & \longrightarrow & K & \rightrightarrows & T & \longrightarrow & R & \longrightarrow & o \\ & & & & \downarrow{\scriptstyle \nu_1} & & \downarrow{\scriptstyle \nu_o} & & \downarrow{\scriptstyle p} & & \\ O & \longrightarrow & B & \rightrightarrows & P & \rightrightarrows & E & \longrightarrow & M & \longrightarrow & o \end{array}$$

where $e^o, e^1\ K \rightrightarrows T$ is the kernel pair of $T \to R$ and $t^o : T \to K$ is the diagonal map. Commutativity of the leftmost square means that each of three distinct squares commutes, i.e. with the upper, middle or lower arrows. Recalling that $E = T/Z(A,T)$ and $P = K/\Delta_Z$

we see that the vertical arrows are onto. Then there is a $\sigma_o : GR \to T$ with $\nu_o \sigma_o = p_o$. Since K is the kernel pair, we have $\sigma_1 : G^2 R \to K$ with $e^i \sigma_1 = \sigma_o \epsilon^i$, $i = o,1$. Then $\nu_1 \sigma_1$ is a possible choice for p_1 and we will assume $p_1 = \nu_1 \sigma_1$. Then $\partial p_2 = (1-s^o d^o) p_1 \epsilon = (1-s^o d^o) \nu_1 \sigma_1 \epsilon = \nu_1 (1-t^o e^o) \sigma_1 \epsilon$. But $e^o (1-t^o e^o) \sigma_1 \epsilon = o$ and $e^1 (1-t^o e^o) \sigma_1 \epsilon = (e^1 - e^o) \sigma_1 \epsilon = \sigma_o (\epsilon^1 - \epsilon^o) \epsilon = \sigma_o \epsilon \epsilon = o$, and since e^o, e^1 are jointly monic, i.e. define a monic $K \to T \times T$, this implies that $\nu_1 (1-t^o e^o) \sigma_1 \epsilon = o$.

Conversely, suppose p, p_o, p_1, p_2 are given and there is a derivation $\tau : G^2 R \to ZA$ such that $\partial p_2 = \tau \epsilon$. Let $\tilde{p}_1 : G^2 R \to p$ be $p_1 - \tau$ where we abuse language and think of τ as taking values in $P \supset ZA$. Then $\tilde{p}_1$ can be easily shown to be an algebra homomorphism above p_o . Choosing $\tilde{p}_2$ above $\tilde{p}_1$ we have new choices $p, p_o, \tilde{p}_1, \tilde{p}_2$ and $\partial \tilde{p}_2 = (1-s^o d^o) d \tilde{p}_2$
$= (1-s^o d^o) \tilde{p}_1 \epsilon = (1-s^o d^o) (p_1 - \tau) \epsilon = (1-s^o d^o) p_1 \epsilon$
$- (1-s^o d^o) \tau \epsilon = \partial p_2 - \tau \epsilon = o$, since $(1-s^o d^o)$ is the identity when restricted to $ZA = \ker d^o \cap \ker d^1$. Thus we can assume that p_o, p_1, p_2 had been chosen so that $\partial p_2 = o$ already.

Let

$$
\begin{array}{ccc}
Q & \xrightarrow{\ q_1\ } & P \\
{\scriptstyle q_2}\big\downarrow & & \big\downarrow{\scriptstyle d^1} \\
GR & \xrightarrow[\ p_o\ ]{} & E
\end{array}
$$

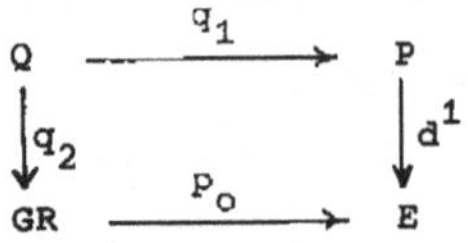

be a pullback. Since the pullback is computed in the underlying module category, d^1 is onto so q_2 is onto. Also the induced map $\ker q_2 \to \ker d^1 = A$ is an isomorphis (this is true in an arbitrary pointed category) and we will identity $\ker q_2$ with a map $a : A \to Q$ such that $q_1 a = \ker d^1$. Now let $u^o, u^1 : G^2 R \to Q$ be defined by the conditions $q_1 u^o = s^o d^o p_1, q_2 u^o = \epsilon^o, q_1 u^1 = p_1, q_2 u^1 = \epsilon^1$. In the commutative diagram

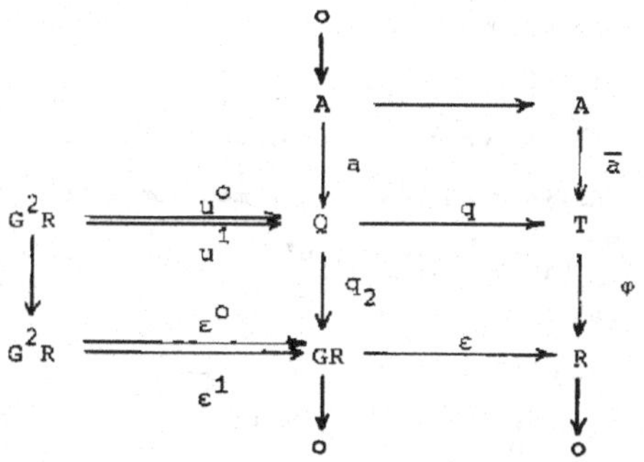

the rows are coequalizers and the columns are exact. The exactness of the right column follows from the commutativity of colimits. We claim that the map $\bar{a}$ is $1 - 1$.

This requires showing that $\mathrm{Im}\ a \cap \mathrm{Ker}\ q = o$. Ker q is the ideal generated by the image of $u = u^o - u^1$. Also $\mathrm{Im}\ a = \ker q_2$. Consequently the result will follow from

<u>**Proposition 2.3.**</u> The image of u is an ideal and $\mathrm{Im}\ u \cap \ker q_2 = o$.

Proof. If $x \in G^2R$, $y \in Q$, let $x' = \delta q_2 y$. We claim that $u(xx') = ux \cdot y$. To prove this it suffices to show that $q_i u(xx') = q_i(ux \cdot y)$ for $i = 1,2$ (because of the definition of pullback). But $q_2 u(xx') = \varepsilon(xx') = \varepsilon^o x \cdot \varepsilon^o x' - \varepsilon^1 x \cdot \varepsilon^1 x'$ $= \varepsilon^o x \cdot q_2 y - \varepsilon^1 x \cdot q_2 y = q_2(u^o x \cdot y) - q_2(u^1 x \cdot y)$ $= q_2(ux \cdot y)$. Next observe that our assumption is that $(1 - s^o d^1)p_1$ is zero on $\mathrm{Im}\ \varepsilon = \mathrm{Ker}\ \varepsilon$. In particular, $(s^o d^1 - 1)p_1\delta = o$. ($\varepsilon\delta = \varepsilon^o\delta - \varepsilon^1\delta = o$.) Also $(s^o d^o - 1)p_1 x \cdot (s^o d^1 - 1)q_1 y \in \ker d^o \cdot \ker d^1 = o$.

Then we have,

$$q_1 u(xx') = (q_1 u^o - q_1 u^1)(xx') = (s^o d^o p_1 - p_1)(xx') = s^o d^o p_1 x \cdot s^o d^o p_1 x' - p_1 x \cdot p_1{}'$$
$$= (s^o d^o p_1 x - p_1 x) s^o d^o p_1 x' + p_1 x \cdot (s^o d^o p_1 x' - p_1 x')$$
$$= (s^o d^o - 1)p_1 x \cdot s^o d^o p_1 \delta q_2 y + p_1 x \cdot (s^o d^o - 1)p_1 \delta q_2 y = (s^o d^o - 1)p_1 x \cdot s^o p_o \varepsilon^o \delta q_2 y$$
$$= (s^o d^o - 1)p_1 x \cdot s^o p_o q_2 y = (s^o d^o - 1)p_1 x \cdot s^o d^1 q_1 y$$
$$= (s^o d^o - 1)p_1 x \cdot q_1 y + (s^o d^o - 1)p_1 x \cdot (s^o d^1 - 1)q_1 y$$
$$= (s^o d^o p_1 x - p_1 x) q_1 y = q_1 ux \cdot q_1 y = q_1(ux \cdot y).$$

Now if $ux \in \ker q_2$, then $o = q_2 ux = \varepsilon x$, $x \in \mathrm{Ker}\,\varepsilon = \mathrm{Im}\ \varepsilon$, and

$o = (s^o d^o - 1)p_1 x = q_1 ux$. But then $ux = o$.

Now to complete the proof of 1.2 we show

Proposition 2.4. There is a $\tau : T \to E$ which is onto, whose kernel is $Z(A,T)$ and such that $p\varphi = \pi\tau$. Let τ be defined as the unique map for which $\tau q = d^o q_1$. This defines a map, for $d^o q_1 u^o = d^o s^o d^o p_1 = d^o p_1 = d^o q_1 u'$. τ is seen to be onto by applying the 5-lemma to the diagram,

$$\begin{array}{ccccccccc} o & \longrightarrow & A & \longrightarrow & T & \longrightarrow & R & \longrightarrow & o \\ & & \| & & \downarrow{\scriptstyle \tau} & & \downarrow{\scriptstyle p} & & \\ & & A & \longrightarrow & E & \longrightarrow & M & \longrightarrow & o \end{array}$$

since p is assumed onto. $\pi\tau q = \pi d^o q_1 = \pi d^1 q_1 = \pi p_o q_2 = p\varepsilon q_2 = p\varphi q$ and q is onto, so $\pi\tau = p\varphi$. Now if we represent elements of Q as pairs $(x,Q) \in GX \times p$ subject to $p_o x = d^1 Q$, $\tau(x,Q) = d^o Q$. Then $\operatorname{Ker} \tau = \{(x,Q) \mid d^o Q = o\}$. That is,

$$\begin{array}{ccc} \operatorname{Ker} \tau & \longrightarrow & \operatorname{Ker} d^o \\ \downarrow & & \downarrow{\scriptstyle d^1} \quad \Big| \ \operatorname{Ker} d^o \\ GX & \xrightarrow{\ p_o\ } & E \end{array}$$

is a pullback. A is represented as $\{(o,Q') \mid d^1 Q' = o\}$. Now $Z(A,T) =$

$= \{(x,Q) \in Q \mid d^1 Q' = o \to QQ' = o\}$

$= \{(x,Q) \in Q \mid Q \in Z(A,P)\}$.

It was observed in §1 that $Z(A,P) = \operatorname{Ker} d^1$. Thus $Z(A,T) = \{(x,Q) \in Q \mid Q \in \operatorname{Ker} d^1\} = \operatorname{Ker} \tau$.

3. The action of H^1 .

This § is devoted to proving the following.

Theorem 3.1. Let $p : R \to M$ be unobstructed. Let $\underline{\Sigma} = \underline{\Sigma}p$ denote the equivalence classes of extensions

$$o \longrightarrow A \longrightarrow T \longrightarrow R \longrightarrow o$$

which induce p . Then the group $H^1(R,ZA)$ acts on $\underline{\Sigma}p$ as a principal homogeneous representation. (This means that for any $\Sigma \in \underline{\Sigma}$, multiplication by Σ is a 1-1 correspondence $H^1(R,ZA) \cong \underline{\Sigma}$.)

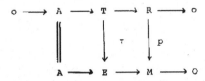

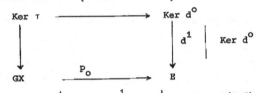

Proof. Let $\underline{\Lambda}$ denote the equivalence classes of singular extensions

$$o \longrightarrow ZA \longrightarrow U \longrightarrow R \longrightarrow o$$

which induce the same module structure on ZA as that given by p (recalling that ZA is always an M-module). Then $\underline{\Lambda} \simeq H^1(R,ZA)$ where the addition in $\underline{\Lambda}$ is by Baer sum and is devoted by $\Lambda_1 + \Lambda_2$, $\Lambda_1, \Lambda_2 \in \underline{\Lambda}$. We will describe operations $\underline{\Lambda} \times \underline{\Sigma} \to \underline{\Sigma}$, denoted by $(\Lambda,\Sigma) \to \Lambda + \Sigma$, and $\underline{\Sigma} \times \underline{\Sigma} \to \underline{\Lambda}$, denoted by $(\Sigma,\Sigma') \to \Sigma - \Sigma'$, such that

a) $(\Lambda_1 + \Lambda_2) + \Sigma = \Lambda_1 + (\Lambda_2 + \Sigma)$

b) $(\Sigma_1 - \Sigma_2) + \Sigma_2 = \Sigma_1$

c) $(\Lambda + \Sigma) - \Sigma = \Lambda$

for $\Lambda, \Lambda_1, \Lambda_2 \in \underline{\Lambda}$, $\Sigma, \Sigma_1, \Sigma_2 \in \underline{\Sigma}$ (proposition 3.2). This will clearly prove theorem 3.1.

We describe $\Lambda + \Sigma$ as follows. Let

$$o \longrightarrow ZA \longrightarrow U \xrightarrow{\psi} R \longrightarrow o \in \Lambda$$
$$o \longrightarrow A \longrightarrow T \xrightarrow{\varphi} R \longrightarrow o \in \Sigma \; .$$

(Here we mean representatives of the equivalence classes.) To simplify notation we assume $ZA < U$ and $A < T$. Let

$$\begin{array}{ccc} V & \longrightarrow & U \\ \downarrow & & \downarrow \\ T & \longrightarrow & R \end{array}$$

be a pullback. This means $V = \{(t,u) \in T \times U \mid \varphi t = \psi u\}$. Then $I = \{(z, -z) \mid z \in ZA\} < V$. Let $T' = V/I$. Map $A \to T'$ by $a \to (a,o) + I$. Map $T' \to R$ by $(t,u) + I \to \varphi t = \psi u$. This is clearly well defined modulo I. Clearly $o \to A \to T' \xrightarrow{\varphi'} R \to o$ is a complex and φ' is onto. It is exact since $\ker(V \to R) = \ker(T \to R) \times o + o \times \ker(U \to R) = A \times o + o \times ZA = A \times o + I$. (since $ZA \subset A$).

$Z(T',A) = \{(t,u) + I \in V/I \mid t \in Z(T,A)\}$. Map $T' \to T/Z(T,A)$ by $(t,u) + I \to t + Z(T,A)$. This is well defined modulo I and its kernel is $Z(T',A)$. Since $U \to R$ is onto, so is $V \to T$, and hence $T' \to T/Z(T,A)$ is also. Thus $T'/Z(T',A) \simeq T/Z(T,A)$ and the isomorphism is coherent with φ and φ' and with the maps $T \leftarrow A \to T'$. Thus

$$o \longrightarrow A \longrightarrow T' \xrightarrow{\varphi'} R \longrightarrow o \quad \in\in \underline{\Sigma} .$$

(This notation means the sequence $\in$ some $\Sigma' \in \underline{\Sigma}$.)

To define $\Sigma_1 - \Sigma_2$ let Σ_i be represent by the sequence

$$o \longrightarrow A \longrightarrow T_i \xrightarrow{\varphi_i} R \longrightarrow o \ , \quad i = 1,2$$

where we again suppose $A < T_i$. We may also suppose $T_1/Z(A,T_1) = E = T_2/Z(A,T_2)$ and $T_1 \overset{\tau_1}{\to} E \overset{\tau_2}{\leftarrow} T_2$ are the projections. Let

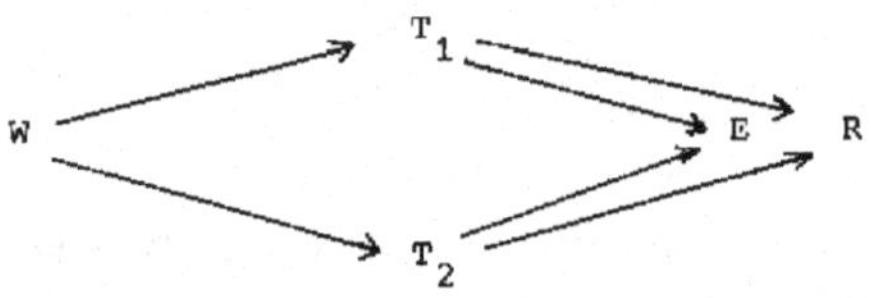

be a limit. This means

$W = \{(t_1,t_2) \in T_1 \times T_2 \mid \tau_1 t_1 = \tau_2 t_2$ and $\varphi_1 t_1 = \varphi_2 t_2\}$.

Then $J = \{(a,a) \mid a \in A\} < W$. Map $ZA \to W/J$ by $z \to (z,o) + J$ and $\varphi : W/J \to R$ by $(t_1,t_2) + J \to \varphi_1 t_1 = \varphi_2 t_2$. If $(t_1,t_2) + J \in \mathrm{Ker}\varphi$, then $\varphi_1 t_1 = o = \varphi_2 t_2$, so $t_1,t_2 \in A$. Then $(t_1,t_2) = (t_1 - t_2,o) + (t_2,t_2)$. But then $\tau_1(t_1 - t_2) = o$, so $t_1 - t_2 \in A \cap Z(A,T_1) = ZA$. Thus $ZA \subset \mathrm{Ker}\varphi$, and clearly $\mathrm{Ker}\varphi \subset ZA$. Now given $r \in R$, we can find $t_i \in T_i$ with $\varphi_i t_i = r$, $i = 1,2$. Then $\pi(\tau_1 t_1 - \tau_2 t_2) = \pi\tau_1 t_1 - \pi\tau_2 t_2 = p\varphi_1 t_1 - p\varphi_2 t_2 = o$, so $\tau_1 t_1 - \tau_2 t_2 = \lambda a$ for some $a \in A$. (Recall $\lambda : A \to E$ is the multiplication map.) But then $\tau_1 t_1 = \tau_2(t_2 + a)$ and $\varphi_1 t_1 = \varphi_2(t_2 + a)$, so $(t_1,t_2 + a) + J \in W/J$ and $\varphi(t_1,t_2 + a) = r$. Thus φ is onto and

$$o \longrightarrow ZA \longrightarrow W/J \longrightarrow R \longrightarrow o \quad \in\in \underline{\Lambda}\,.$$

Note that the correct R-module structure is induced on ZA because p is the same.

<u>Proposition 3.2.</u> For any $\Lambda, \Lambda_1, \Lambda_2 \in \underline{\Lambda}$, $\Sigma, \Sigma_1, \Sigma_2 \in \underline{\Sigma}$,

a) $(\Lambda_1 + \Lambda_2) + \Sigma = \Lambda_1 + (\Lambda_2 + \Sigma)$,

b) $(\Sigma_1 - \Sigma_2) + \Sigma_2 = \Sigma_1$,

c) $(\Lambda + \Sigma) - \Sigma = \Lambda$.

Proof. a) Let

$$o \longrightarrow ZA \longrightarrow U_i \xrightarrow{\psi_i} R \longrightarrow o\,, \quad i = 1,2,$$
$$o \longrightarrow A \longrightarrow T \xrightarrow{\varphi} R \longrightarrow o$$

represent Λ_1, Λ_2, Σ respectively. An element of $(\Lambda_1 + \Lambda_2) + \Sigma$ is represented by a triple (u_1,u_2,t) such that $\psi(u_1,u_2) = \varphi t$ where $\psi(u_1,u_2) = \psi_1 u_1 = \psi_2 u_2$. An element of $\Lambda_1 + (\Lambda_2 + \Sigma)$ is represented by a triple (u_1,u_2,t) where $\psi_1 u_1 = \varphi'(u_2,t)$ and $\varphi'(u_2,t) = \psi_2 u_2 = \varphi t$. Thus each of them is the limit

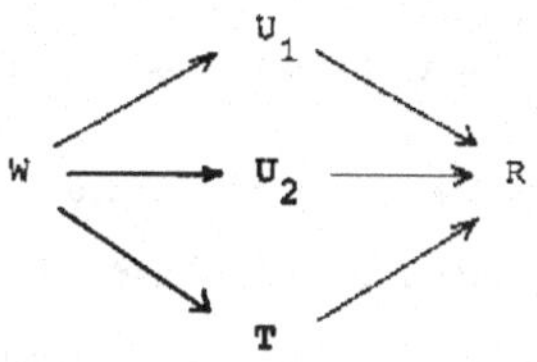

modulo a certain ideal which is easily seen to be the same in each case, namely $\{(z_1,z_2,z_3) \mid z_i \in Z \text{ and } z_1 + z_2 + z_3 = o\}$.

b) Let Σ_1 and Σ_2 be represented by sequences $o \to A \to T_i \xrightarrow{\varphi_i} R \to o$. Let $\tau_i : T_i \to E$ as above for $i = 1,2$. Let $(\Sigma_1 - \Sigma_2) + \Sigma_2$ be represented by

$$o \longrightarrow A \longrightarrow T \longrightarrow R \longrightarrow o\,.$$

Then an element of T can be represented as a triple (t_1,t_2,t_2') subject to the condition $\tau_1 t_1 = \tau_2 t_2$, $\varphi_1 t_1 = \varphi_2 t_2 = \varphi_2 t_2'$. These conditions imply that $t_2'-t_2 \in A$ and we can map $\sigma : T \to T_2$ by $\sigma(t_1,t_2,t_2') = t_1 + (t_2' - t_2)$.

To show that σ is an algebra homomorphism, recall that $\tau_1 t_1 = \tau_2 t_2$ implies that t_1 and t_2 act the same on A. Now if $(t_1,t_2,t_2^!)$, $(s_1,s_2,s_2^!) \in T$,

$\sigma(t_1,t_2,t_2^!)\cdot\sigma(s_1,s_2,s_2^!) = (t_1 + (t_2^! - t_2))(s_1 + (s_2^! - s_2))$

$= t_1s_1 + t_1(s_2^! - s_2) + (t_2^! - t_2)s_1 + (t_2^! - t_2)(s_2^! - s_2)$

$= t_1s_1 + t_2(s_2^! - s_2) + (t_2^! - t_2)s_2 + (t_2^! - t_2)(s_2^! - s_2)$

$= t_1s_1 + t_2^!s_2^! - t_2s_2 = \sigma(t_1s_1,t_2s_2,t_2s_2,t_2^!s_2^!)$

$= \sigma((t_1,t_2,t_2^!)(s_1,s_2,s_2^!)$. Also the diagram

$$\begin{array}{ccccccccc} 0 & \longrightarrow & A & \longrightarrow & T & \longrightarrow & R & \longrightarrow & 0 \\ & & \downarrow & & \downarrow & & \downarrow & & \\ 0 & \longrightarrow & A & \longrightarrow & T_2 & \longrightarrow & R & \longrightarrow & 0 \end{array}$$

commutes and the sequences are equivalent.

c) Let Λ and Σ and $(\Lambda + \Sigma) - \Sigma$ be represented by sequences

$$0 \longrightarrow ZA \longrightarrow U \xrightarrow{\psi} R \longrightarrow 0 ,$$

$$0 \longrightarrow A \longrightarrow T \xrightarrow{\varphi} R \longrightarrow 0 ,$$

$$0 \longrightarrow ZA \longrightarrow U' \xrightarrow{\psi} R \longrightarrow 0 , \quad \text{respectively.}$$

An element of U' is represented by a triple (t,u,t') subject to $\varphi t = \psi u = \varphi t'$ and $\tau t = \tau t'$. The equivalence relation is generated by all $(z,a-z,a)$, $a \in A$, $z \in ZA$. The relations imply that $t - t' \in ZA$, so the map $\sigma : U' \to U$ which takes $(t,u,t') \to u + (t-t')$ makes sense and is easily seen to be well defined. For $s,s',t,t' \in T$, $u,v \in U$, we have $\sigma(t,u,t')\,\sigma(s,v,s') = (u+t-t')(v+s-s')$

$= uv + u(s-s') + (t-t')v + (t-t')(s-s')$

$= uv + t(s-s') + (t-t')s'$

$= uv + ts - t's' = \sigma(ts,uv,t's')$

$= \sigma((t,u,t')(s,v,s'))$.

Since $ZA \to U'$ takes $z \to (z,0,0)$ and $\psi'(t,u,t') = \psi u = \psi u + \psi(t-t')$, the diagram

$$\begin{array}{ccccccccc} 0 & \longrightarrow & ZA & \longrightarrow & U' & \longrightarrow & R & \longrightarrow & 0 \\ & & \downarrow & & \downarrow & & \downarrow & & \\ 0 & \longrightarrow & ZA & \longrightarrow & U & \longrightarrow & R & \longrightarrow & 0 \end{array}$$

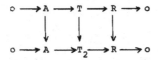

commutes and gives the equivalence.

4. Every element of H^2 is on obstruction.

The title of this § means the following. Given an R-module Z and a class $\xi \in H^2(R,Z)$, it is possible to find an algebra A and an $E \in EA$ of the form

$$o \longrightarrow ZA \longrightarrow A \longrightarrow E \longrightarrow M \longrightarrow o$$

and a surjection $p : R \to M$ such that $Z \simeq ZA$ as an R-module (via p) and $[p] = \xi$. It is clear that this statement together with theorem 2.2 characterizes H^2 completely. No smaller group contains all obstructions and no factor group is fine enough to test whether a p comes from an extension. In particular, this shows that in degrees 1 and 2 these groups nust coincide with those of Harrison (renumbered) (see [Harr]) and Lichtenbaum and Schlessinger (see [LS]). In particular those coincide. See also Gerstenhaber ([G],[G']) and Barr ([B]).

Theorem 4.1. Every element of H^2 is an obstruction.

Proof. Represent ξ by a derivation $\varrho : G^3R \to Z$. This derivation has the property that $\varrho\varepsilon = o$ and by the simplicial normalization theorem we may also suppose $\varrho\delta^o = p\delta^1 = o$. Let $V = \{(x,z) \in G^2R \times Z \mid \varepsilon^1 x = o\}$. (Here Z is given trivial mul-tiplication.) Let $I = \{(\varepsilon^o y,-\varrho y) \mid y \in G^3R,\ \varepsilon^1 y = \varepsilon^2 y = o\}$. $I \subset V$ for $\varepsilon^1\varepsilon^o y = \varepsilon^o\varepsilon^2 y = o$. I claim that $I \triangleleft V$. In fact for $(x,z) \in V$, $(\varepsilon^o y,-\varrho y) \in I$, $(x,z)(\varepsilon^o y,-\varrho y) = (x\cdot\varepsilon^o y,o)$. Now $\delta^o x\cdot y \in G^3R$ satisfies $\varepsilon^o(\delta^o x\cdot y) = x\cdot\varepsilon^o y$, $\varepsilon^i(\delta^o x\cdot y) = \varepsilon^i\delta^o x\ \varepsilon^i y = o$, $i = 1,2$. Moreover $\varrho(\delta^o x\cdot y) = \varrho\delta^o x\cdot y + \delta^o x\cdot\varrho y$. Now $\varrho\delta^o = o$ by assumption and the action of G^3R on Z is obtained by applying face operators into R (any composite of them is the same) and then multiplying. In particular, $\delta^o x\cdot\varrho y = \varepsilon^1\varepsilon^1\delta^o x\cdot\varrho y = \varepsilon^1 x\cdot\varrho y = o$, since $\varepsilon^1 x = o$. Thus $(x,z)(\varepsilon^o y,-\varrho y) = (\varepsilon^o(\delta^o x\cdot y),-\varrho(\delta^o x\cdot y))$ and I is an ideal. Let $A = V/I$. I claim that the composite $Z \to V \to V/I$ is 1-1 and embeds Z as ZA. For if $(o,z) = (\varepsilon^o y,-\varrho y)$, then $\varepsilon^o y = \varepsilon^1 y = \varepsilon^2 y = o$ so that y is a cycle and hence a boundary, $y = \varepsilon z$. But then

$\varrho y = \varrho\varepsilon z = o$. This shows $Z \cap I = o$. If $(x,z) + I \in ZA$, $(x,z)(x',z') = (xx',o) \in I$ for all $(x',z') \in V$. In particular $\varepsilon(xx') = \varepsilon^o(xx') = \varepsilon^o x \cdot \varepsilon^o x' = o$ for all x' with $\varepsilon^1 x' = o$. By the simplicial normalization theorem this means $\varepsilon^o x \cdot \ker \varepsilon = o$. Let $w \in GR$ be the basis element corresponding to $o \in R$. Then w is not a zero divisor, but $w \in \operatorname{Ker} \varepsilon$. Hence $\varepsilon^o x = o$ and $x = \varepsilon y$ and by the normalization theorem we may suppose $\varepsilon^1 y = \varepsilon^2 y = o$. Therefore $(x,z) = (\varepsilon^o y, -\varrho y) + (o, z+\varrho y) \equiv (o, z+\varrho y) \pmod{I}$. On the other hand $Z + I \subset ZA$.

Let GR operate on V by $y(x,z) = (\delta y \cdot x, yz)$ where GR operates on Z via $p\varepsilon$. I is a GR-submodule for $y'(\varepsilon^o y, -\varrho y) = (\delta y' \cdot \varepsilon^o y, -y' \cdot \varrho y) = (\varepsilon^o(\delta\delta y' \cdot y), -\varrho(\delta\delta y' \cdot y))$, since $\varrho(\delta\delta^1 y \cdot y) = \delta\delta y' \cdot \varrho y + \varrho\,\delta\delta y' \cdot y = y' \cdot \varrho y$. Hence A is a GR-algebra.

Let E be the algebra of endomorphisms of A which is generated by the multiplications from GR and the inner multiplications. Let $p_o : GR \to E$ and $\lambda : A \to E$ be the indicated maps. Then $E = \operatorname{Im} p_o + \operatorname{Im} \lambda$. This implies that πp_o is onto where $\pi : E \to M$ is coker λ .

Now we wish to map $p : R \to M$ such that $p\varepsilon = \pi p_o$. In order to do this we must show that for $x \in G^2R, p_o\varepsilon^o x$ and $p_o\varepsilon^1 x$ differ by an inner multiplication. First we show that if $(x',z) \in V$, then $(x \cdot x' - \delta\varepsilon^o x \cdot x', o) \in I$. In fact let
$y = (1-\delta^o\varepsilon^1)(\delta^1 y \cdot \delta^o x)$. Then $\varepsilon^1 y = o$ and $\varepsilon^2 y = o$ also, since $\varepsilon^2\delta^o x' = \delta\varepsilon^1 x' = o$.
$\varepsilon^o y = (\varepsilon^o - \varepsilon^1)(\delta^1 x \cdot \delta^o x') = \delta\varepsilon^o y \cdot x - x \cdot x'$. Finally $\varrho y = o$ because of the assumption we made that $\varrho\delta^i = o$. Now $(p_o\varepsilon^o x - p_o\varepsilon^1 x)(x',z)$

$= ((\delta\varepsilon^o x - \delta\varepsilon^1 x)x', xz - xz)$

$\equiv ((x - \delta\varepsilon^1 x)x', o) \pmod{I}$

$= (x - \delta\varepsilon^1 x, o)(x',z)$ where $(x - \delta\varepsilon^1 x, o) \in V$. Thus we have shown

<u>Lemma 4.2</u> $p_o\varepsilon^o x - p_o\varepsilon^1 x$ is the inner multiplication $\lambda((x-\delta\varepsilon^1 x, o)+I)$. Then map $p : R \to M$ as indicated. Now $\pi p_o = p\varepsilon$ is a surjection and so is p .

P is constructed as pairs $(e,a), e \in E, a \in A$ with multiplication $(e,a)(e',a') = (ee', ea' + e'a + aa')$. Map $p_1 : G^2R \to E$ by $p_1 x = (p_o\varepsilon^1 x, (x - \delta^o\varepsilon^1 x, o) + I)$. Then $d^o p_1 x = p_o\varepsilon^1 x + \lambda((x - \delta^o\varepsilon^1 x, o) + I) = p_o\varepsilon^1 x + p_o\varepsilon^o x - p_o\varepsilon^1 x = p_o\varepsilon^o x$ by lemma 4.2.

Also $d^1p_1x = p_o\varepsilon^1x$ and thus p_1 is a suitable map. If $p_2 : G^3R \to B$ is chosen as prescribed, then for any $x \in G^3R$, $(1-s^od^1)dp_2x = (1-s^od^1)p_1\varepsilon x$ $= (1-s^od^1)(p_o\varepsilon^1\varepsilon x,(\varepsilon x-\delta^o\varepsilon^1\varepsilon x,o)+I) = (p_o\varepsilon^1\varepsilon x,\varepsilon x-\delta^o\varepsilon^1\varepsilon x,o)+I) - (p_o\varepsilon^1\varepsilon x,o)$ $= (o,(\varepsilon x-\delta^o\varepsilon^1\varepsilon x,o)+I)$. The proof is completed by showing that $(\varepsilon x-\delta^o\varepsilon^1\varepsilon x,o) \equiv (o,\varrho x)$ (mod I). Let $y = (1-\delta^o\varepsilon^1)(1-\delta^1\varepsilon^2)x$. Then $\varepsilon^1y = \varepsilon^2y = o$ clearly and $\varepsilon^oy = (\varepsilon^o-\varepsilon^1)(1-\delta^1\varepsilon^2)x = (\varepsilon^o-\varepsilon^1+\varepsilon^2-\delta^o\varepsilon^1\varepsilon^o)x = (\varepsilon^o-\varepsilon^1+\varepsilon^2-\delta^o\varepsilon^1(\varepsilon^o-\varepsilon^1+\varepsilon^2))x = (\varepsilon-\delta^o\varepsilon^1\varepsilon x)$, while $\varrho y = \varrho x$, since we have assumed that $\varrho\delta^i = o$. Thus $\partial p = \varrho$ and $[p] = \xi$. This completes the proof.

REFERENCES

E-M S. Eilenberg and S. Mac Lane, Cohomology theory in abstract groups. II. Group extensions with a non-abelian kernel. Ann. of Math. 48, (1947), 326-341.

Hoch G. Hochschild, Cohomology and representations of associative algebras, Duke Math J. 14, (1947), 921-948.

Hoch' ______________, Lie algebra kernels and cohomology, Amer. J. Math. 76, (1954), 698-716.

Mac S. Mac Lane, Extensions and obstructions for rings, Ill. J. Math. 2, (1958), 316-345.

Shuk U. Shukla, Cohomologie des algèbres associatives, Ann. Sci. Ecole Norm. Sup. 78, (1961), 163-209.

Harr D.K. Harrison, Commutative algebras and cohomology, Trans. Amer. Math. Soc. 98, (1962), 191 - 204.

Hu P.J. Huber, Homotopy theory in general categories, Math. Ann. 144, (1961), 361 - 385.

B - B M. Barr and J. Beck, Homology and standard constructions, this volume.

L - S S. Lichtenbaum and S. Schlessinger, The cotangent complex of a morphism, Trans. Amer. Math. Soc. 128, (1967), 41 - 70.

B M. Barr, A note on commutative algebra cohomology, Bull. Amer. Math. Soc. (to appear).

G M. Gerstenhaber, On the deformation of rings and algebras. II, Ann. of Math (2) 84, (1966), 1 - 19.

G' ________________, The third cohomology group of a ring and the commutative cohomology theory, Bull. Amer.Math. Soc. 73, (1967) 950 - 954.

D - P A. Dold and D. Puppe, Homologie nicht-additiver Funktoren; Anwendungen, Ann. Inst. Fourier, 11 , (1961), 201 - 312.

ON COTRIPLE AND ANDRÉ (CO)HOMOLOGY, THEIR RELATIONSHIP WITH CLASSICAL HOMOLOGICAL ALGEBRA

by Friedrich Ulmer [1)]

This summary is to be considered as an appendix to André [1], Beck [3] and Barr-Beck [2], [4]. Details will appear in another Lecture Notes volume later on [27]. When André and Barr-Beck presented their non-abelian derived functors in seminars at the Forschungsinstitut during the winter of 1965-66 and the summer of 1967, I noticed some relationship to classical homological algebra.[2)] On the level of functor categories, their non-abelian derived functors A_* and H_* turn out to be abelian derived functors. The aim of this note is to sketch how some of the properties of A_* and H_* can be obtained within the abelian framework and also to indicate some new results. Further applications are given in the detailed version [27]. The basic reason why abelian methods are adequate lies in the fact that the simplicial resolutions André and Barr-Beck used to construct A_* and H_* become acyclic resolutions in the functor category in the usual sense. However, it should be noted that the abelian approach is not always really different from the approach of André and Barr-Beck. In some cases the difference lies rather in looking at the same phenomena from two different standpoints, namely, from an elementary homological algebra view instead of from the view of a newly developed machinery which is simplicially oriented. In other cases, however, the abelian viewpoint leads to simplifications of proofs and generalizations of known facts as well as to new results and insights. An instance

---

1) Part of this work was supported by the Forschungsinstitut für Mathematik der E.T.H. and the Deutsche Forschungsgemeinschaft.

2) In the meantime some of the results presented here were found independently by several authors. Among them are M. Bachmann (in a thesis under the supervision of B. Eckmann), E. Dubuc [9], U. Oberst [20] and D. Swan (unpublished). The paper of U. Oberst, which is to appear in the Math. Zeitschrift, led me to revise part of this note (or rather [27]) to include some of his results. Part of the material herein was first observed during the winter of 1967-68 after I had received an earlier version of [2]. The second half of [2] Ch. X was also developed during this period and illustrates the mutual influence of the material presented there and of the corresponding material here and in [27].

for the latter is the method of acyclic models which turns out to be standard procedure in homological algebra to compute the left derived functors of the Kan extension E_J by means of projectives [3] or, more generally, by E_J-acyclic resolutions.[4] We mostly limit ourselves to dealing with homology and leave it to the reader to state the corresponding (i.e. dual) theorems for cohomology. To make this work in practice, we try to avoid exactness conditions (i.e. AB5)) on the coefficient category $\underline{A}$. However, a few results depend on AB5) and are provably false without it. This reflects the known fact that certain properties hold only for homology, but not for cohomology.

Morphism sets, natural transformation and functor categories are denoted by brackets $[-,-]$, comma categories by parentheses $(-,-)$. The categories of sets and abelian groups are denoted by $\underline{S}$ and $\underline{Ab}.\underline{Gr}$. . The phrase "Let $\underline{A}$ be a category with direct limits" always means that $\underline{A}$ has direct limits over small index categories. However, we sometimes also consider direct limits of functors $F : \underline{D} \to \underline{A}$, where $\underline{D}$ is not necessarily small. Of course we then have to prove that this specific limit exists.

I am indebted to Jon Beck and Michael Barr for many stimulating conversations without which the paper would not have its present form.

Our approach is based on the notions of Kan extension [14], [26] and generalized representable functor [24], which prove to be very useful in this context. We recall these definitions.

(1) A generalized representable functor from a category $\underline{M}$ to a category $\underline{A}$ (with sums) is a composite [5]

$$A \otimes [M,-] : \underline{M} \to \underline{Sets} \to \underline{A}$$

where $[M,-] : \underline{M} \to \underline{S}$ is the hom-functor associated with $M \in \underline{M}$ and $A \otimes$ the left

3) This is the case for the original version of Eilenberg-MacLane.

4) This is the case for the version of Barr-Beck [2] Ch. XI.

5) In the following we abbreviate "$\underline{Sets}$" to "$\underline{S}$" .

adjoint of $[A,-] : \underline{A} \to \underline{S}$, where $A \in \underline{A}$. Recall that $A \otimes : \underline{S} \to \underline{A}$ assigns to a set Λ the Λ-fold sum of A (cf. [24] intro.).[6]

(2) Let $\underline{M}$ be a full subcategory of $\underline{C}$ and $J : \underline{M} \to \underline{C}$ be the inclusion. The (right) Kan extension [7] of a functor $t : \underline{M} \to \underline{A}$ is a functor $E_J(t) : \underline{C} \to \underline{A}$ such that for every functor $S : \underline{C} \to \underline{A}$ there is a bijection $[E_J(t),S] \cong [t,S\cdot J]$, natural in S . Clearly $E_J(t)$ is unique up to equivalence. One can show that $E_J(t)\cdot J \cong t$, in other words $E_J(t)$ is an extension of t . If $E_J(t)$ exists for every t , then $E_J : [\underline{M},\underline{A}] \to [\underline{C},\underline{A}]$ is left adjoint to the restriction $R_J : [\underline{C},\underline{A}] \to [\underline{M},\underline{A}]$. The functor E_J is called the Kan extension. We show below that it exists if either $\underline{M}$ is small and $\underline{A}$ has direct limits or if $\underline{M}$ consists of "projectives" in $\underline{C}$ and $\underline{A}$ has coequalizers. Necessary and sufficient conditions for the existence of E_J can be found in [26].

The notions of generalized representable functors and Kan extensions are closely related.[8] Before we can illustrate this, we have to recall the two basic properties of generalized representable functors which illustrate that they are a useful substitute for the usual hom-functors in an arbitrary functor category $[\underline{M},\underline{A}]$.[9]

(3) Yoneda Lemma. For every functor $t : \underline{M} \to \underline{A}$ there is a bijection

$$[A \otimes [M,-],t] \cong [A,tM]$$

natural in t , $M \in \underline{M}$ and $A \in \underline{A}$.

This is an immediate consequence of the usual Yoneda lemma and the induced adjoint pair $[\underline{M},\underline{A}] \to [\underline{M},\underline{S}]$, $t \rightsquigarrow [-,A]\cdot t$ and $[\underline{M},\underline{S}] \to [\underline{M},\underline{A}]$, $r \rightsquigarrow A \otimes \cdot r$. (Note that

---

6) For the notion of a corepresentable functor $\underline{M} \to \underline{A}$ we refer the reader to [24]. Note that a corepresentable is also covariant. The relationship between representable and corepresentable functors $\underline{M} \to \underline{A}$ is entirely different from the relationship between covariant and contravariant hom-functors.

7) There is the dual notion of a left Kan extension. Here we will only deal with the rightKan extension and call it the Kan extension.

8) The same holds for the left Kan extension and corepresentable functors.

9) In view of this, we abbreviate "generalized representable functor" to "representable functor".

$[A\otimes[M,-],t] \cong [[M,-],[A,-]\cdot t] \cong [A,tM]$.)

(4) Theorem. Every functor $t : \underline{M} \to \underline{A}$ is canonically a direct limit of representable functors. In other words, like the usual Yoneda embedding, the Yoneda functor $\underline{M}^{opp} \times \underline{A} \longrightarrow [\underline{M},\underline{A}]$, $M\times A \rightsquigarrow A\otimes[M,-]$ is dense (cf. [25]) or adequate in the sense of Isbell. Moreover, there is a direct limit representation

$$(5) \qquad t = \varinjlim td\alpha\otimes[r\alpha,-]$$

where α runs through the morphisms of $\underline{M}$ and $d\alpha$ and $r\alpha$ denote the domain and range of α respectively.[10] Note that $\underline{M}$ need not be small. For a proof we refer the reader to [24] 2.15, 2.12 or [27].

(6) Corollary. If $\underline{M}$ is small, then there is for every functor $t : \underline{M} \to \underline{A}$ a canonical epimorphism

$$\varphi(t) : \bigoplus_{\alpha} (td\alpha\otimes[r\alpha,-]) \longrightarrow t$$

which is objectwise split. This can also be established directly using the Yoneda lemma (3). (Note that $\varphi(t)$ restricted on a factor $td\alpha\otimes[r\alpha,-]$ corresponds to $t\alpha$ under the Yoneda isomorphism $[td\alpha\otimes[r\alpha,-],t] \cong [td\alpha,tr\alpha]$.)

(7) From the Yoneda lemma (3) it follows immediately that the Kan extension of a representable functor $A\otimes[M,-] : \underline{M} \to \underline{A}$ is $A\otimes[JM,-] : \underline{C} \to \underline{A}$. This is because $[A\otimes[M,-],S\cdot J] \cong [A,S(JM)] \cong [A\otimes[JM,-],S]$. Since $J : \underline{M} \to \underline{C}$ is full and faithful, $A\otimes[JM,-]$ is an extension of $A\otimes[M,-]$. Since the Kan extension $E_J : [\underline{M},\underline{A}] \to [\underline{C},\underline{A}]$ is a left adjoint, it is obvious from (5) and the above that the Kan extension of a functor $t : \underline{M} \to \underline{A}$ exists iff $\varinjlim td\alpha\otimes[Jr\alpha,-]$ exists and that $E_J(t) = \varinjlim td\alpha\otimes[Jr\alpha,-]$ is valid. (Hence $E_J(t)$ is an extension of t .) This also shows that $E_J : [\underline{M},\underline{A}] \to [\underline{C},\underline{A}]$ exists if $\underline{M}$ is small and $\underline{A}$ has direct limits.[11]

---

10) More precisely, the index category for this representation is the subdivision of $\underline{M}$ in the sense of Kan [14].

11) Kan [14] gave a different proof of this. He constructed $E_J(t) : \underline{C} \to \underline{A}$ objectwise using the category $(\underline{M},C)$ over $C \in \underline{C}$ (also called comma category; its objects are morphisms $M \to C$ with $M \in \underline{M}$). He showed that the direct limit of the functor $(\underline{M},C) \to \underline{A}$, $(M \to C) \rightsquigarrow tM$, is $E_J(t)(C)$. The relation between the two constructions is discussed in [26]. It should be noted that besides small-

(8) If $\Psi : t \to t'$ is an objectwise split natural transformation of functors from $\underline{M}$ to $\underline{A}$ (i.e. $\Psi(M) : tM \to tM'$ admits a section $\sigma(M)$ for every $M \in \underline{M}$, but σ need not be a natural transformation), then every diagram

(9)
$$\begin{array}{ccc} & & t' \\ & \nearrow & \downarrow \Psi \\ A \otimes [M,-] & \longrightarrow & t \end{array}$$

can be completed in the indicated way. This is a consequence of the Yoneda lemma (3). Thus if $\underline{A}$ is abelian,[12] then the representable functors are projective relative to the class $\mathcal{P}$ of short exact sequences in $[\underline{M},\underline{A}]$ which are objectwise split exact. From (6) it follows that there are enough relative projectives in $[\underline{M},\underline{A}]$ if $\underline{M}$ is small.

(10) If however A is a projective in $\underline{A}$, then it follows form the Yoneda lemma that the above diagram can be complete only assuming that $\Psi : t \to t'$ is epimorphic.[12] This shows that $A \otimes [M,-]$ is projective in $[\underline{M},\underline{A}]$, provided $A \in \underline{A}$ is projective. One easily deduces from this and (6) that $[\underline{M},\underline{A}]$ has enough projectives, if $\underline{A}$ does and $\underline{M}$ is small.

By standard homological algebra we obtain the following:

(11) Theorem. Let $J : \underline{M} \to \underline{C}$ be the inclusion of a small subcategory of $\underline{C}$ and $\underline{A}$ be an abelian category with sums. Then the Kan extension $E_J : [\underline{M},\underline{A}] \to [\underline{C},\underline{A}]$ and its left derived functors $\mathcal{P}\text{-}L_*E_J$ relative to $\mathcal{P}$ exist. If either $\underline{A}$ has enough projectives or $\underline{A}$ satisfies Grothendieck's axiom AB4) [13], then the absolute derived functors L_*E_J also exist. In the latter case the functors L_nE_J and $\mathcal{P}\text{-}L_nE_J$ coincide for $n \geq 0$. In the following we denote $L_*E_J(-)$ by $A_*(\ ,-)$ and call it the André homology.

---

ness there are other conditions on $\underline{M}$ which guarantee the existence of the Kan extension $E_J : [\underline{M},\underline{A}] \to [\underline{C},\underline{A}]$ (for instance, if $\underline{M}$ consists of "projectives" of $\underline{C}$). One can also impose conditions which imply the existence of $E_J(t)$ for a particular functor t. This is the case one meets mostly in practice.

12) From now on we always assume that $\underline{A}$ is abelian.

13) I.e. sums are exact in $\underline{A}$.

__Proof.__ The only thing to prove is that the functors $\mathcal{P}\text{-}L_*E_J$ are the absolute derived functors of E_J if $\underline{A}$ is AB4). For every $t \in [\underline{M},\underline{A}]$ the epimorphism $\varphi(t)$ in (6) gives rise to a relative projective resolution

$$(12) \quad P_*(t) : \cdots \bigoplus_\alpha (t_n d\alpha \otimes [r\alpha,-]) \cdots \bigoplus_\alpha (t_1 d\alpha \otimes [r\alpha,-]) \xrightarrow{\varphi(t_1)} \bigoplus_\alpha (t_0 d\alpha \otimes [r\alpha,-]) \xrightarrow{\varphi(t)} t \to 0$$

where $t_0 = t$ and $t_1 = \ker \varphi(t)$, etc.[14] Using (7) and the property AB4) of $\underline{A}$, one can show that a short exact sequence of functors $0 \to t' \to t \to t'' \to 0$ in $[\underline{M},\underline{A}]$ gives rise to a short exact sequence $0 \to E_J P_*(t') \to E_J P_*(t) \to E_J P_*(t'') \to 0$ of chain complexes in $[\underline{C},\underline{A}]$. The long exact homology sequence associated with it makes $\mathcal{P}\text{-}L_*E_J$ into an absolute exact connected sequence of functors. Since $\mathcal{P}\text{-}L_nE_J$ vanishes for $n > o$ on sums[15] $\bigoplus_\alpha (td\alpha \otimes [r\alpha,-])$, it follows by standard homological algebra that $\mathcal{P}\text{-}L_*E_J$ is left universal. In other words, the functors $\mathcal{P}\text{-}L_nE_J$ are the (absolute) left derived functors L_nE_J of the Kan extension $E_J : [\underline{M},\underline{A}] \to [\underline{C},\underline{A}]$.

(13) A comparison with André's homology theory $H_*(\ ,-) : [\underline{M},\underline{A}] \to [\underline{C},\underline{A}]$ in [1] p.14, shows that $H_0(\ ,-)$ agrees with the Kan extension E_J on sums of representable functors.[16] Since both $H_0(\ ,-)$ and E_J are right exact, it follows from the exactness of (12) that they coincide. Since $H_n(\ ,-) : [\underline{M},\underline{A}] \to [\underline{C},\underline{A}]$ vanishes for $n > o$ on sums of representable functors, it follows by standard homological algebra that the functors $H_*(\ ,-)$ are the left derived functors of E_J. Hence $H_*(\ ,-) \cong A_*(\ ,-) = L_*E_J(-)$ is valid. It may seem at first that this is "by chance" because André constructs H_* in an entirely different way (cf. [1] p. 3). This however is not so. Recall that he associates with every functor $t : \underline{M} \to \underline{A}$ a complex of functors $C_*(t) : \underline{C} \to \underline{A}$ and defines $H_n(-,t)$ to be the n-th homology of $C_*(t)$ (cf. [1] p.3). It is not difficult to show that the restriction of $C_n(t)$ on $\underline{M}$ is a sum of representable functors and that the Kan extension of $C_n(t) \cdot J$ is $C_n(t)$. Moreover, $C_*(t) \cdot J$ is an objectwise split exact resolution of t. Thus André's construction

---

14) $P_*(t)$ denotes the non-augmented complex, i.e. without t.

15) Recall that $td\alpha \otimes [r\alpha,-]$ is a relative projective in $[\underline{M},\underline{A}]$.

16) In the notation of André, $\underline{C}$ should be replaced by $\underline{N}$. Note that in view of (7), a "foncteur élémentaire" of André is the Kan extension of a sum of representable functors $\underline{M} \to \underline{A}$.

turns out to be the standard procedure in homological algebra to compute the left derived functors E_J . Namely: choose an E_J-acyclic [17] resolution of t , apply E_J and take homology. The same is true for his computational device [1] prop. 1.5 (i.e. the restriction of the complex S_* on $\underline{M}$ is an E_J-acyclic resolution of $S\cdot J$ and $E_J(S_*\cdot J) = S_*$ is valid).[18] We now list some properties of $A_*(,-)$ which in part generalize results of André [1]. They are consequences of (11), (12) and the nice behaviour of the Kan extension on representable functors.

(14) Theorem.

(a) For every functor t the composite $A_p(J-,t) : \underline{M} \to \underline{C} \to \underline{A}$ is zero for $p > o$.

(b) Assume that $\underline{A}$ is an AB5) category and let C be an object in $\underline{C}$ such that the comma category $(\underline{M},C)$ is directed.[19] Then $A_p(C,t)$ vanishes for $p > o$.

(c) Assume moreover that for every $M \in \underline{M}$ the hom-functor $[JM,-] : \underline{C} \to \underline{S}$ preserves direct limits over directed index categories. Then $A_*(-,t) : \underline{C} \to \underline{A}$ also preserves direct limits. (In most examples this assumption is satisfied if the objects of $\underline{M}$ are finitely generated.)

If $\underline{M}$ has finite sums, it follows from a) and c) that $A_p(-,t) : \underline{C} \to \underline{A}$ vanishes on arbitrary sums $\bigoplus_\nu M_\nu$, where $M_\nu \in \underline{M}$. As for applications, it is of great importance to establish this without assuming that $\underline{A}$ is AB5) . We will sketch later on how this can be done.

The properties a) - c) are straightforward consequences of (12), (11), (7), footnote 11) and the fact that in an AB5) category direct limits over directed index categories are exact.

(15) A change of models gives rise to a spectral sequence (cf. [1] prop. 8.1). For this, let $\underline{M}'$ be a small full subcategory of $\underline{C}$ containing $\underline{M}$. Denote by $J' : \underline{M}' \to \underline{C}$

---

17) A functor is called E_J-acyclic if $L_n E_J : [\underline{M},\underline{A}] \to [\underline{C},\underline{A}]$ vanishes on it for $n > o$.

18) We will show later that this computational method is closely related with acyclic models.

19) A category $\underline{D}$ is called directed if for every pair D, D' of objects in $\underline{D}$ there is a $D'' \in \underline{D}$ together with morphisms $D \to D''$, $D' \to D''$ and if for a pair of morphisms $\lambda, \mu : D_0 \rightrightarrows D_1$ there is a morphism $\gamma : D_1 \to D_2$ such that $\gamma\lambda = \gamma\mu$

and $J'' : \underline{M} \to \underline{M}'$ the inclusions and by A'_* and A''_* the associated André homologies. Since $J = J' \cdot J''$, it follows that $E_J : [\underline{M},\underline{A}] \to [\underline{C},\underline{A}]$ is the composite of $E_{J''} : [\underline{M},\underline{A}] \to [\underline{M}',\underline{A}]$ with $E_{J'} : [\underline{M}',\underline{A}] \to [\underline{C},\underline{A}]$. Thus by standard homological algebra there is for every functor $t : \underline{M} \to \underline{A}$ a spectral sequence

(16) $$A'_p(\; ,A''_q(-,t)) \Rightarrow A_{p+q}(-,t)$$

provided $\underline{A}$ is AB4) or $\underline{A}$ has enough projectives (cf. [13] 2.4.1). One only has to verify that $E_{J''}$ takes E_J-acyclic objects into $E_{J'}$-acyclic objects. But this is obvious from (7), because the Kan extension of a representable functor is again representable. The same holds for projective representable functors (cf. (10)).[20]
The "Hochschild-Serre" spectral sequence of André [1] p. 33 can be obtained in the same way.

Likewise a composed coefficient functor gives rise to a universal coefficient spectral sequence.

(17) Theorem. Let $t : \underline{M} \to \underline{A}$ and $F : \underline{A} \to \underline{A}'$ be functors, where $\underline{A}$ and $\underline{A}'$ are either AB4) categories or have enough projectives. Assume that the left derived functors L_*F exist and that F has a right adjoint. Then there is a spectral sequence

$$L_pF \cdot A_q(-,t) \Rightarrow A_{p+q}(-,F \cdot t)$$

provided the values of t are F-acyclic (i.e. $L_qF(tM) = 0$ for $q > o$).

(18) Corollary. If for every $p > o$ the functor $A_p(-,t) : \underline{C} \to \underline{A}$ vanishes on an object $C \in \underline{C}$, then $A_*(C,F \cdot t) \cong L_*F(A_o(C,t))$. This gives rise to an infinite coproduct formula, provided $A_o(-,t) : \underline{C} \to \underline{A}$ is sum-preserving. For, let $C = \bigoplus_\nu C_\nu$ be an arbitrary sum with the property $A_p(C_\nu,t) = 0$ for $p > o$. Then the canonical map

(19) $$\bigoplus_\nu A_*(C_\nu,F \cdot t) \xrightarrow{\cong} A_*(\bigoplus_\nu C_\nu,F \cdot t)$$

is an isomorphism because $\bigoplus_\nu L_*F(A_o(C_\nu,F \cdot t)) \cong L_*F(\bigoplus_\nu A_o(C_\nu,F \cdot t)) \cong L_*F(A_o(\bigoplus_\nu C_\nu,F \cdot t))$

20) The assumption that $\underline{M}'$ is small can be replaced by the following: The Kan extension $E_{J'} : [\underline{M}',\underline{A}] \to [\underline{C},\underline{A}]$ and its left derived functors $L_*E_{J'}$ exist and $L_nE_{J'}$ vanishes on sums of representable functors for $n > o$.

holds.[21]

(20) <u>Corollary.</u> Assume that $\underline{A}$ is AB5) (but not $\underline{A}'$). Then every finite coproduct formula for $A_*(-,t) : \underline{C} \to \underline{A}$ gives rise to an infinite coproduct formula for $A_*(-,F\cdot t) : \underline{C} \to \underline{A}'$. In other words, $A_*(\bigoplus_\nu C_\nu, F\cdot t) \cong \bigoplus_\nu A_*(C_\nu, F\cdot t)$ holds if (14c) holds and $A_*(\bigoplus_i C_{\nu_i}, t) \cong \bigoplus_i A_*(C_{\nu_i}, t)$ is valid for every finite subsum $\bigoplus_i C_{\nu_i}$ of $\bigoplus_\nu C_\nu$.[22]

<u>Proof of (17) and (20).</u> (Sketch) By $E_C : [\underline{C},\underline{A}] \to \underline{A}$ and $E'_C : [\underline{C},\underline{A}] \to \underline{A}'$ we denote the evaluation functors associated with $C \in \underline{C}$. The assumptions on F and t imply that the diagram

$$\begin{array}{ccccc} & & \underline{A} & & \\ & \nearrow^{E_C \cdot E_J} & & \searrow^{F} & \\ [\underline{M},\underline{A}] & \xrightarrow{E'_J(F\cdot -)} & [\underline{C},\underline{A}'] & \xrightarrow{E'_C} & \underline{A}' \end{array}$$

is commutative. The derived functors of $E'_C E'_J(F\cdot -) : [\underline{M},\underline{A}] \to \underline{A}'$ can be identified with $t \rightsquigarrow A_*(C,F\cdot t)$. As above in (15) and (16), the spectral sequence arises from the decomposition of $E'_C E'_J(F\cdot -)$ into $E_C \cdot E_J$ and F . The infinite coproduct formula for $A_*(-,t) : \underline{C} \to \underline{A}$ can be established by means of (14c). Thus it also holds for the E_2-term of the spectral sequence (17). One can show that the direct sum decomposition of the E_2-term is compatible with the differentials and the associated filtration of the spectral sequence. In this way one obtains an infinite coproduct formula for $A_*(-,F\cdot t) : \underline{C} \to \underline{A}'$.

(21) An abelian interpretation of André's non-abelian resolutions and neighborhoods is contained in a forthcoming paper of U. Oberst [20]. We include here a somewhat improved version of this interpretation and use it to solve a central problem which remained open in Barr-Beck [2] Ch. X. For this we briefly review the tensor product $\otimes$ between functors, which was investigated by D. Buchsbaum [5], J. Fisher [10],

---

21) This applies to the categories of groups and semi groups and yields infinite coproduct formulas for homology and cohomology of groups and semi groups without conditions. For it can be shown that they coincide with A_* and A^* if $t = \mathrm{Diff}$ and F is tensoring with or homming into some module (cf. [2] Ch. 1 and Ch. X).

22) This applies to all finite coproduct theormes established in Barr-Beck [2] Ch.7 and André [1] with t and F as in footnote 21).

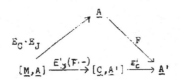

P. Freyd [11], D. Kan [14], U. Oberst [19] [20], C. Watts [28], N. Yoneda [29] and the author. The tensor product is a bifunctor

(22) $$\underline{\otimes} : [\underline{M}^{opp}, \underline{Ab.Gr.}] \times [\underline{M}, \underline{A}] \longrightarrow \underline{A}$$

defined by the following universal property. For every $s \in [\underline{M}^{opp}, \underline{Ab.Gr.}]$, $t \in [\underline{M}, \underline{A}]$ and $A \in \underline{A}$ there is an isomorphism

(23) $$[s \underline{\otimes} t, A] \cong [s, [t-, A]]$$

natural in s, t and A. It can be constructed like the tensor product between Λ-modules, namely stepwise: 1) $\Lambda \otimes Y = Y$; 2) $(\bigoplus_\nu \Lambda_\nu) \otimes Y = \bigoplus_\nu Y_\nu$, where $\Lambda_\nu = \Lambda$ and $Y_\nu = Y$; 3) for an arbitrary module X choose a presentation $\bigoplus_\nu \Lambda_\nu \to \bigoplus_\mu \Lambda_\mu \to X \to 0$ and define $X \otimes Y$ to be the cokernel of the induced map $\bigoplus_\nu Y_\nu \to \bigoplus_\mu Y_\mu$, where $Y_\nu = Y = Y_\mu$.

The rôle of Λ is played by the family of contravariant representable functors $\mathbf{Z} \otimes [-, M] : \underline{M}^{opp} \to \underline{Ab.Gr.}$, where $M \in \underline{M}$ and $\mathbf{Z}$ denotes the integers.23) Thus we define

(24) $$(\mathbf{Z} \otimes [-, M]) \underline{\otimes} t = tM$$

and continue as above. The universal property (23) follows from the Yoneda lemma (3) in the following way: $[\mathbf{Z} \otimes [-, M] \underline{\otimes} t, A] \cong [tM, A] \cong [\mathbf{Z}, [tM, A]] \cong [\mathbf{Z} \otimes [-, M], [t-, A]]$. One can show by means of (24), (3) and the classical argument about balanced bifunctors that the derived functors of $\underline{\otimes} t : [\underline{M}^{opp}, \underline{Ab.Gr.}] \to \underline{A}$ and $s \underline{\otimes} : [\underline{M}, \underline{A}] \to \underline{A}$ have the property $(L_* s \underline{\otimes})(t) \cong L_*(\underline{\otimes} t)(s)$, provided $\underline{A}$ is AB4) (resp. AB5)) and the values of s are free (resp. torsion free) abelian groups. Under these conditions the notion $\mathrm{Tor}_*(s, t)$ makes sense and has its usual properties, e.g. $\mathrm{Tor}_*(s, t)$ can be computed by projective or flat resolutions in either variable. We remark without proof that every representable functor $A \otimes [M, -] : \underline{M} \to \underline{A}$ is flat. It should be noted that for this and the following (up to (31)) one cannot replace the condition AB4) by the assumption that $\underline{A}$ has enough projectives.

(25) U. Oberst [20] considers the class $\mathcal{P}$ of short exact sequences in $[\underline{M}^{opp}, \underline{Ab.Gr.}]$

---

23) Note that the functors $\mathbf{Z} \otimes [-, M]$, $M \in \underline{M}$ are projective and form a generating family in $[\underline{M}^{opp}, \underline{Ab.Gr.}]$. This follows easily from the Yoneda lemma (3) and (10).

and $[\underline{M},\underline{A}]$ which are objectwise split exact. He shows that the derived functors of s and t relative to $\mathcal{P}$ have the property $\mathcal{P}\text{-}(L_*s\otimes)(t) \cong \mathcal{P}\text{-}(L_*\otimes t)(s)$ without any conditions on s and t. Thus the notion $\mathcal{P}\text{-}\operatorname{Tor}_*(s,t)$ makes sense. With every object $C \in \underline{C}$ there is associated a functor $\mathbb{Z}\otimes[J-,C] : \underline{M}^{opp} \to \underline{Ab}.\underline{Gr}.$ the values of which are free abelian groups. (Recall that $J : \underline{M} \to \underline{C}$ is the inclusion.) He establishes an isomorphism

$$(26) \qquad A_*(C,t) = \mathcal{P}\text{-}\operatorname{Tor}_*(\mathbb{Z}\otimes[J-,C],t)$$

for every functor $t : \underline{M} \to \underline{A}$ and points out that a non-abelian resolution of C in the sense of André [1] p.17 is a relative projective resolution of $\mathbb{Z}\otimes[J-,C]$. Thus André's result that $A_*(C,t)$ can be computed either by the complex $C_*(t) : \underline{C} \to \underline{A}$ evaluated at C (cf. (13) and [1] p.3) or a non-abelian resolution of C turns out to be a special case of the well known fact that $\mathcal{P}\text{-}\operatorname{Tor}_*(-,-)$ can be computed by a relative projective resolution of either variable. U. Oberst also observes that a neighborhood ("voisinage") of C (cf. [1] p.38) gives rise to a relative projective of $\mathbb{Z}\otimes[J-,C]$. Therefore it is obvious that $A_*(C,t)$ can also be computed by means of neighborhoods.

(27) The notion of relative $\mathcal{P}\text{-}\operatorname{Tor}_*(-,-)$ is somewhat difficult to handle in practice. For instance, the spectral sequences (16) and (17) and the coproduct formulas (19) and (20) cannot be obtained with it because of the misbehaviour of the Kan exten- on relative projective resolutions. Moreover, André's computational method [1] prop. 1.8 (cf. also (13)) cannot be explained by means of $\mathcal{P}\text{-}\operatorname{Tor}_*(\mathbb{Z}\otimes[J-,C],t)$ because the resolution of the functor t in question need not be relative projective. It appears that our notion of an absolute Tor_* does not have this disadvantage. The basic reason for the difference lies in the fact that relative projective resolutions of s and t are always flat resolutions of s and t but the converse is not true.[24] The properties (26) etc. of the relative $\mathcal{P}\text{-}\operatorname{Tor}_*$ can be established similarly for the absolute Tor_* using the techniques of U. Oberst [20]. We now We now sketch a different way to obtain these.

---

24) Note that a projective resolution of $\mathbb{Z}\otimes[J-,C]$ is also relative projective and vice versa.

The fundamental relationship between $\underline{\otimes}$ and the Kan extension $E_J : [\underline{M},\underline{A}] \to [\underline{C},\underline{A}]$ is given by the equation

$$(28) \qquad (Z\otimes[J-,C])\underline{\otimes}t = E_J(t)(C)$$

where t and C are arbitrary objects of $[\underline{M},\underline{A}]$ and $\underline{C}$ respectively. To see this, let $[J-,C] = \varinjlim [-,M_\nu]$ be the canonical representation of $[J-,C] : \underline{M}^{opp} \to \underline{S}$ as a direct limit of contravariant hom-functors (cf. [24] 1.10). Note that the index category for this representation is isomorphic with the comma category $(\underline{M},C)$ (cf. footnote 11)). Hence $Z\otimes[J-,C] = \varinjlim Z\otimes[-,M_\nu]$ and it follows from (24) and Kan's construction (cf. footnote 11)) that $(Z\otimes[J-,C])\underline{\otimes}t = \varinjlim tM_\nu = E_J(t)(C)$. Since $A_*(-,t)$ is the homology of the complex $E_J P_*(t)$, where $P_*(t)$ is the flat resolution (12) of t , it follows from (28) that

$$(29) \qquad A_*(C,t) \cong Tor_*(Z\otimes[J-,C],t)$$

Thus $A_*(C,t)$ can be computed either by projective resolutions of $Z\otimes[J-,C]$ (e.g. non-abelian resolutions and neighborhoods [25]) or flat resolutions of t (e.g. $P_*(t)$ or André's resolutions $C_*(t)\cdot J$ and S_* , cf. (13)).

The above methods prove very useful in establishing the theorem below which is basic for many applications.

(30) Theorem. Let $\underline{M}$ be a full small subcategory of a category $\underline{C}$ which has sums. $\underline{M}$ need not have finite sums. However, if a sum $\bigoplus_i M_i \in \underline{C}$ is already in $\underline{M}$, it is assumed that every subsum of $\bigoplus_i M_i$ is also in $\underline{M}$. Moreover, assume that for every pair of objects $M \in \underline{M}$ and $\bigoplus_\nu M_\nu \in \underline{C}$ every morphism $M \to \bigoplus M_\nu$ factors through a subsum belonging to $\underline{M}$, where $M_\nu \in \underline{M}$. Let $\underline{A}$ be an AB4) category and $t : \underline{M} \to \underline{A}$ a sum-preserving functor.[26]

---

25) Further examples are provided by the simplicial resolutions of Barr-Beck [2] Ch.5, the projective simplicial resolutions of type (X,o) of Dold-Puppe [8] and the pseudo-simplicial resolutions of Tierney-Vogel [22]. A corollary of this is that the André (co)homology coincides with the theories developed by Barr-Beck [2], Dold-Puppe [8] and Tierney-Vogel [22] when $\underline{C}$ and $\underline{M}$ are defined appropriately.
Note that there are many more projective resolutions of $Z\otimes[J-,C]$ than the ones described so far (for instance, the resolutions used in the proof of (30)).

26) The meaning is that t has to preserve the sums which exist in $\underline{M}$.

Then for $p > o$ the functor $A_p(-,t) : \underline{C} \to \underline{A}$ vanishes on arbitrary sums $\oplus_{\mu} M_{\mu}$, where $M_{\mu} \in \underline{M}$. 27)

The idea of the proof is to construct a projective resolution of $Z \otimes [J-, \oplus_{\mu} M_{\mu}]$ which remains exact when tensored with $\otimes t : [\underline{M}^{opp}, \underline{Ab}.\underline{Gr}.] \to \underline{A}$. It is a subcomplex of the complex $Z \otimes \bar{\underline{M}}_*(-, \oplus_{\mu} M_{\mu})$ André associated with the object $\oplus_{\mu} M_{\mu} \in \underline{C}$ and the full subcategory $\bar{\underline{M}}$ of $\underline{M}$ consisting of those subsums of $\oplus_{\mu} M_{\mu}$ which belong to $\underline{M}$ (cf. [1] p.38). In dimension n the resolution consists of a sum $\oplus(Z \otimes [-, M_n])$ with $M_n \in \bar{\underline{M}}$; more precisely, for every ascending chain of subsums $M_n \to M_{n-1} \to \dots M_o \to \oplus_{\mu} M_{\mu}$ in $\oplus_{\mu} M_{\mu}$ such that $M_i \in \bar{\underline{M}}$ for $n \geq i \geq o$ there is a summand $Z \otimes [-, M_n]$. The conditions on the inclusion $J : \underline{M} \to \underline{C}$ imply that the subcomplex evaluated at each $M \in \underline{M}$ has a contraction and hence it is a resolution of $Z \otimes [J-, \oplus_{\mu} M_{\mu}]$. The condition on t implies that the resolution, when tensored with $\otimes t$, also has a contraction. For deatils see [27].

(31) <u>Corollary.</u> Let $\underline{M}'$ be the full subcategory of $\underline{C}$ consisting of sums of objects in $\underline{M}$. Assume that the Kan extension $E_{J'} : [\underline{M}', \underline{A}] \to [\underline{C}, \underline{A}]$ and its left derived functors $L_* E_{J'} = A'_*$ exist and that $L_n E_{J'}$ vanishes on representable functors for $n > o$. Then by (30) the spectral sequence (16) collapses and one obtains an isomorphism

$$A_*(-,t) \xrightarrow{\cong} A'_*(-, A''_o(-,t))$$

where $t : \underline{M} \to \underline{A}$ is a functor as in (30) and $A''_o(-,t) : \underline{M}' \to \underline{A}$ is its Kan extension on $\underline{M}'$. 28)

The value of (31) lies in the fact that $A'_* : \underline{C} \to \underline{A}$ can be identified with the homology associated with a certain cotriple in $\underline{C}$ (the model induced cotriple (cf. (42), (43)). In this way every André homology can be realized as a cotriple homology and all information about the latter carries over to the former and vice versa.

---

27) In an earlier version of this theorem I assumed in addition that a certain s.s. set satisfies the Kan condition, which is the case in the examples I know. M. André then pointed out to me that this condition is redundant. This led to some simplifications in my original proof. He also found a different proof which is based on the methods he developed in [1].

28) If $\underline{A}$ is AB5), then the theorem is also true if $\underline{M}'$ is an arbitrary full subcategory of $\underline{C}$ such that for every $M' \in \underline{M}'$ the associated category $(\underline{M}, M')$ is directed. This follows easily from (14b) and (16).

It has become apparent in several places that the smallness of $\underline{M}$ is an unpleasant restriction which should be removed. André did this by requiring that every $C \in \underline{C}$ has a neighborhood in $\underline{M}$. Another way of expressing the same condition is to assume that every functor $Z \otimes [J-,C] : \underline{M}^{opp} \to \underline{Ab.Gr}$. admits a projection resolution. It is then clear from the above that there is an exact connected sequence of functors $A_*(\ ,-) : [\underline{M},\underline{A}] \to [\underline{C},\underline{A}]$ with the properties $A_0(\ ,-) = E_J$ and $A_n(-,A \otimes [-,M]) = 0$ for $n > 0$. Since $\underline{M}$ is not small, not every functor $t : \underline{M} \to \underline{A}$ need be a quotient of a sum of representable functors and one cannot automatically conclude that $L_*E_J = A_*$. In many examples this is however the case, e.g. if $\underline{M}$ consists of the G-projectives of a cotriple G in $\underline{C}$.

(32) So far we have only dealt with André homology with respect to an inclusion $J : \underline{M} \to \underline{C}$ and not with the homology associated with a cotriple G in $\underline{C}$ (for the definition of a cotriple we refer to Barr-Beck [2] intro.). One reason for this is that the corresponding model category $\underline{M}$ is not small. Another is that the presence of a cotriple is a more special situation in which theorems often hold under weaker conditions and proofs are easier. The additional information is due to the simple behavior of the Kan extension on functors of the form $\underline{M} \xrightarrow{G} \underline{M} \xrightarrow{t} \underline{A}$, where G is the restriction of the cotriple on $\underline{M}$. We now outline how our approach works for cotriple homology.

(33) Let G be a cotriple in $\underline{C}$ and denote by $\underline{M}$ any full subcategory of $\underline{C}$, the objects of which are G-projectives and include every GC, where $C \in \underline{C}$. (Recall that an object $X \in \underline{C}$ is called G-projective if $\varepsilon(X) : GX \to X$ admits a section, where $\varepsilon : G \to id_{\underline{C}}$ is the co-unit of the cotriple. The objects GC are called free.) With every functor $t : \underline{C} \to \underline{A}$ Barr-Beck [2] associate the cotriple derived functors $H_*(-,t)_G : \underline{C} \to \underline{A}$, also called cotriple homology with coefficient functor t. Their construction of $H_*(-,t)_G$ only involves the values of t on the free objects of $\underline{C}$. Thus $H_*(-,t)_G$ is also well defined when t is only defined on $\underline{M}$.

(34) __Theorem.__ The Kan extension $E_J : [\underline{M},\underline{A}] \to [\underline{C},\underline{A}]$ exists. It assigns to a functor $t : \underline{M} \to \underline{A}$ the zeroth cotriple derived functor $H_0(-,t)_G : \underline{C} \to \underline{A}$. In particular

$E_J(t\cdot G) = t\cdot G$ is valid. (Note that $\underline{A}$ need not be AB3) or AB4) for this.)

<u>Proof.</u> According to (2), we have to show that for every $S : \underline{C} \to \underline{A}$ the restriction map $[H_o(-,t)_G, S] \to [t, S\cdot J]$ is a bijection. We limit ourselves to giving a map in the opposite direction and leave it to the reader to check that they are inverse to eachother. A natural transformation $\varphi : t \to S\cdot J$ gives rise to a diagram

$$\begin{array}{ccccc}
tG^2C & \overset{t\varepsilon(GC)}{\underset{tG(\varepsilon C)}{\rightrightarrows}} & tGC & \longrightarrow & H_o(C,t)_G \\
\downarrow{\scriptstyle \varphi(G^2C)} & & \downarrow{\scriptstyle \varphi(GC)} & & \\
SG^2C & \overset{S\varepsilon(GC)}{\underset{SG(\varepsilon C)}{\rightrightarrows}} & SGC & \xrightarrow{S\varepsilon(C)} & SC
\end{array}$$

for every $C \in \underline{C}$. The top row is by construction of $H_o(C,t)_G$ a coequalizer. Thus there is a unique morphism $H_o(C,t)_G \to SC$ which makes the diagram commutative. In this way one obtains a natural transformation $H_o(-,t)_G \to S$. $\square$

The properties established in [2] Ch. 1 imply that $H_*(\ ,-)_G : [\underline{M},\underline{A}] \to [\underline{C},\underline{A}]$ is an exact connected sequence of functors. Since $H_n(-\ ,t\cdot G) = 0$ for every functor $t : \underline{M} \to \underline{A}$ and since the canonical natural transformation $tG \to t$ is an (objectwise split) epimorphism, we obtain by standard homological algebra the following:

(35) <u>Theorem.</u> The left derived functors of the Kan extension $E_J : [\underline{M},\underline{A}] \to [\underline{C},\underline{A}]$ exist and $L_*E_J(-) \cong H_*(\ ,-)_G$ is valid. Moreover $\mathcal{P}\text{-}L_*E_J(-) \cong H_*(\ ,-)_G$ holds, where $\mathcal{P}$ denotes the class of short exact sequences in $[\underline{M},\underline{A}]$ which are objectwise split exact. We remark without proof that for $n > o$ $H_n(\ ,-)_G$ vanishes on sums of representable functors.

(36) <u>Corollary.</u> The cotriple homology depends only on the G-projectives.[29] In particular, two cotriples G and G' in $\underline{C}$ give rise to the same homology if their

29) In many cases the cotriple homology depends only on the finitely generated G-projectives. An object $X \in \underline{C}$ is called finitely generated if the hom-functor $[X,-] : \underline{C} \to \underline{S}$ preserves filtered unions.

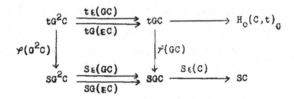

projectives coincide. One can show that the converse is also true.

(37) It is obvious from the above that the axioms of Barr-Beck [2] Ch.3 for $H_*(\ ,-)_G$ are the usual acyclicity criterion for establishing the universal property of an exact connected sequence of functors. As in (13) the construction of $H_*(-,t)_G$ in [2] by means of a s.s. resolution $tG_* : \underline{C} \to \underline{A}$ is actually the standard procedure in homological algebra. This is because the restriction of tG_* on $\underline{M}$ is an E_J-acyclic resolution of t and because $E_J(tG_*\cdot J) = tG_*$ holds.

(38) It also follows from (35) that cotriple homology $H_*(-,t)_G$ and André homology $A_*(-,t)$ coincide, provided the models for the latter are $\underline{M}$. One might be tempted to deduce this from the first half of (35), but apparently it can only be obtained from the second half. The reason is a set theoretical difficulty. For details we refer to [27].

From this it is obvious that the properties previously established for the André homology carry over to the cotriple homology. We list below some useful modifications which result from direct proofs of these properties.

(39) The assumption in (14c), which is seldom present in examples when $\underline{M}$ is not small, can be replaced by the following: The cotriple $G : \underline{C} \to \underline{M}$ and the functor $t : \underline{M} \to \underline{A}$ preserve directed direct limits.

In (17) - (20) the functor F need not have a right adjoint. It suffices instead that F is right exact. For (20) G has to preserve directed direct limits.

From (35), (16) and footnote 20) we obtain for a small subcategory $\bar{\underline{M}}$ of $\underline{M}$ a spectral sequence

$$(40) \qquad H_p(-,\bar{A}_q(-,t))_G \Longrightarrow A_{p+q}(-,t)$$

where $\underline{M}$ is as in (33), and $\bar{A}_*$ and A_* denote the left derived functors of the Kan extensions $[\bar{\underline{M}},\underline{A}] \to [\underline{M},\underline{A}]$ and $[\bar{\underline{M}},\underline{A}] \to [\underline{C},\underline{A}]$ respectively.

The tensor product $\otimes : [\underline{M}^{opp},\underline{Ab}.\underline{Gr}.] \times [\underline{M},\underline{A}] \longrightarrow \underline{A}$, $s,t \rightsquigarrow s\otimes t$ is defined as in (23) but may not exist for every $s \in [\underline{M}^{opp},\underline{Ab}.\underline{Gr}.]$. However (28) and

(41) $$\mathrm{Tor}_*(Z\otimes[J-,C],t) \cong H_*(C,t)_G \cong \mathcal{P}-\mathrm{Tor}_*(Z\otimes[J-,C],t)$$

hold. The first isomorphism shows that $H_*(C,t)_G$ can be computed by either a projective resolution of $Z\otimes[J-,C]$ or an E_J-acyclic resolution of t. The former is a generalization of the main result in [2] 5.1, because a G-resolution is a projective resolution of $Z\otimes[J-,C]$; the latter generalizes the acyclic model argument in [4] (cf. also (53)).

(42) With every small subcategory $\underline{\bar{M}}$ of a category $\underline{C}$ there is associated a cotriple G, called the model induced cotriple (cf. [2] Ch. 10). Recall that its functor part $G : \underline{C} \to \underline{C}$ assigns to an object $C \in \underline{C}$ the sum $\bigoplus_{f\to C} df$, indexed by morphisms $f : df \to C$ the domain df of which belongs to $\underline{\bar{M}}$. The co-unit $\varepsilon(C) : \bigoplus df \to C$ restricted on a summand df is $f : df \to C$. The theorems (35) and (31) enable us to compare the André homology $A_*(-,-)$ of the inclusion $\underline{\bar{M}} \to \underline{C}$ with the homology of the model induced cotriple G in $\underline{C}$. Since every sum $\bigoplus_\nu \bar{M}_\nu$ of objects $\bar{M}_\nu \in \underline{\bar{M}}$ is G-projective, we obtain the following:

(43) <u>Theorem.</u> Assume that the inclusion $\underline{\bar{M}} \to \underline{C}$ satisfies the conditions in (30). Then for every sum preserving functor $t : \underline{\bar{M}} \to \underline{A}$ the canonical map

$$A_*(-,t) \xrightarrow{\cong} H_*(-,A_0(-,t))_G$$

is an isomorphism, provided $\underline{A}$ is an AB4) category.
If t is contravariant and takes sums into products, we obtain likewise for cohomology

$$A^*(-,t) \xleftarrow{\cong} H^*(-,A^0(-,t))_G$$

provided $\underline{A}$ is an AB4)* category.
The theorem asserts that André (co)homology can be realized under rather weak conditions as (co)homology of a cotriple, even of a model induced cotriple. In this way, the considerations in Barr-Beck [2] 7.1,7.2 (coproduct formulas), [2] 8.1 (homology sequence of a map) and [2] 9.1, 9.2 (Mayer Vietoris) also apply to André (co)homology. Moreover, the fact that cotriple cohomology tends to classify extensions (cf. [3]) carries over to André cohomology. We illustrate the use of this realization with some examples.

(44) Examples.

a) Let $\underline{C}$ be a category of algebras with rank$(C) = \alpha$ in the sense of Linton [16]. Recall that if $\alpha = \aleph_0$ then $\underline{C}$ is a category of universal algebras in the classical sense (cf. Lawvere [15]). By means of (43) and (36) one can show that (co)homology of the free cotriple in $\underline{C}$ coincides with the André (co)homology associated with the inclusion $\underline{\bar{M}} \to \underline{C}$, where the objects of $\underline{\bar{M}}$ are free algebras on less than α generators.

b) Let $\underline{C} = \underline{Ab}.\underline{Gr}.$ and let $\underline{\bar{M}}$ be the subcategory of finitely generated abelian groups. Using (43), one can show that the first André cohomology group $A^1(C,[-,Y])$ is isomorphic to the group of pure extensions of $C \in \underline{Ab}.\underline{Gr}.$ with kernel $Y \in \underline{Ab}.\underline{Gr}.$ in the sense of Harrison. The same holds if $\underline{C}$ is a category of Λ-modules, where Λ is a ring with a unit.

c) Let $\underline{C} = \Lambda$-algebras and let $\underline{\bar{M}}$ be the subcategory of finitely generated tensor algebras. Let C be a Λ-algebra and Y be a Λ-bimodule. Then $A^1(C,[-,Y])$ classifies singular extensions $E \to C$ with kernel Y such that the underlying Λ-module extension is pure in the sense of Harrison (cf. b)).[30]

The cases b) and c) set the tone for a long list of similar examples, which indicate that Harrison's theory of pure group extensions can be considerably generalized.

We conclude the summary by establishing a relationship between acyclic models and elementary homological algebra. The generalization of acyclic models [31] in [2] Ch.XI and the fact that the cotriple derived functors are the left derived functors of the Kan extension $E_J : [\underline{M},\underline{A}] \to [\underline{C},\underline{A}]$ (cf. (35)) make it fairly obvious that acyclic models and Kan extensions are closely related. Roughly speaking, the technique of acyclic models (à la Eilenberg-MacLane) turns out to be the standard procedure in homological algebra to compute the left derived functors of the Kan extension by means of projectives.[32] Eilenberg-MacLane showed that two augmented complexes $T_* \to T_{-1}$

---

30) I am indebted to M. Barr for correnting an error I had made in this example.

31) The method of acyclic models was introduced by Eilenberg-MacLane. Barr-Beck [4] gave a different version by means of cotriples.

32) In a recent paper of Dold, MacLane, Oberst [7] a relationship between acyclic

and $\bar{T}_* \to \bar{T}_{-1}$ of functors from $\underline{C}$ to $\underline{A}$ with the property $T_{-1} \cong \bar{T}_{-1}$ are homotopically equivalent, provided they are acyclic on the models $\underline{M}$ and the functors $T_n, \bar{T}_n$ are "representable" for $n \geq 0$.[33)] The homotopy equivalence between T_* and $\bar{T}_*$ can be obtained in the following way [34)]: There are projective resolutions t_* and $\bar{t}_*$ of $T_{-1} \cdot J$ and $\bar{T}_{-1} \cdot J$ which are mapped onto T_* and $\bar{T}_*$ by the Kan extension $E_J ; [\underline{M},\underline{A}] \to [\underline{C},\underline{A}]$. By standard homological algebra t_* and $\bar{t}_*$ are homotopically equivalent. Hence so are T_* and $\bar{T}_*$.

(45) In more detail, let $J : \underline{M} \to \underline{C}$ be the inclusion of a full small subcategory (referred to as models) and let $\underline{A}$ be an abelian category with sums and enough projectives. As shown in (10), a representable functor $P \otimes [M,-] : \underline{M} \to \underline{A}$ is projective iff P is projective in $\underline{A}$. Choose for every functor $t : \underline{M} \to \underline{A}$ and every $M \in \underline{M}$ an epimorphism $P_M \to tM$, where P_M is projective.[35)] In view of the Yoneda isomorphism $[P_M \otimes [M,-],t] \cong [P_M,tM]$ (cf. (3)), the family of epimorphisms determines a natural transformation $\bigoplus_{M \in \underline{M}} (P_M \otimes [M,-]) \to t$ which can easily be shown to be epimorphic. Thus t is projective in $[\underline{M},\underline{A}]$ iff it is a direct summand of a sum of projective representables. If t is the restriction of a functor $T : \underline{C} \to \underline{A}$, then the family and the Yoneda isomorphisms $[P_M \otimes [JM,-],T] \cong [P_M, TJM]$ determine also a natural transformation $\varphi(T) : \bigoplus_{M \in \underline{M}} (P_M \otimes [JM,-]) \to T$.

(46) <u>Theorem.</u> A functor $T : \underline{C} \to \underline{A}$ is "representable" iff $T \cdot J$ is projective in

---

models and projective classes was also pointed out. As in André [1] and Barr-Beck [2], the Kan extension does not enter into the picture of [7], and all considerations are carried out in $[\underline{C},\underline{A}]$, the range of the Kan extension. The reader will notice that the results of this chapter are based on the improved version of acyclic models in [2] Ch. XI and not on [7]. It seems to me that the use of the Kan extension establishes a much closer relationship between acyclic models and homological algebra than the one in [7]. Moreover, it gives rise to a useful generalization of acyclic models which cannot be obtained by the methods of [7].

33) In order not to confuse Eilenberg-MacLane's notion of a "representable" functor with ours, we use quotation marks for the former.

34) This was also observed by D. Swan (unpublished).

35) If $\underline{A}$ is the category of abelian groups, one can choose P_M to be the free abelian group on tM. It is instructive for the following to have this example in mind. It links our approach with the original one of Eilenberg-MacLane.

$[\underline{M},\underline{A}]$ and $E_J(T\cdot J) = T$ holds.[36)]

<u>Proof.</u> Eilenberg-MacLane call a functor $T : \underline{C} \to \underline{A}$ "representable" if the above natural transformation $\varphi(T)\colon \bigoplus_M (P_M \otimes [JM,-]) \to T$ admits a section σ (in other words, T is a direct summand of $\bigoplus_M (P_M \otimes [JM,-])$. Since $E_J(P_M \otimes [M,-]) = P_M \otimes [JM,-]$ (cf. (7)) and J is a full inclusion, the composite $E_J \cdot R_J : [\underline{C},\underline{A}] \to [\underline{M},\underline{A}] \to [\underline{C},\underline{A}]$ maps the sum $\bigoplus_M (P_M \otimes [JM,-])$ on itself. Obviously the same holds for a direct summand T of $\bigoplus_M (P_M \otimes [JM,-])$, i.e. $E_J(T\cdot J) = T$ is valid. The theorem follows readily from this and the above (45).

As a corollary we obtain

(47) <u>Theorem.</u> Let $T_* : \underline{C} \to \underline{A}$ be a positive complex [37)] of functors together with an augmentation $T_* \to T_{-1}$. The following are equivalent:

i) The functors T_n are "representable" for $n \geq 0$ and the augmented complex $T_* \cdot J \to T_{-1} \cdot J \to 0$ is exact.

ii) $T_* \cdot J$ is a projective resolution of $T_{-1} \cdot J$ and $E_J(T_n \cdot J) = T$ holds for $n \geq 0$.

(48) <u>Corollary.</u> If $\bar{T}_* \to \bar{T}_{-1}$ is another augmented complex satisfying i) such that $T_{-1} \cdot J \cong \bar{T}_{-1} \cdot J$ holds, then T_* and $\bar{T}_*$ are homotopically equivalent. Every chain map $T_* \to \bar{T}_*$ is a homotopy equivalence, provided its restriction on $\underline{M}$ is compatible with the augmentation isomorphism $T_{-1} \cdot J \cong \bar{T}_{-1} \cdot J$. Moreover, the n-th homology of T_* (and $\bar{T}_*$) is the value of the n-th left derived functor of E_J at t, where $T_{-1} \cdot J \cong t \cong \bar{T}_{-1} \cdot J$ (i.e. $H_n(T_*) \cong L_n E_J(t)$).

<u>Proof of (48).</u> By (47) the complexes $T_* \cdot J$ and $\bar{T}_* \cdot J$ are projective resolutions of t, and hence there is a homotopy equivalence $T_* \cdot J \simeq \bar{T}_* \cdot J$. Applying the Kan extension yields $T_* \simeq \bar{T}_*$. Clearly the restriction of every chain map $f_* : T_* \to \bar{T}_*$ on $\underline{M}$ is a homotopy equivalence $f_* J : T \cdot J \simeq \bar{T}_* \cdot J$, provided $f_* J$ is compatible with the augmentation isomorphism $T_{-1} \cdot J \cong \bar{T}_{-1} \cdot J$. Applying the Kan extension E_J on $f_* J$

---

36) This shows that the notions of a "representable" functor (Eilenberg-MacLane) and a representable functor (Ulmer [24]) are closely related. Actually the functor T is also projective in $[\underline{C},\underline{A}]$, but this is irrelevant in the following.

37) This means that $T_n = 0$ for negative n. In the following we abbreviate "positive complex" to "complex".

yields again f_* . Hence f_* is also a homotopy equivalence. By standard homological algebra $L_nE_J(t) \cong H_nE_J(T_*\cdot J) = H_nT_*$ holds.

(49) Remark. Roughly speaking, the above shows that the method of acyclic models is the standard procedure in homological algebra to compute the left derived functors of the Kan extension by means of projectives. It is well known that the left derived functors of E_J can be computed not only with projectives but, more generally, by E_J-acyclic resolutions. This leads to a useful generalization of acyclic models. For the considerations below, one can drop the assumption that $\underline{A}$ has enough projectives and sums. Call a functor $T : \underline{C} \to \underline{A}$ "weakly representable" iff $E_J(T\cdot J) = T$ and $L_nE_J(T\cdot J) = 0$ for $n > o$. Clearly a "representable" functor is "weakly representable". By standard homological algebra we obtain the following:

(50) Theorem. Let $T_*, \bar{T}_* : \underline{C} \to \underline{A}$ be complexes of "weakly representable" functors together with augmentations $T_* \to T_{-1}$ and $\bar{T}_* \to \bar{T}_{-1}$ such that $T_{-1}\cdot J \cong \bar{T}_{-1}\cdot J$ is valid and the augmented complexes $T_*\cdot J \to T_{-1}\cdot J \to 0$ and $\bar{T}_*\cdot J \to \bar{T}_{-1}\cdot J \to 0$ are exact Then $H_n(T_*) \cong L_nE_J(t) \cong H_n(\bar{T}_*)$ holds for $n \geq o$, where t is a functor isomorphic to $T_{-1}\cdot J$ or $\bar{T}_{-1}\cdot J$.

(51) Moreover, one can show that every chain map $f_* : T_* \to \bar{T}_*$, the restriction of which on $\underline{M}$ is compatible with $T_{-1}\cdot J \cong \bar{T}_{-1}\cdot J$, induces a homology isomorphism. In general, there is no homotopy equivalence between T_* and $\bar{T}_*$. In practice this lack lack can be compensated by the following: Let $F : \underline{A} \to \underline{A}'$ be an additive functor with a right adjoint, such that the objects T_nC and $\bar{T}_nC$ are F-acyclic for $C \in \underline{C}$ and $n \geq o$ [38] ($\underline{A}'$ abelian). Then $Ff_* : F\cdot T_* \to F\cdot \bar{T}_*$ is still a homology isomorphism. If in addition $\underline{A}$ and $\underline{A}'$ are Grothendieck AB4) categories, then $H_n(F\cdot T_*) \cong H_n(F\cdot \bar{T}_*)$ and every chain map $g_* : F\cdot T_* \to F\cdot \bar{T}_*$ is a homology isomorphism, provided its restriction on $\underline{M}$ is compatible with $F\cdot T_{-1}\cdot J \cong F\cdot \bar{T}_{-1}\cdot J$.
This can be proved by means of (18) and the mapping cône technique of Dold [6].

(52) Remark. The isomorphism $H_n(T_*) \cong H_n(\bar{T}_*)$ in (51) can also be obtained from a result of André (cf. [1] p.7), provided $\underline{A}$ is AB4). Since $A_o(\ ,-) : [\underline{M},\underline{A}] \to [\underline{C},\underline{A}]$

---

38) In the examples, T_nC and $\bar{T}_nC$ are usually projective, $C \in \underline{C}$, $n \geq o$.

coincides with E_J (for $A_0(,-)$, see (11)), it follows from (13) that a functor $\underline{C} \to \underline{A}$ is "weakly representable" iff it satisfies André's conditions in [1] p.7. Hence it follows that $H_n(T_*) \cong A_n(-,T_{-1}\cdot J) \cong H_n(\bar{T}_*)$. This shows that André's computational device is actually a generalization of acyclic models, the notion "representable" being replaced by "weakly representable". Barr-Beck [2] Ch. XI used this computational device to improve their original version of acyclic models in [4]. Their presentation in [2] Ch.XI made me realize the relationship between acyclic models and Kan extensions. We conclude this summary with an abelian interpretation of their version of acyclic models.

(53) Let $\underline{A}$ be an abelian category and $\mathbb{G}$ be a cotriple in a category $\underline{C}$. Let $\underline{M}$ be the full subcategory of $\underline{C}$ consisting of objects GC , where $C \in \underline{C}$, and denote by $J : \underline{M} \to \underline{C}$ the inclusion. Let $T_*' : \underline{C} \to \underline{A}$ be a complex of functors together with an augmentation $T_* \to T_{-1}$. Their modified definition of "representability" is : $H_0(-,T_n)_{\mathbb{G}} = T_n$ and $H_j(-,T_n)_{\mathbb{G}} = 0$ for $j > o$ and $n \geq o$; and of acyclicity : $T_*M \to T_{-1}M \to 0$ is an exact complex for every $M \in \underline{M}$. Note that these conditions are considerably weaker than their original ones in [4]. To make the connection between this version of acyclic models and homological algebra, we first recall that $H_0(,-)_{\mathbb{G}} : [\underline{C},\underline{A}] \to [\underline{C},\underline{A}]$ is the composite of the restriction $R_J : [\underline{C},\underline{A}] \to [\underline{M},\underline{A}]$ with the Kan extension $E_J : [\underline{M},\underline{A}] \to [\underline{C},\underline{A}]$ (cf. (35)). Moreover, $H_*(,-)_{\mathbb{G}} : [\underline{C},\underline{A}] \to [\underline{C},\underline{A}]$ is the composite of R_J with $L_*E_J : [\underline{M},\underline{A}] \to [\underline{C},\underline{A}]$. This shows that the modified notions of acyclicity and representability of Barr-Beck [2] Ch. XI coincide with "acyclic" and "weakly representable" as defined in (49). Hence their method of acyclic models is essentially the standard procedure in homological algebra to compute the left derived functor of the Kan extension $E_J : [\underline{M},\underline{A}] \to [\underline{C},\underline{A}]$ by means of E_J-acyclic resolutions. We leave it to the reader to state theorems analogous to (50) and (51) for this situation.

Bibliography

[1] André, M., Méthode simpliciale en algèbre homologique et algèbre commutative, Lecture Notes, No. 32, Springer 1967.

[2] Barr, M. and Beck, J., Homology and standard constructions, this volume.

[3] Beck, J., Triples, algebras and cohomology, Dissertation, Columbia 1964 - 1967.

[4] Barr, M. and Beck, J., Acyclic models and triples, Coca (La Jolla), Springer 1966.

[5] Buchsbaum, D., Homology and universal functors, Lecture Notes No.61,Springer 1968.

[6] Dold, A., Zur Homotopietheorie der Kettenkomplexe, Math. Annalen 140, 1960.

[7] Dold, A., MacLane, S. and Oberst, U., Projective classes and acyclic models, Lecture Notes 47, 1967.

[8] Dold, A., Puppe,D. Homologie nicht-additiver Funktoren,Ann.Inst. Fourier 11, 1961.

[9] Dubuc, E., Adjoint triangles, Lecture Notes, No. 61,Springer, 1968.

[10] Fisher, J., The tensor product of functors, satellites and derived functors, to appear in J. of Algebra.

[11] Freyd, P., Abelian Categories, Harper and Row, New York, 1964.

[12] Freyd, P., Functor categories and their application to relative homology, Mimeographed notes, University of Pennsylvania, 1962.

[13] Grothendieck,A. Sur quelques points d'algèbre homologique, Tõhoku, Math.J.9, 1957.

[14] Kan, D., Adjoint functors, Trans. Amer. Math. Soc. 87, 1958.

[15] Lawvere, F., Functorial semantics and algebraic theories, Proc. Nat. Acad. Sci. U.S.A. 50, 1963

[16] Linton, F., Some aspects of equational categories, Coca (La Jolla), Springer 1966.

[17] MacLane, S., Homology, Springer 1963.

[18] Mitchell, B., Theory of categories, Academic Press, New York 1965.

[19] Oberst, U., Basiserweiterung in der Homologie kleiner Kategorien, Math. Zeitschrift 100, 1967.

[20] Oberst, U., Homology of categories and exactness of direct limits, to appear.

[21] Röhrl, H., Satelliten halbexakter Funktoren, Math. Zeitschrift 79, 1962.

[22] Tierney, M. and Vogel, W., Simplicial resolutions and derived functors, Mimeographed Notes, Eidg. Techn. Hochschule, 1966.

[23] Ulmer, F., Satelliten und derivierte Funktoren I. Math. Zeitschrift 91, 1966.

[24] Ulmer, F., Representable functors with values in arbitrary categories, J. of Algebra 8, 1968.

[25] Ulmer,F. Properties of dense and relative adjoint functors, J. of Algebra 8, 1968.

[26] Ulmer, F., Properties of Kan extensions, Mimeographed Notes, Eidg. Techn. Hochschule, 1966.

[27] Ulmer, F., On cotriple and André (co)homology, their relationship with classical homological algebra, to appear in Lecture Notes.

[28] Watts, C.E., A homology theory for small categories, Coca (La Jolla) 1966.

[29] Yoneda, N., On Ext and exact sequences, J. Fac. Sci. Tokyo, Sec. I, 8, 1961.

Offsetdruck: Julius Beltz, Weinheim/Bergstr

Lecture Notes in Mathematics

Bisher erschienen/Already published

Bitte wenden / Continued